D0710499

ECOLOGICAL ENGINEERING AND ECOSYSTEM RESTORATION

William J. Mitsch and Sven E. Jørgensen, July 2001, Copenhagen, Denmark

ECOLOGICAL ENGINEERING AND ECOSYSTEM RESTORATION

William J. Mitsch

Sven Erik Jørgensen

WILEY

JOHN WILEY & SONS, INC.

Library of Congress Cataloging-in-Publication Data:

Mitsch, William J.
  Ecological engineering and ecosystem restoration / William J. Mitsch, Sven Erik Jørgensen.
    p. cm.
  Includes bibliographical references and index.
  ISBN 0-471-33264-X (cloth)
    1. Restoration ecology. 2. Ecological engineering. I. Jørgensen, Sven Erik, 1934– II. Title.

QH541.15.R45M58 2003
628—dc21

2003041105

Printed in the United States of America

10 9 8 7 6 5 4 3 2

To

Howard T. Odum (1924–2002) and Eugene P. Odum (1913–2002), pioneers of the ecology that is fundamental to both this book and our careers

H. T. and Gene Odum, First Annual Meeting of American Ecological Engineering Society, Athens, Georgia, May 2001

CONTENTS

PREFACE   ix

I INTRODUCTION   1

1 Why Ecological Engineering and Ecosystem Restoration?   3

2 Definitions   23

3 Classification of Ecological Engineering   40

4 Ecosystems   56

5 Ecological Design Principles   94

II APPLICATIONS OF ECOLOGICAL ENGINEERING   103

6 Lake and Reservoir Restoration   105

7 Stream and River Restoration   125

8 Wetland Creation and Restoration   163

9 Coastal Restoration   195

10 Treatment Wetlands   230

11 Bioremediation: Restoration of Contaminated Soils   263

12 Mine and Disturbed Land Restoration   287

13 Ecological Engineering in China   309

III   ECOLOGICAL ENGINEERING TOOLS   337

14   **Modeling in Ecological Engineering and
     Ecosystem Restoration**   **339**

References   **367**

ORGANISM INDEX   397

SUBJECT INDEX   401

PREFACE

Fourteen years ago the two authors of this volume edited a book entitled *Ecological Engineering: An Introduction to Ecotechnology* (John Wiley & Sons, Inc., 1989). That multiauthored book had contributions and ideas from a number of authors and gave many early examples of the use of ecological engineering to solve typical environmental problems in both the West and East. Our present book, *Ecological Engineering and Ecosystem Restoration,* is *not* a second edition of that volume; it is a completely new book—the fields of ecological engineering and ecosystem restoration have advanced that much. Over a decade ago (1992), the international peer-reviewed journal *Ecological Engineering* (Elsevier) began publication, enhancing the maturation of the field by presenting both experimental results and case studies of ecological engineering. About the same time (1993), the journal *Restoration Ecology* (Blackwell) began publication and was soon joined by *Ecological Restoration,* the renamed and updated *Restoration and Management Notes.* Today, therefore, it is possible to write this volume as a textbook containing the basic information that is needed to select and design ecological solutions. Approaches include case studies and ecosystem models that are applied to design, manage, and modify ecosystems.

This book is divided into three parts. Part I covers the basic concepts needed for the application of ecological engineering and ecosystem restoration. The aim of the first chapter is to place the fields in the context of other environmental disciplines and concepts. The next chapter gives definitions of ecological engineering and ecosystem restoration, their history, and basic concepts. Ecological engineering and ecosystem restoration are used widely today to solve a number of environmental problems, so Chapter 3 provides an overview of applications and presents a classification based on these applications. In Chapter 4 we give the concepts in systems ecology that are fundamental for ecological engineering and restoration. As chemistry is the basic for chemical engineering, so ecology is the basis for ecological engineering and ecosystem restoration. In Chapter 5 we expand the 13 principles of ecological engineering that we published in 1989 and now have 19 principles that should serve as a starting point for governing ecological engineering and ecosystem restoration. It is recommended that these principles be used as guidelines for all appropriate projects. If a project violates several of these principles, a sustainable ecological solution may not occur.

Part II includes eight chapters on applications of ecological engineering and ecosystem restoration, mostly in an ecosystem approach; this is the heart and soul of the book. Chapters 6 to 9 and Chapter 12 cover fields that are often described as ecosystem restoration: restoration of lakes and reservoirs, restoration of rivers and streams, creation and restoration of wetlands, restoration of coastal areas, and restoration of severely altered terrestrial landscapes, such as mined land. Chapters 10 and 11 encompass the use of ecological engineering methods to solve pollution problems: how wetlands can be employed in treating point and nonpoint water pollution, and how contaminants can be removed from soil in situ by bioremediation. The last chapter in this part, Chapter 13, covers the history and application of ecological engineering methods in China, including how agriculture can be developed more sustainably. Throughout the second part, case studies are provided to illustrate a wide variety of practical ecological management techniques that use ecological restoration and engineering.

In Part III we present a major tool that should be used in almost every ecological engineering and ecological restoration project: ecological modeling. The tool has been developed substantially over the past 30 years and is the one technique we have for predicting ecosystem structure and function, a necessity in any ecological engineering or restoration project.

We expect this book to be a catalyst to the continued development of ecological engineering and ecosystem restoration. We believe that ecological engineering and ecosystem restoration are basically the same thing on a spectrum of ecosystem management approaches. We strongly resist the idea that these are two separate fields, one for engineers and the other for everybody else. Our collection of principles and case studies here is meant to provide a demonstration of the breadth of application of ecological engineering and ecosystem restoration and to suggest the significance of these applications. It is up to environmental scientists and engineers in academia, government, and industry to continue to find innovative solutions to environmental problems through this "partnership with nature." Some day, ecologists in our society may have both "ecological" and "engineering" in their titles and not be subjected to the criticism of not knowing ecology or not knowing engineering. We are suggesting that the time has come.

This book could not have come about with a significant amount of assistance from others. Ruthmarie H. Mitsch assisted with a major amount of editing and permission requests. Anne Mischo kindly drew several of the illustrations. Photos, illustrations, and other assistance were provided by Ken Strait and Jeff Pantazes, PSEG, New Jersey; John Day, Louisiana State University; Li Zhang, The Ohio State University; Akira Miyawki, Japanese Center for International Studies in Ecology, Yokahama; Christopher Henry, ASCONIT Consultants, Lyon, France; Naiming Wang, Lou Toth, and Paul Whalen, South Florida Water Management District; Tim Nightengale, Academy of Natural Sciences; Karen Bushaw-Newton, American University; Mike Weinstein, Rutgers University; Mark Brown, University of Florida; and Carlos Hoffmann, Danish Environmental Survey. Elsevier Science kindly provided permission for many illustrations from our previous books and from *Ecological Engineering*.

Note that we have dedicated the book to the Odum brothers—Howard T. Odum and Eugene P. Odum—both of whom died in 2002. They are, in our opinion, the two founders of modern ecology. We have stood on their broad shoulders and seen farther.

WILLIAM J. MITSCH
Columbus, Ohio

SVEN ERIK JØRGENSEN
Copenhagen, Denmark

I

INTRODUCTION

1

WHY ECOLOGICAL ENGINEERING AND ECOSYSTEM RESTORATION?

We are now in a position to make a substantial contribution to the "greening" of the planet through ecological engineering and ecosystem restoration. We find ourselves in a retrospective period of human history, both politically and ecologically, where although not necessarily questioning all we have built and engineered to date, we are determining (1) whether to continue practices as usual (and whether we can afford to do so) and (2) what new approaches are available for restoring the "bodily functions" of nature, on which we depend. Signs all around us confirm that a paradigm shift is taking place, both within and outside the ecological and engineering professions, to accommodate ecological approaches to what was formerly done through rigid engineering and a general avoidance of any reliance on natural systems.

Engineers, ecologists, resource managers, and even politicians are now completely redesigning, at a cost of almost $8 billion, the plumbing in the southern Florida Everglades to provide a more ecologically integrated system (Figure 1.1). As part of the effort in the Florida Everglades, the Kissimmee River in Florida is being "restored"—at an enormous cost—to something resembling its former self before it was canalized 30 years ago (Figure 1.2). Ecological approaches are being investigated to reduce nonpoint-source pollution from reaching the Baltic Sea, where extensive eutrophication is occurring. The Gulf of Mexico continues to have annual "dead zones" that now spread well over 20,000 km², approaching the size of the state of Massachusetts. Discussions are being held not on whether to restore the Mississippi River Basin to a more natural state by removing levees and restoring wetlands and riparian forests (Figure 1.3), but when and how that restoration will occur. In a related effort, the Louisiana delta and coastline are disappearing into the sea, and major efforts are under way to reduce land loss along that coastline.

3

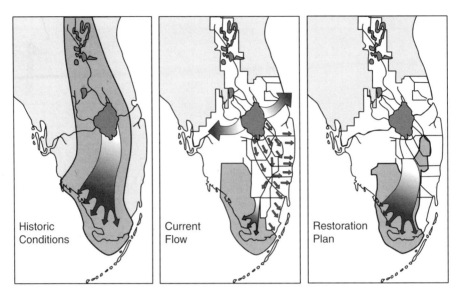

Figure 1.1 The largest ecological redesign ever attempted is the $8 billion, 20-year restoration of the Everglades in southern Florida. The historic flow of water as seasonal overflow from Lake Okeechobee through the Everglades has been altered dramatically by a system of drainage canals and pump stations built in the twentieth century that now divert a large amount of the water east and west to the Atlantic Ocean and Gulf of Mexico. The restoration plan is to restore historical hydrologic flow of water through the Everglades and to Florida Bay to the south. (Redrawn from U.S. Army Corps of Engineers.)

Figure 1.2 Part of the Everglades area restoration has involved the remeandering and restoration of the Kissimmee River that flows from central Florida to Lake Okeechobee. At approximately 10 times the cost of the stream straightening that occurred in the 1960s, the river is being partially restored by low-head dams and reattachment of backwaters and old meanders to the stream channel. (Courtesy of Lou Toth; photo by Paul Whalen, South Florida Water Management District, West Palm Beach, Florida; reprinted with permission.)

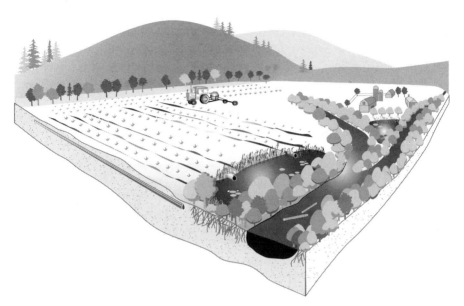

Figure 1.3 Creating and restoring wetlands and riparian ecosystems in an agricultural landscape are being proposed as means of controlling nonpoint-source pollution in Scandinavia to protect the Baltic Sea and in the United States to protect the Gulf of Mexico and the Chesapeake Bay. (From Mitsch and Gosselink, 2000; copyright 2000; reprinted with permission from John Wiley & Sons, Inc.)

Denmark is bringing back its largest river, the Skern River, to its old meandering course to prevent continued deterioration of coastal waters (Figure 1.4). Treating wastewater with constructed wetlands (Figure 1.5), an approach just begun with experiments in the 1960s and 1970s, is now an accepted approach for wastewater treatment throughout the world. Thousands of hectares of coastal marshes are being restored along the Delaware Bay in New Jersey from hay farms to a status they have not seen since the eighteenth century by removing dikes and carving tidal creeks (Figure 1.6).

The planners of the expensive and controversial Biosphere 2 enclosed ecosystems in Arizona (Figure 1.7) have perhaps unwittingly illustrated the great value of natural ecosystems. In an ecological engineering of the most extreme kind, a group of ecosystems were designed in a 1.25-ha glass-enclosed system to illustrate potential use of such enclosed systems for human support in future space colonies. But fans were used for air movement instead of natural air movement and pumps for water movement to replace the free water flow from the hydrologic cycle, illustrating exactly the point of this book: Ecosystems, running on the natural energies of sunlight, wind, and water, are our real support systems, providing a great variety of free public service functions that we do not realize are important until they are gone. Costanza et al.'s (1997) classic answer to the question "what is nature

Figure 1.4 A major river restoration project in Europe is being carried out on the Skern River in Jutland, Denmark, where Denmark's largest river is being restored to its former meandering course. When the restoration is complete, 2200 ha of meadows and wetlands—half of the area drained in the 1960s—will be restored. A reason for the restoration is coastal pollution in a fjord adjacent to the North Sea caused by the straightened river and pollution transport from agriculture. The Danish public funds earmarked for this project are 8.5 times the cost of the original river drainage project, most of which occurred in the mid-twentieth century. (Photo by W. J. Mitsch.)

worth?" suggested that ecosystems are providing services equivalent to about $64,000 per square kilometer per year. By using the cost of construction and maintenance of Biosphere 2 as an indicator of what it would take to produce ecosystems devoid of our climate, air, water, and winds, the value of an ecosystem escalates to about $1 billion per square kilometer per year (Mitsch,

Figure 1.5 A 3-ha treatment wetland adjacent to a wastewater treatment plant in central Ohio. The treatment plant delivers secondarily treated wastewater to the wetland that has, as its main function, tertiary treatment of the wastewater to prevent nitrogen and phosphorus from reaching the adjacent north fork of the Licking River. (Courtesy of W. J. Mitsch.)

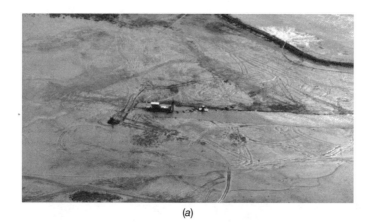

(a)

(b)

Figure 1.6 One of the largest coastal wetland restoration projects in the eastern United States involves the restoration, enhancement, and preservation of 5000 ha of coastal salt marshes on Delaware Bay in New Jersey and Delaware. One part of that project involves reflooding salt hay farms by breeching coastal dikes and constructing "starter" tidal creeks to create a high percentage cover of desirable vegetation such as *Spartina alterniflora,* a relatively low percent of open water, and the absence of the invasive reed grass *Phragmites australis.* Tidal restoration construction work was completed at the 460-ha Maurice River site in New Jersey in early 1988. (a) Tidal creek construction activity in 1995; (b) vegetation recovery of the site by *Spartina alterniflora* by June 2001. Restoration of the site by *Spartina* and other desirable vegetation occurred on 71 percent of the site after four growing seasons [(a) Courtesy of Ken Strait, PSEG; reprinted with permission from PSEG, Salem, New Jersey; (b) photo by W. J. Mitsch.]

(a)

(b)

(c)

Figure 1.7   Biosphere 2, a 1.25-ha glass-enclosed mesocosm created in the Arizona desert, an example of ecological engineering in its extreme: (a) savannah, ocean, wetland, and desert containment; (b) intensive agriculture containment; (c) human living quarters. A 10-MW power plant (not pictured) is necessary to provide sufficient power to create sufficient water flow and air movement for various ecosystems to flourish in this high-CO_2 environment. Based on this project, it was estimated that to create such artificial ecological systems in place of nature providing the same services free would cost $1 billion per square kilometer per year (Mitsch, 1999). (Photos by W. J. Mitsch.)

1999). This is the cost we would incur to create our natural Earth in space, forgetting the cost to get there. The message of these estimates is of course clear: We should protect and enhance our Biosphere 1, allowing it to provide the services it provides and, in some cases, enhancing its ability to provide more.

Ecologists are now refining the techniques of restoring function in degraded ecosystems, and countless ecologists now call themselves restoration ecologists. Agricultural engineers, known for the efficiency with which they drained the landscape, are changing their names and their actions in many locations by restoring ditches to stream channels and farmlands to wetlands. Civil engineers, the nation's top river straighteners, are busy removing dams and restoring river meanders. The U.S. Army Corps of Engineers is now "greening" its mission to specifically include ecological restoration; some in that organization see themselves not only as the nation's water resource managers, but also as the nation's ecological engineers. Restoration and creation of ecosystems is now an industry.

1.1 FORTY YEARS OF ENVIRONMENTAL PROTECTION AND RESTORATION

We are approaching an age of diminishing resources where the growth of the human population continues and we have not yet found the proper means to solve local, regional, and global pollution and shortage of renewable resources. The first green wave that appeared in the middle and late 1960s was thought to offer feasible ways to solve pollution problems completely. Visible problems were mostly limited to point sources of air and water pollution and a comprehensive "end-of-the-pipe technology" (i.e., environmental technology) was developed and refined in that time to solve pollution problems. It was even seriously forecast in the early 1970s that what was called *zero discharge* could be attained for water pollution. For example, the U.S. Congress declared in the Clean Water Act in 1972 that all waters of the nation should be fishable and swimmable by 1983. The year came and went, yet half the rivers in the country were not "fishable and swimmable" as the act had stipulated just because the politicians said it should be. There were complications far beyond controlling industrial and municipal wastewater.

It became clear that zero-discharge or similar policies would be too expensive and that we should also rely on the self-purification ability of ecosystems. That outlook called for the development of environmental and ecological models to assess the self-purification capacity of ecosystems and to set up emission standards reflecting the relationship between impacts and effects in the ecosystems (Figure 1.8). In this case, models were used to relate an emission to its effect on the ecosystem, and toxicological studies were used to determine the effects on its components (e.g., fish). Those relationships were then used to determine a good solution to the environmental prob-

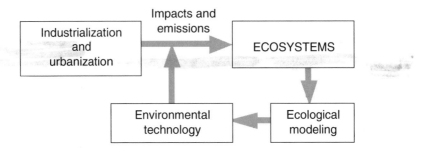

Figure 1.8   Strategy applied in environmental management in the early 1970s. An ecological model is used to relate an emission or discharge to its effect on the ecosystem and its components. The relationship is applied to the selection of a good solution to the environmental problems by application of environmental technology. (From Jørgensen and Bendoricchio, 2001; copyright 2001; reprinted with permission from Elsevier Science.)

lems by application of environmental technology (e.g., wastewater treatment systems).

Meanwhile, we have found that what we could call the environmental crisis is much more complex than thought. We could, for instance, remove heavy metals from wastewater, but where should we dispose the sludge containing the heavy metals? Resource management pointed toward recycling instead of removal. Nonpoint sources of toxic substances and nutrients, originating primarily from agriculture, emerged as new threatening environmental problems in the late 1970s. The focus on global environmental problems such as acid deposition, the greenhouse effect, and the decomposition of the ozone layer in the 1980s added to the complexity of the situation. It was revealed that we use as many as 100,000 chemicals that may threaten the environment, due to their more-or-less toxic effects on plants, animals, humans, and entire ecosystems. In most industrialized countries, comprehensive environmental legislation was introduced to regulate the wide spectrum of different pollution sources. Trillions of dollars have been invested in pollution abatement on a global scale, but it seems that two or more new problems emerge for each problem that we solve. Our society seems not to be geared to solving environmental problems, or is there perhaps another explanation?

The complexity and additional options of today's environmental management includes simultaneous application of environmental technology, cleaner technology, environmental legislation, ecological engineering, and ecosystem restoration (Figure 1.9). Traditional environmental technology offers a broad range of methods that are able to remove pollutants from water, air, and soil, and these methods are particularly applicable to cope with point sources. Cleaner technology explores the possibilities of recycling by-products or final waste products or attempts to change the entire production technology to obtain reduced emission. It attempts to answer the pertinent question: Couldn't we produce our product by a more environmentally friendly method?

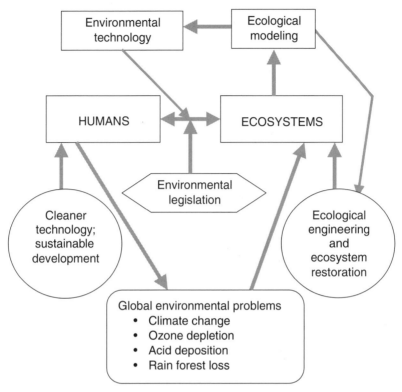

Figure 1.9 Environmental management in the twenty-first century is much more complex than the early approach shown in Figure 1.8. Models are still used in environmental management, but the management tools are more numerous and complex and include the approaches of environmental technology, cleaner technology and sustainable development (sometimes called *industrial ecology*), and ecological engineering/ecosystem restoration. Models can be used to select the best environmental management strategy. In addition, global environmental problems, which also require the use of models as synthesizing tools, have become important environmental issues with little solution except adaptation and pollution source reduction. (Redrawn from Jørgensen and Bendoricchio, 2001.)

To a great extent, the answer will be based on environmental risk assessment, life-cycle analysis, and environmental auditing. The ISO 14000 series and risk reduction techniques are among the most important tools in the application of cleaner technology. Environmental legislation and green taxes may be used in addition to the classes of technology. The fourth option—ecological engineering and ecosystem restoration combined—is the subject of this book.

Figure 1.10 shows the flows of material (and energy) in the history of a product, from raw materials to final disposal as waste. The exact number of products in modern technological society is not known, but it is probably on the order of 10^7 to 10^8. All these products emit pollutants to the environment

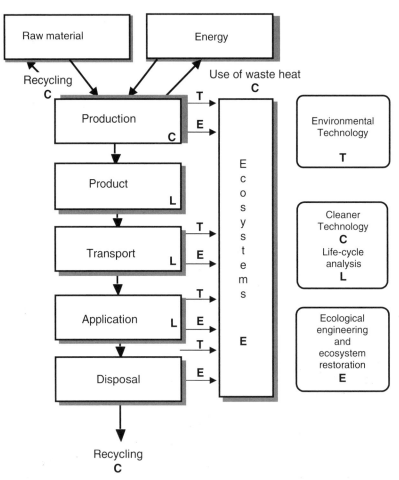

Figure 1.10 Pollution flow and possible approaches to its control. The large arrows cover mass flows and the thin arrows symbolize control possibilities. Production includes both industrial and agricultural production, the latter being mostly responsible for nonpoint pollution. Environmental technology (T) is applied on point pollution, while ecological engineering and ecosystem restoration (E) are needed for the solution of many ecological and environmental problems. Clean technology (C) explores the possibilities of changing the present production methods to obtain a reduction in pollution, either through recycling by-products and waste products or by a more-or-less radical change in the production technology. Life-cycle analyses (L) are used to determine where in the history of a product the pollution actually takes place, with the view to change the product, its transport, and/or its application. (Redrawn from Jørgensen, 2000.)

during their production, transportation from producer to use, application, and final disposal as waste. The core problems in environmental management are how to control pollutants properly and how to manage our ecological systems in this less-than-perfect world. The answer is that we must have a wide spectrum of methods. Notice that Figure 1.10 is based on the principles of conservation of matter and energy and on our wider use of ecosystem properties in environmental management. Environmental legislation and green taxes are not included in Figure 1.10, as they may in principle be used as regulating instruments in every step of the flow from raw materials and energy to final waste disposal of the product used.

From this short introduction to environmental management and the wide spectrum of methods that can be implemented to solve environmental problems, we can conclude that environmental management is a very complex issue. A local environmental problem may be solved by selection of another raw material or energy source, by a partial or complete change in the method of production, by increased use of recycling, by selection of the proper combination of technological methods taken from any of the four aforementioned classes of technologies, by a slight change in the properties of the product, by a combination of environmental technology with recovery of the ecosystem affected, and so on. The number of possible solutions is enormous, yet the environmental management strategy should attempt to find the optimum solution from an economic–ecological point of view.

Two major forces can lead to apathy: naive technological optimism—the idea that some technological wonder will always save us regardless of what we do, and gloom and doom pessimism—the idea that nothing will work and our destruction is assured. Ecological and environmental problems are very complex and difficult to solve. The best starting point to solving the complex and difficult problems must be an understanding of the nature of the problems. Ecological principles and ecological methods can then be employed to solve some of them.

1.2 WHY ECOLOGICAL ENGINEERING AND RESTORATION ARE NEEDED

The state of our environment, combined with a dwindling in nonrenewable natural resources available to solve environmental problems, suggests that the time has come for a new paradigm that involves ecosystem- and landscape-scale questions and solutions. There are a great number of environmental and resource problems that need an ecosystem approach, not just a standard technological solution. We have finally recognized that we cannot achieve complete elimination of pollutants, owing to a number of factors, and that we need new approaches better attuned to our natural ecosystems. In our attempts to control our own environment, we have also seen that we have tried to control nature too much at times, with disastrous consequences such as enormous floods, invasive species, air and water pollution being transported hundreds and thousands of kilometers instead of a few kilometers, and the production of massive quantities of solid wastes that need disposal or use somewhere. But why now, and why do we suggest this new field, especially to include engineers, whom many blame for the difficult situation in which we now find ourselves?

Limited Resources

There is a finite quantity of resources to address to the problems of pollution control and natural resource disappearance. This is particularly true for developing countries that wish to have the standard of living and technology of developed countries but currently must deal with pollution problems often more serious than those in the developed world. The limited resources and the high and increasing human population force us to find a trade-off between the two extremes of pollution and totally unaffected ecosystems. We cannot and must not accept a situation of no environmental control, but neither can we afford zero-discharge policies, knowing that we do not provide one-third of the world's population with sufficient food and housing.

Three pronounced developments have caused the environmental crisis that we are now facing: the growth in population, industrialization, and urbanization. Figure 1.11 illustrates world population growth, past and projected. From the graph it can be seen that population growth has experienced decreasing doubling time, which implies that growth is more than exponential (exponential growth corresponds to a constant doubling time). The growth of population from 1 billion to 2 billion people took about 100 years, whereas the next doubling in population took only 45 years. The net birth rate at present is about 370,000 people per day, while the death rate is 150,000 per day. The population growth is determined by the differences between the two: population increase = birth rate − death rate. The present growth rate implies that the world's population is increasing by more than 200,000 people per day, or about 1.5 million per week, corresponding to more than 80 million per year.

Renewable resources are those that can maintain themselves or be replenished continuously if managed wisely. Food crops, animals, wildlife, air, water, forest, and so on, belong to this class. Land and open space can also be considered renewable, but they shrink as the population increases. Although we cannot run out of these resources, we can use them faster than they can be regenerated, or by using them unwisely, we can affect the environment. Other resources, such as fossil fuels and minerals, are nonrenewable resources, whose finite supplies can be depleted. Theoretically, some of these resources are renewable, but only over hundreds of millions of years, whereas the time scale of concern to humans is only hundreds of years. When we talk about finite supplies of resources, we should qualify this by discussing finite supplies of substances presently considered to be resources. Often, our most

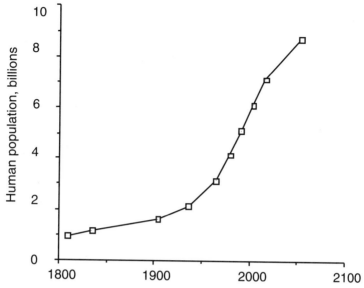

Figure 1.11 Human population through the nineteenth and twentieth centuries and projection to the year 2050. An optimistic prognosis is applied which predicts that population growth rate will begin to decrease after the year 2015. In the period 1805–1975, population growth was more than exponential. Even with a slowing growth rate in the first half of the twenty-first century, human population is expected to almost double from its level in 2000 to 9 billion by the year 2050.

important consideration is whether the pollution costs of extraction and use of a resource outweigh its benefits as population or per capita consumption increases.

The Shell Game ✳✳✳ Changing pollutants from 1 form to another

When we control pollution through technological means, we are often playing a shell game with the pollution. There are many examples of how traditional environmental technology removes pollutants from one medium and sends them to another. Toxic substances present in municipal wastewater cannot be biodegraded in a mechanical–biological wastewater treatment plant but will depend on the water solubility found either in the treated water or in the sludge. If the sludge is used as a soil conditioner in agriculture, the toxic substance will contaminate the soil. If the sludge is incinerated, it may cause air pollution or be found in the ash.

We use scrubbers to prevent sulfur emissions from power plants and are then faced with enormous solid waste storage problems from the sludge left behind. We build solid waste facilities and water pollution control systems, and the result can be atmospheric emissions of the greenhouse gas methane. We use industrial wastewater treatment methods to remove heavy metals from a factory and are left with a metal-rich sludge. We burn sludge and solid wastes and create air pollution problems. We are moving materials around in a shell game—if they are not under one shell, they are under another.

Climate Change and the Secondary Pollution Effect

During the last few decades we have observed a distinct increase in different types of pollution. The concentrations of carbon dioxide and other greenhouse gases in the atmosphere have increased uniformly and throughout the world in the past century. Concentrations of many toxic substances have increased in soil and water, and the ecological balance has been changed in our ecosystems. In many major river systems, oxygen depletion has been recorded more often, and many recreational lakes and coastlines are suffering from eutrophication as a result of high concentrations of nutrients—mainly nitrogen and phosphorus.

We are now confronted with the fact that seemingly inert carbon dioxide (CO_2)—the principal gas emitted by our own respiratory system—has increased by 20 percent in the twentieth century, due primarily to the increased use of fossil fuels (Figure 1.12). This continuing increase is now considered to be the chief cause of change in our climate. International efforts such as the Kyoto Treaty are now attempting to limit the burning of fossil fuels to minimize future climate effects, although many fear that climate changes are already inevitable, even with drastic fuel use reductions. Environmental technology is often fossil fuel–based. When expensive environmental technology is used to solve a pollution problem, we are solving a problem of one type and may be contributing to another through an increase in the global emissions of CO_2. Often, we have used a great amount of fossil fuel in the economy to develop that technology. The amount of energy used is generally proportional to the cost of the technology, and the CO_2 emission is proportional to the amount of fossil fuel used. The use of alternative "nature-based" technologies and low-cost technologies is therefore vital in overall pollution abatement.

Figure 1.12 also illustrates a more dramatic change in our environment that is slowly leading to eutrophication of the planet. Bioavailable nitrogen essentially doubled in the twentieth century, due to nitrogen fertilizer production and use, increased nitrogen fixing in cultured plants, and release of nitrogen from long-term soil storage because of land drainage. Humans are essentially eutrophying the planet, causing dramatic shifts in the ecology of both natural ecosystems such as estuaries and restored systems such as grasslands, forests, and wetlands. Restoration of ecosystems to systems remembered from publications and maps from the nineteenth century may be impossible because of this change; restoration that includes adapting to this change is a better approach.

What has caused this sudden increase in pollution? The answer is not simple, but the growth in population is obviously one factor that influences

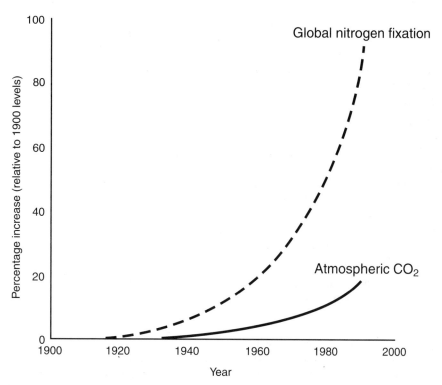

Figure 1.12 Pattern of atmospheric carbon dioxide and bioavailable nitrogen in the twentieth century. (Redrawn from Vitousek et al., 1997.)

our environment. Other factors include our rate of consumption, the type and amount of waste we produce, and the fact that consumption of resources in the production of environmental technology leads to an even greater production of secondary pollution. The economy that supports this technology is consuming fossil fuels while producing CO_2 and emitting nitrogen to the biosphere.

A Sustainable Society

Since we cannot solve all our environmental problems through high-technology solutions alone, and since our energy future appears quite clouded, we must investigate alternative means of cleaning the environment. Ecotechnology and ecological engineering will play a significant role in a sustainable society. *Sustainability* has become one of the buzzwords of our time, used again and again in the environmental debate—sometimes in the wrong context. It is therefore important to provide a clear definition here to avoid misunderstandings later in the book. The Brundtland report (World Commission on Environment and Development, 1987) produced the following definition:

Sustainable development is development that meets the needs of the present without compromising the ability of future generations to meet their own needs. Note that this definition includes no reference to environmental quality, biological integrity, ecosystem health, or biodiversity. Klostermann and Tukker (1998) discuss sustainability as based on product innovation and introduce the concept of *ecoefficiency,* the reciprocal of the weighted sum of the environmental claims, including ecological impacts and draws on renewable and nonrenewable resources.

Conservation philosophy has been divided into two schools: *resourcism* and *preservationism.* These are understood, respectively, as seeking maximum sustained yield of renewable resources, and excluding human inhabitation and economic exploitation from remaining areas of an undeveloped nature. These two philosophies of conservation are mutually incompatible. They are both reductive and ignore nonresources, and seem not to answer the core issue: how to achieve sustainable development, although preservationism has been retooled and adapted to conservation biology.

Lemons et al. (1998) gave a more down-to-earth solution by formulating the following rules:

1. *Output rule.* Waste emission from a project should be within the assimilative capacity of the local environment to absorb without unacceptable degradation of its future waste absorptive capacity or other important services.
2. *Input rule.* Harvest rates of renewable resources inputs should be within the regenerative capacity of the natural system that generates them, and depletion rates of nonrenewable resource inputs should be equal to the rate at which renewable substitutes are developed by human invention and investment.

An Overdue Alliance

Engineering and ecology are ripe for integration into one field, rather than following separate approaches, which are often adversarial. Ecology as a science is not routinely integrated in engineering curricula, even in environmental engineering programs. Engineers thus miss the one science that could help them most in environmental matters. Similarly, environmental scientists and managers are lacking an important element in their profession—problem solving. Although tremendously competent in describing problems and perhaps even in managing ecosystems one species at a time, ecologists are not well versed in prescribing solutions to problems. The basic science of ecological engineering is ecology, a field that has now matured to the point where it needs to have a prescriptive—rather than just a descriptive—aspect.

Nature Needs Help, Too

Most of the discussion above is on solving human problems, with little discussion of the impact on nature of what we do. We solve human problems and create problems for nature. That has been the history of humankind, at least in the Western world. We need to adopt approaches to solving environmental problems not only for our human well-being but also to protect streams, river, lakes, wetlands, forests, and savannahs. We need to work symbiotically with nature where we use her public service functions, but recognize the need to conserve nature as well. The idea of nature conservation is so important that it needs to become a goal of engineering, not just one of its possible outcomes. We must seek additional approaches to reducing the adverse effects of pollution while preserving our natural ecosystems and conserving our nonrenewable energy resources. Ecotechnology and ecological engineering offer such additional means for coping with some pollution problems, by recognizing of the self-designing properties of natural ecosystems. The prototype machines for ecological engineers are the ecosystems of the world.

Ecological Engineering Is Already Happening Anyhow

Ecological engineering is now, in effect, being practiced by many professions under a great variety of names, including *ecotechnology, ecosystem restoration, artificial ecology, biomanipulation, ecosystem rehabilitation, nature engineering* (in the Netherlands), *hydroecology* (in eastern Europe), and *bioengineering.* But little theory backs these practices. Engineers are building wetlands, lakes, and rivers with little understanding of the long-term biological integrity of these systems. Ecologists and landscape architects now design ecosystems with homespun methodologies that must be relearned each time. Engineers and ecologists who design and construct ecosystems use cookbooks that do not always work and often do not publish their successes and failures in the open literature. Theory has not yet connected with the practice.

Some of the ecotechnological methods presented in the book are not new; some have in fact, been practiced for centuries. In earlier times these methods were considered good empirical approaches. Today, ecology has developed sufficiently to understand the scientific background of ecological engineering, to formalize use of these approaches, and to develop new ones. We must understand not only how we can influence the processes in the ecosystem and how the ecosystem components are linked together, but also how changes in one ecosystem can produce changes in neighboring ecosystems.

1.3 APPROPRIATE TIMING FOR ECOLOGICAL ENGINEERING

We must acknowledge that there are now 2 billion more people on Earth than there were 20 years ago, and that the nonrenewable resources are more limited. We therefore need to find new ways to approach environmental concerns. We have attempted to solve problems through the use of available technology. Unfortunately, those attempts have partially failed. Therefore, we must think more ecologically and consider additional means. If applied properly, ecological engineering and ecosystem restoration are based on ecological considerations and attempt to optimize ecosystems (including limited resources) and human-made systems for the benefit of both. It should therefore afford additional opportunities for solving the crisis. We have had several energy crises during the past 30 years, and we know that new crises will appear in the future. Therefore, we have to rely more on solar-based ecosystems, which are the bases for ecological engineering.

In the short term, ecological engineering could bring immediate attention to the importance of "designing, building, and restoring ecosystems" as a logical extension of the field of ecology as it applies directly to solving environmental problems. In the long term, it will provide the basic and applied scientific results needed by environmental regulators and managers to control some types of pollution while reconstructing the landscape in an ecologically sound way. Formalization of the idea that natural ecosystems have values for humans other than directly commercial ones is also a benefit of ecotechnology and will go a long way toward enhancing even further a global ecological conservation ethic.

1.4   ECOLOGICAL ENGINEERING AND OUR EDUCATIONAL SYSTEM

Environmental problems are by definition multidisciplinary, but our educational systems are not. There is a need in environmental management for integration of disciplines, particularly between ecology and engineering, as emphasized in this book. Ecological training in schools and universities—at least at the ecosystem scale needed for solving many of our ecological problems—is often out of date and not taught well in biology-dominated programs. Ecology is considered by many to be a subset of biology, but the opposite is true; it is often taught far better in environmental science or similar programs. Aside from a recent frenzy in restoration ecology that is still not focused, it is unlikely that one can get a proper education in restoring and creating ecosystems in traditional biology programs. Similarly, engineers have not taken the demands and value of ecosystems into consideration in their development of technologies and their planning of production. Biology and ecology have scarcely been mentioned in undergraduate engineering education, partially because most engineers view ecology as what is portrayed in the popular press rather than as a substantially quantifiable science. An improperly trained engineer is a danger in ecological restoration.

A far better integration of disciplines is urgently needed for repairing the planet. Our higher educational system has for many years encouraged specialization. Plaudits have been given to scientists or experts who fully mastered a very narrow problem or topic, and analyses have been more appreciated than syntheses. But this system has created isolation from other problems and topics and has paralyzed interaction among disciplines. The result has been that ecologists have often not understood the need for quantifying ecosystems and dealing with applications of ecology in a technological framework. Therefore, we have been slowed by our educational system in solving environmental problems properly. Engineers want to quantify everything and lose appreciation for the many ecological pathways and feedbacks that need to be considered. A much more integrated education is urgently needed in the future if we are to find the right environmental solutions, asking the relevant questions and considering human society and ecosystems as an entity.

This does not imply that we should develop educational systems and curricula that give all students a little knowledge about everything. That type of "cafeteria" educational approach has been discredited in many circles, and we concur. Rather, we need to educate fundamentally sound generalists as well as specialists. Specialists must be taught to work together on multidisciplinary projects. Such collaboration will force specialists to understand other languages, it will prevent isolation, and it will encourage cooperation and coordination. Future students must devote some time to learning what other disciplines can offer.

The science of ecology is expected to continue its progress, and owing to the rapid growth of computer technology, we shall be able to describe and solve more and more complex ecological problems. However, a better general ecological understanding is needed by our politicians and the entire population, too, if we are to use ecological engineering successfully in the future. Such understanding will not come with a more complex technology if our educational system lags behind. Therefore, better ecological education with multidisciplinary aspects is needed urgently. Ecology should be introduced as a basic and compulsory subject at all school levels, including the elementary. It should be as required in our basic college education together with the humanities, social sciences, and chemistry.

1.5 FUTURE OF ECOLOGICAL ENGINEERING

Our experience with the formal fields of ecological engineering and ecosystem restoration is limited, but our experience is growing exponentially. We have had one decade of formal peer-reviewed experience in the journals *Ecological Engineering* and *Restoration Ecology*, two professional societies, and a few academic settings in the world where the fields are beginning to be taught. Although current results look very promising, we need to integrate the application of ecological engineering and ecosystem restoration much more in our pollution control and environmental planning in the future. Thus, continuous development of systems ecology, applied ecology, ecological modeling, and ecological engineering is much needed to complement the basic science of ecology. Ecological engineering and ecosystem restoration offer us a very useful tool for better planning in the future. It will be a challenge to humankind to use properly the tools that develop in this field. If we do, the twenty-first century can be an ecological century to which we can very much look forward.

2

DEFINITIONS

2.1 ECOLOGICAL ENGINEERING

Ecological engineering is "the design of sustainable ecosystems that integrate human society with its natural environment for the benefit of both" (Mitsch, 1996, 1998). This is a slight variation of the definition we gave in the late 1980s (Mitsch and Jørgensen, 1989) when we defined ecological engineering as "the design of human society with its natural environment for the benefit of both."

In a word, ecological engineering involves creating and restoring sustainable ecosystems that have value to both humans and nature. Ecological engineering combines basic and applied science for the restoration, design, and construction of aquatic and terrestrial ecosystems. The goals of ecological engineering are (1) the restoration of ecosystems that have been substantially disturbed by human activities such as environmental pollution or land disturbance, and (2) the development of new sustainable ecosystems that have both human and ecological value. This is engineering in the sense that it involves the design of this natural environment using quantitative approaches grounded in the basic science of ecology. It is a technology where the primary tools are self-designing ecosystems. It is biology and ecology in the sense that the components are all the biological species of the world. And as the goals suggest, it is often ecological restoration.

2.2 ECOSYSTEM RESTORATION

Restoration ecology was described as "full or partial placement of structural or functional characteristics that have been extinguished or diminished and the substitution of alternative qualities or characteristics than the ones originally present with the proviso that they have more social, economic, or ecological value than existed in the disturbed or displaced state" (Cairns, 1988a). A definition of *ecological restoration* established as part of a National Academy of Science study in the early 1990s was "the return of an ecosystem to a close approximation of its condition prior to disturbance" [National Research Council (NRC), 1992]. Several restoration fields have developed somewhat independently, and all appear to have the design of ecosystems as their theme. Although related to ecological engineering or even a part of it, several of these approaches seem to lack one of the two important cornerstones of ecological engineering: (1) recognizing the self-designing ability of ecosystems, or (2) basing the approaches on a theoretical base, not just empiricism.

The concepts of ecological engineering and ecosystem restoration are intertwined. Further, there are a host of words and terms that refer to parts of these fields. In some ways ecological engineering is not a new field but an amalgam of several fields, some almost homespun, that deal with restoration and creation of ecosystems. Some of the many synonyms and subsets of ecological engineering are illustrated in Table 2.1, a list updated from the original one that we developed a decade ago (Mitsch, 1993). Early work in Europe used the concept of *bioengineering,* or using plants as engineering materials. Much has been written about the fields of *restoration ecology* (Aber and Jordan, 1985; Jordan et al., 1987; Buckley, 1989) and *ecosystem rehabilitation* (Cairns, 1988b; Wali, 1992). An ecological approach has been applied to *river and stream restoration* (Gore, 1985), agriculture as *agroecosystems* (Lowrance et al., 1984), *riparian restoration,* and *wetland creation and restoration.*

Some of these terms, such as *bioengineering* (Schiechtl, 1980), *ecotechniques, biomanipulation* (related to fishery management for eutrophication control; Hosper and Jagtman, 1990), *ecotechnology, ecotech* (Straskraba, 1993; Moser, 1994, 1996), and even *nature engineering* (Aanen et al., 1991), convey a sense of ecological engineering. Restoration ecology and the restoration fields (terrestrial, aquatic, wetland, etc.) have many features in common with ecological engineering. In fact, Bradshaw (1997) called ecosystem

Table 2.1   Terms that are synonyms, subdisciplines, or fields similar to ecological engineering

• Synthetic ecology	• Biomanipulation
• Restoration ecology	• River and lake restoration
• Bioengineering	• Wetland restoration
• Sustainable agroecology	• Reclamation ecology
• Habitat reconstruction	• Nature engineering
• Ecohydrology	• Ecotechnology
• Ecosystem rehabilitation	• Engineering ecology
• Biospherics	• Solar aquatics

restoration "ecological engineering of the best kind" because we are putting back ecosystems that used to exist, not creating new combinations of populations or systems that are subjected to excessive stresses, such as pollution. Another term, *ecohydrology*, integrates the science of stream and river ecology with the hydrologic nature of the streams, rivers, and watersheds. It has been defined as the use of ecosystem properties as management tools for enhancing the absorptive capacity of aquatic ecosystems against human impact (Zalewski, 2000a).

2.3 HISTORY OF ECOLOGICAL ENGINEERING

The term *ecological engineering* was coined by Howard T. Odum in the 1960s and has since been used extensively in North America, Europe, and China. Odum first defined ecological engineering as "those cases in which the energy supplied by man is small relative to the natural sources, but sufficient to produce large effects in the resulting patterns and processes " (Odum, 1962) and "environmental manipulation by man using small amounts of supplementary energy to control systems in which the main energy drives are still coming from natural sources" (Odum et al., 1963). Odum (1971) elaborated on the breadth of ecological engineering in his book *Environment, Power, and Society* by stating that "the management of nature is ecological engineering, an endeavor with singular aspects supplementary to those of traditional engineering. A partnership with nature is a better term." He later stated in *Systems Ecology* (Odum, 1983): "The engineering of new ecosystem designs is a field that uses systems that are mainly self-organizing."

Concurrent but separate from development of ecological engineering concepts in the West was the development of the term *ecological engineering* in China (see Chapter 13). Under the leadership of Ma Shijun, known as "the father of ecological engineering in China," ecologists in China began using the term *ecological engineering* in the 1960s, with much of their work written in Chinese-language publications. In one of the first publications in Western literature, Ma (1985) described the application of ecological principles in the concept of ecological engineering in China. Much of the approach to environmental management in China started as an art, but in the past two decades there has been explicit use of the term *ecological engineering* in China, first to describe a formal "design with nature" philosophy for wastewater. Ma (1988) later defined ecological engineering as "a specially designed system of production process in which the principles of the species symbiosis and the cycling and regeneration of substances in an ecological system are applied with adopting the system engineering technology and introducing new technologies and excellent traditional production measures to make a multi-step use of substance." He suggested that ecological engineering was first proposed in China in 1978 and is now used throughout the entire country, with about 500 sites at the time of publication (1988) practicing *agroecological engi-*

neering, defined as an "application of ecological engineering in agriculture" (Ma, 1988). That figure was updated to about 2000 applications of ecological engineering in China by the early 1990s (Yan and Zhang, 1992; Yan et al., 1993). At a symposium on agroecological engineering in Beijing, Qi and Tian (1988) suggested that "the objective of ecological research [in China] is being transformed from systems analysis to system design and construction," stating that ecology now has a great knowledge base from observational and experimental ecology and is in the position to meet global environmental problems through ecosystem design, the main task of ecological engineering. Yan and Yao (1989) described integrated fish culture management as it is practiced in China as ecological engineering because of its attention to waste utilization and recycling.

Meanwhile, there was a similar development of the field of ecotechnology in central Europe in the mid-1980s. Uhlmann (1983), Straskraba (1984, 1985), and Straskraba and Gnauek (1985) defined *ecotechnology* as the "use of technological means for ecosystem management, based on deep ecological understanding, to minimize the costs of measures and their harm to the environment." Straskraba (1993) further elaborated on this point and called ecotechnology "the transfer of ecological principles into ecological management." In this book we consider ecological engineering and ecotechnology as generally synonymous but agree that the former term involves primarily the creation and restoration of ecosystems, whereas the latter term involves managing ecosystems. Which is the more encompassing term is difficult to say, but it could be, as Straskraba (1993) suggested, that "ecotechnology is in a sense broader [than ecological engineering] . . . being that environmental management [ecotechnology] is considered not only the creation and restoration of ecosystems."

The U.S. marine scientist John Todd also was a leader in applying both the term and the concepts of ecological engineering to wastewater treatment, first at his New Alchemy Institute and later at his Ocean Ark Center in New England. The term *ecological engineering* was applied to the treatment of wastewater and septage in ecologically based "green machines," with indoor greenhouse applications built both in Europe and North America in the 1980s and continuing through the present. Here the applications are described as "environmentally responsible technology [which] would provide little or no sludge, generate useful byproducts, use no hazardous chemicals in the process chain and remove synthetic chemicals from the wastewater" (Guterstam and Todd, 1990). All applications within this subset of ecological engineering have the commonality of using ecosystems for treatment of human wastes, with an emphasis on truly solving problems with an ecological system rather than simply shifting the problem to another medium. Following this wastewater treatment direction, one of the first meetings with the title *ecological engineering* was held in Trosa, Sweden, in March 1991; the papers at that meeting were published in two editions (Etnier and Guterstam, 1991, 1997).

Table 2.2   Workshops and subsequent special issue publications of a SCOPE[a]

Workshop Title	Workshop Location and Date	Special Issue Publication
Ecological engineering in central and eastern Europe: remediation of ecosystems damaged by environmental contamination	Tallin, Estonia November 6–8, 1995	Mitsch and Mander, 1997
Ecological engineering in developing countries	Beijing, China October 7–11, 1996	R. Wang et al., 1998
Ecological engineering applied to river and wetland restoration	Paris, France July 29–31, 1998	Lefeuvre et al., 2002
Ecology of postmining landscapes	Cottbus, Germany March 15–19, 1999	Hüttl and Bradshaw, 2001

[a]Scientific Committee on Problems of the Environment, project Ecological Engineering and Ecosystem Restoration.

Finally, our 1989 book entitled *Ecological Engineering: An Introduction to Ecotechnology* (Mitsch and Jørgensen, 1989) and subsequent initiation of the scientific journal *Ecological Engineering: The Journal of Ecotechnology* in 1992 have together brought ecological engineering principles and practice to a wide audience. A 1993 workshop in Washington, DC, sponsored by the U.S. Committee to SCOPE (Scientific Committee on Problems in the Environment), led to an international SCOPE project on Ecological Engineering and Ecosystem Restoration. The SCOPE committee developed several journal special issues based on workshops held around the world (Table 2.2). As a result of the great interest in ecological engineering that developed in the 1990s, the International Ecological Engineering Society (IEES) was established in Utrecht in the Netherlands, in 1993, and the American Ecological Engineering Society (AEES) had its inaugural meeting in Athens, Georgia, in 2001.

2.4   BASIC CONCEPTS IN ECOLOGICAL ENGINEERING

There are a few basic concepts that collectively distinguish ecological engineering from more conventional approaches to solving environmental problems through engineering approaches. These include the following concepts about ecological engineering:

1. It is based on the self-designing capacity of ecosystems.
2. It can be the acid test of ecological theories.

3. It relies on system approaches.
4. It conserves nonrenewable energy sources.
5. It supports ecosystem conservation.

These concepts are discussed in more detail below.

Self-Design

Self-design and the related concept of self-organization must be understood as important properties of ecosystems in the context of their creation and restoration. In fact, their application may be the most fundamental concept of ecological engineering. Self-organization is the property of systems in general to reorganize themselves given an environment that is inherently unstable and nonhomogeneous. Although somewhat vague and possibly nonapplicable at the species level (where the major self-organization may be evolution itself), self-organization is a systems property that applies very well to ecosystems where species are continually introduced and deleted, species interactions (e.g., predation, mutualism) change in dominance, and the environment itself changes. All of these activities go on at one degree or another all the time. In a sense, the organization is not derived from some outside force but from within the system itself. Self-organization manifests itself in microcosms and newly created ecosystems "showing that after the first period of competitive colonization, the species prevailing are those that reinforce other species through nutrient cycles, aids to reproduction, control of spatial diversity, population regulation, and other means" (Odum, 1989a). Ecological engineering often involves the development of new ecosystems as well as the use of pilot-scale models such as mesocosms to test ecosystem behavior; the self-organizing capacity of ecosystems remains both an enigma to ecologists and an important concept for ecological engineering.

All systems have some level of organization, but Pahl-Wostl (1995) argues that there are two types of ways that systems can be organized: by rigid top-down control or external influence (imposed organization), or by self-organization. These two types of organization are contrasted in Table 2.3. Imposed organization, such as that implemented in many conventional engineering approaches, results in rigid structures and little potential for adapting to change. This is, of course, desirable for engineering design, where predictability of safe and reliable structures are necessary, such as for bridges, furnaces, and sulfur scrubbers. Self-organization, on the other hand, develops flexible networks with a much higher potential for adaptation to new situations. It is thus the latter property that is desirable for solving many of our ecological problems. When biological systems are involved, the ability of the ecosystems to change, adapt, and grow according to their forcing functions and internal feedbacks is most important.

We define *self-design* as "the application of self-organization in the design of ecosystems." The presence and survival of species in ecosystems after their

Table 2.3 Systems categorized by type of organization

Characteristic	Imposed Organization	Self-Organization
Control	Externally imposed, centralized control	Endogenously imposed, distributed control
Rigidity	Rigid networks	Flexible networks
Potential for adaptation	Little potential	High potential
Application	Conventional engineering	Ecological engineering
Examples	Machine	Organism
	Fascist or socialist society	Democratic society
	Agriculture	Natural ecosystem

Source: After Pahl-Wostl (1995).

introduction by nature or humans is more up to nature than to humans (Figure 2.1). "Many are called but few are chosen" is the essence of the functional development of any ecosystem (see also Chapter 4). Self-design is an ecosystem function in which the chance introduction of species is analogous to the chance development of mutations necessary for evolution to proceed (Mitsch, 1998). Multiple seeding of species into ecologically engineered systems is a means of speeding the selection process in this self-organization or self-design (Odum, 1989b). In the context of ecosystem development, self-design means that if an ecosystem is open to allow "seeding" of enough species and their propagules through human or natural means, the system itself will optimize its design by selecting for the assemblage of plants, microbes, and animals that is best adapted for existing conditions. The ecosystem then "designs a mix of man-made and ecological components [in a] pattern that maximizes performance, because it reinforces the strongest of alternative pathways that are provided by the variety of species and human initiatives" (Odum, 1989a).

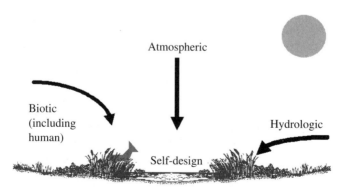

Figure 2.1 Species introduction to an ecosystem enhances the ability of that ecosystem to self-design its assemblage of plants, animals, and microbes. In the wetland example shown here, species propagule introduction occurs through the atmosphere, by animal movement and activity (including human intervention), and through the inflow of surface water.

Self-design in ecological engineering provides a new approach to the fundamental thing that conventional engineers do—design. In a typical engineering project, the engineer tries to anticipate everything that might happen and provides a predictable and reliable system that will perform its desired function for the given lifetime of the system, be it an airplane, a building, a microchip, or a dam on a river (Figure 2.2*a*). Biology is usually not involved, and thus rarely is an ecosystem involved. Rigidity and predictable structures that have safety factors designed into them result. The ecological engineer, on the other hand, relies as much or more on nature's ability to self-design a self-sustaining ecosystem than on the ecological engineer's ability to know perfectly what species to add at what time (Figure 2.2*b*). Nature contributes to the final ecosystem design as much or more than the engineer does. Nature is the chief contractor: human engineers provide the initial conditions and possibly initial propagules that collectively enhance an ecosystem's ability to self-design.

Ecological engineering depends on the self-designing capability of ecosystems and nature. When changes occur, natural systems shift, species are substituted for each other, and food chains reorganize. As individual species sort, with some selected and others not, a new system ultimately emerges that is much better suited to the environment that is superimposed on it. Humans participate in self-design by providing choices of initial species, matching species with the environment. Nature does the rest. For example, in designing a wetland, we may want to introduce dozens of different plants at different water depths because of our inability to predict exactly where certain plants will survive and even whether they will survive at all. Nature then takes over and chooses the plants that meet certain water depths, soil conditions, and grazing pressures.

In contrast to the self-design approach, the approach that is more commonly used today by many biologists for much of what we call ecological restoration may be even closer to conventional engineering than is ecological engineering. Plants are planted with the expectation that they will survive. Forestry and horticulture could easily have been called "botanical engineering" or "tree engineering" if engineers had been around before the Industrial Revolution. Our ecosystems are created, restored, or managed to provide habitat for either a specific animal species (e.g., an endangered species) or a group of species (e.g., waterfowl). This could easily be called "zoological engineering." In these cases, the introduction of specific organisms is the goal, and the survival of these organisms becomes the measure of success of a project. Systems are being designed in both cases, perhaps without the precision of systems that engineers design with physics and chemistry as their main sciences. Biology adds to the variability of the systems, but otherwise, design and predictable structures are carried out and desired. This forceful "design" of ecosystems has sometimes been called the *designer approach* and the ecosystems produced are termed *designer ecosystems* (van der Valk, 1998). There is more honesty in these terms than many are willing to ac-

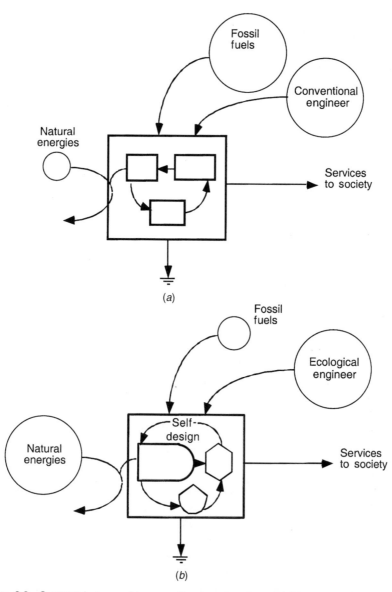

Figure 2.2 Contrast between (a) conventional engineering and (b) ecological engineering. Both provide services to society. Ecological engineering depends less on fossil fuel–based energy and more on the self-design capacity of nature. Conventional engineering generally provides more rigid structures and depends more on the products of a fossil fuel–based economy. (From Mitsch, 1998; copyright 1998; reprinted with permission from Elsevier Science.)

knowledge. The designer approach, while understandable because of the natural human tendency to want to control events, is less sustainable than an approach that relies more on nature's capacity to self-design. Pahl-Wostl (1995) would call it more ecofascism than ecodemocracy.

As we described in an earlier book (Mitsch and Jørgensen, 1989): "Ecological engineering is engineering in the sense that it involves the design of this natural environment using quantitative approaches and basing our approaches on basic science. It is technology with the primary tool being self-designing ecosystems. The components are all of the biological species of the world."

Ecological engineering involves an acceptance of the concept of the self-designing capability of ecosystems and nature. We may even go so far as to say that the concept of a polluted ecosystem is an anthropogenic view, one that may not recognize the beauty of natural systems shifting, substituting species, reorganizing food chains, adapting as individual species, and ultimately designing a system that is ideally suited to the environment that is superimposed on it. The ecosystem also performs another wonderful function—it begins to manipulate its physical and chemical environment to make it a little more palatable! It is this self-designing capability of ecosystems that ecological engineering recognizes as a significant feature, because it allows nature to do some of the "engineering." We participate as the choice generator and as a facilitator of matching environments with ecosystems, but nature does the rest. Thus, ecological engineering is neither a license to pollute nor an open door to invasive species, but human society *and* ecosystems are viewed as vitally connected.

The Acid Test

Restoration ecologists have long suggested the tie between basic research and ecosystem restoration in making an analogy that the best way to understand a system, whether a car, a watch, or an ecosystem, is to "attempt to reassemble it, to repair it, and to adjust it so that it works properly" (Jordan et al., 1987). Ecological engineering will be the ultimate test of many of our ecological theories. Bradshaw (1987), who has described the restoration of a disturbed ecosystem as the "acid test of our understanding of that system," has stated that because we cannot prove that a restored ecosystem proves an ecological theory, we will "learn more from our failures than from our successes since a failure clearly reveals the inadequacies in an idea, while a success can only corroborate and support, and can never absolutely confirm, an assertion." Cairns (1988a) has been more direct: "One of the most compelling reasons for the failure of theoretical ecologists to spend more time on restoration ecology is the exposure of serious weaknesses in many of the widely accepted theories and concepts of ecology."

Ecological theories that have been put forward in the scholarly literature over the past 100 years should serve as the basis of the language and the practice of ecological engineering and ecosystem restoration. But just as there is the possibility of these theories providing the basis for engineering design of ecosystems, there is also a possibility of finding that some of these ecological theories are wrong. Thus, ecological engineering is really a technique for fundamental ecological research and advancement of the field of ecology.

A Systems Approach

Pahl-Wostl (1995) argues that just as self-organization is a property of a system as a whole, it is meaningless at the level of the parts. Ecological engineering requires a more holistic viewpoint than we are used to bringing to many ecosystem management strategies. Ecological engineering emphasizes, as does ecological modeling for systems ecologists, the need to consider the entire ecosystem, not just individual species. Restoration ecology, the subfield of ecological engineering, has been described as a field in which "the investigator is forced to study the entire system rather than components of the system in isolation from each other" (Cairns, 1988a).

As a result, the practice of ecological engineering cannot be supported completely by reductive, analytic experimental testing and relating. Approaches such as modeling and whole-ecosystem experimentation are more important, as ecosystem design and prognosis cannot be predicted by summing the parts to make a whole. One must also be able to synthesize a great number of disciplines to understand and deal with the design of ecosystems.

All applications of technologies, whether of biotechnology, chemical technology, or ecotechnology, require quantification. Because ecosystems are complex systems, the quantification of their reactions becomes complex. Systems tools such as ecological modeling represent well-developed approaches for analyzing ecosystems, their reactions, and the linkage of their components. Ecological modeling is able to synthesize the various pieces of ecological knowledge, which must be put together to solve a certain environmental problem as it takes a holistic view of environmental systems. Optimization of subsystems does not necessarily lead to an optimal solution of the entire system. There are many examples in environmental management where optimal management of one or two aspects of a resource separately does not optimize management of the resource as a whole. Although they usually have one or more specific goals, ecological engineering projects should try to balance between the good of human beings and the good of nature.

Nonrenewable Resource Conservation

Because most ecosystems are primarily solar-based systems, they are self-sustaining. Once an ecosystem is constructed, it should be able to sustain itself indefinitely through self-design with only a modest amount of intervention. This means that the ecosystem, running on solar energy or the products of solar energy, should not need to depend on technological fossil energies as much as it would if a traditional technological solution to the same problem were implemented. If the system does not sustain itself, it does not mean that the ecosystem has failed us (its behavior is ultimately predictable). It means that the ecological engineering has not facilitated the proper interface between nature and the environment. Figure 2.2 illustrates the difference in reliance on renewable and nonrenewable energies in conventional engineering and ecological engineering. For the most part, modern technology and environmental technology are based on an economy supported by nonrenewable (fossil fuel) energy; ecotechnology is based on the use of some nonrenewable energy expenditure at the start (the design and construction work by the ecological engineer), but subsequently on dependence on solar energy.

A corollary to the fact that ecological engineering systems use less nonrenewable energy is that they generally cost less than conventional means of solving pollution and resource problems, particularly in systems maintenance and sustainability. One interesting point is that because of the reliance on solar-driven ecosystems, a larger part of land or water is needed than would be with technological solutions. Therefore, if property purchase (which is, in a way, the purchase of solar energy) is involved in regions where land prices are high, ecological engineering approaches may not be feasible. It is in the daily and annual operating expenses in which the work of nature provides subsidies and thus lower costs for ecological engineering alternatives.

Ecosystem Conservation

Nature provides many valuable functions for humans. Of course, many of these values come from the harvest of plants and animals for food and fiber. The identification of these as well as many nonmarket values of nature, as illustrated by Costanza et al. (1997), for example, has led to an increased emphasis on conservation by illustrating that nature has value. The development of new sustainable ecosystems and their values by ecological engineering will have the same effect. Ecological engineering involves identifying those biological systems that are most adaptable to the human needs and those human needs that are most adaptable to existing ecosystems.

Ecological engineers have in their toolboxes all the ecosystems, communities, populations, and organisms that the world has to offer. Therefore, a direct consequence of ecological engineering is that it would be counterproductive to eliminate or even disturb natural ecosystems unless absolutely necessary. This is analogous to the conservation ethic that is shared by many farmers even though they may till the landscape. Therefore, ecological engineering may lead to a greater environmental conservation ethic than has been realized up to now. For example, when wetlands began to be recognized for their ecosystem values of flood control and water quality enhancement, wetland protection efforts gained a much wider degree of acceptance and even enthusiasm than they had before, despite their long understood values as habitat for fish and wildlife (Mitsch and Gosselink, 2000). In short, recognition of ecosystem values provides greater justification for the conservation of ecosystems and their species.

Aldo Leopold, the great midwestern U.S. conservationist, stated this concept much more eloquently. As compiled by Aldo's son, Luna Leopold, after his father's death (Leopold, 1972): "The last word in ignorance is the man who says of an animal or plant: 'What good is it?' If the land mechanism as a whole is good, then every part is good, whether we understand it or not. If the biota, in the course of eons, has built something we like but do not understand, then who but a fool would discard seemingly useless parts? *To keep every cog and wheel is the first precaution of intelligent tinkering.*" The ecological engineer is nature's tinker.

2.5 CONTRASTS WITH OTHER FIELDS

Environmental Engineering

It may also be useful in discussing ecological engineering and ecotechnology to define what they are not. Ecological engineering is not the same as *environmental engineering*, a field that has been well established in universities and the workplace since the 1960s, and that was around for decades before under the name of *sanitary engineering*. The environmental engineer is certainly involved in the application of scientific principles to clean up or prevent pollution problems, and the field is a well-honored one. Environmental engineers are taught the basics of *unit processes* and how to apply these in valuable environmental technologies (called *unit operations*), such as sedimentation basins, scrubbers, sand filters, and flocculation tanks.

Ecological engineering, by contrast, is involved in identifying those ecosystems that are most adaptable to human needs and in restoring, creating, or redesigning such systems to meet human needs. As do other forms of engineering and technology, ecological engineering uses the basic principles of science (in this case it is mainly the multifaceted science of ecology) to design better living conditions for human society. However, unlike other forms of engineering and technology, ecological engineering has as its raison d'être the design of ecosystems for the benefit of humanity and nature. Ecological engineering has in its toolbox all the ecosystems, communities, and organisms that the world has to offer.

The focus on, and utilization of, biological species, communities, and ecosystems with a reliance on self-design is the feature that most distinguishes ecological engineering from the traditional environmental technologies, which rely on devices and facilities to remove, transform, or contain pollutants, but which do not consider direct manipulation of ecosystems.

Biotechnology

Ecological engineering should not be confused with the fields of *biotechnology* and *bioengineering* that involve genetic manipulation to produce new strains and organisms to carry out specific functions. In this chapter, *biotechnology* is defined in the narrow sense as the development of new species and varieties through alteration of genetic structure. We recognize that the term *biotechnology* is used in a broad sense also to include manipulation of biological systems, without genetic changes. We include the latter changes at the species level and higher as ecological engineering or ecotechnology.

Some of the contrasts between ecotechnology and biotechnology related to basic principles, control, design, and costs to society are shown in Table 2.4. By its very design, biotechnology involves the manipulation of the genetic structure of the cell to produce new strains and organisms capable of carrying out certain functions. Ecotechnology does not manipulate at the genetic level but considers an assemblage of species and their abiotic environment as a self-designing system that can adapt to changes brought about by outside forces—by the changes introduced by humans or the natural forcing functions. Biotechnology began with much fanfare but now is beginning to realize the enormous costs and concerns arising from manipulation at this micro level. In contrast, ecotechnology introduces no new species that nature has not dealt with before, nor does it involve major laboratory development costs except for microcosm studies of new combinations of organisms.

Ecology

Ecological engineering is a new field with its roots in the science of ecology. It can be viewed as the design, restoration, and creation ecosystems according to ecological principles that we have learned over the past century (Figure 2.3). Ecology, often described as a subdiscipline within the biological sciences, has a strong history of development over the past century, dating back to the coining of the term *ecology* by the German biologist Ernst Hackle in the mid-nineteenth century. The principles of the field have been developed by scientists such as Cowles, Shelford, Clements, Gleason, Latke, Elton, Heinemann, Fore, Lineman, Likens, Hutchinson, the Odum brothers, and oth-

Table 2.4 Comparison of ecotechnology and biotechnology

Characteristic	Ecotechnology	Biotechnology
Basic unit	Ecosystem	Cell
Basic principles	Ecology	Genetics, cell biology
Control	Forcing functions, organisms	Genetic structure
Design	Self-design with some human help	Human design
Biotic diversity	Protected	Changed
Maintenance and development costs	Reasonable	Enormous
Energy basis	Solar based	Fossil fuel–based

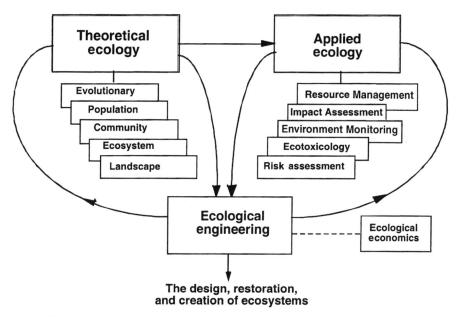

Figure 2.3 Relationships among theoretical ecology, applied ecology, and ecological engineering in the general field of ecology. Ecological economics is shown as the field that can best be used as the approach for reviewing the costs and benefits of ecological engineering projects. (Adapted from Mitsch, 1993.)

ers into a well-defined science that is actually much broader than a subfield of biology. As with any science, discussion abounds as to which theories are correct, particularly with concepts such as succession and energy strategies, but a strong science has developed at the population, community, and ecosystem levels.

Applied ecology, as an extension of these ecological theories, has become popular since the 1960s, particularly in light of our concern for environmental matters. But it has usually been limited to monitoring and assessing environmental impacts or managing natural resources (i.e., it has principally remained *descriptive* in nature). Good examples of recent applied fields in ecology are ecotoxicology and landscape ecology, both of which are descriptive of humanity's effects on the environment.

Both basic and applied ecology provide fundamental concepts for ecological engineering (Figure 2.3) but do not define it completely. Ecological engineering should have its roots in the science of ecology, just as chemical engineering is close to chemistry and biochemical engineering is close to biochemistry. Logically, it should be considered both a branch of ecology and a new field of engineering.

Also shown in Figure 2.3 in the feedbacks from ecological engineering to both theoretical and applied ecology is the fact that there is a high probability of advancing the understanding of ecological systems in ecological engineering. The success or failure of ecologically engineered ecosystems may cause us to reconsider whether time-honored ecological principles are correct and may also show us that some ecosystem management approaches or risk assessment methods need to be reassessed. Furthermore, ecologically engineered systems give us another data point for establishing reliable design standards. Created and restored ecosystems provide unique research opportunities to scientists that natural ecosystems do not provide, particularly in subjects such as self-organization and ecosystem development.

Figure 2.3 also shows that the newly developing metrics of ecosystem value—generally discussed in the new field of *ecological economics*—are the means for us to keep score on whether these projects make both ecological and economic sense. Neither simple cost–benefit approaches nor cost comparisons illustrate the benefits of nature sufficiently.

The integration of the science of ecology into the field of engineering into a formal application is progressing perhaps more rapidly in Europe and China than in the United States; yet the recognition that a prescriptive field such as ecological engineering may be needed for ecology has been slow to take hold. Most ecologists still remain low on the learning curve compared to engineers when confronted after their formal education with real-life issues of wetland construction, river restoration, habitat reconstruction, or mine land rehabilitation. Although they may rise to the occasion quickly with homespun ecotechnology and novel approaches, the integration is not there, the theory does not support the technique, and techniques are relearned each time. Ecological theory may be mentioned, but it has usually not been integrated into a framework within ecology that ecological engineering could provide.

Ecotechniques/Cleaner Technology

There is a great deal of interest in developing green technology in buildings and human structures (Johansson, 1992). Sometimes these approaches are referred to as *ecotechniques* (Thofelt and Englund, 1996). In early stages, there was discussion as to whether these approaches, which range from solar water heaters on rooftops to composting vegetable and animals wastes, are by themselves ecological engineering. What sets ecological engineering apart from most ecotechniques is the use or design of ecosystems in the former. Thus, solar heaters and the splitting of gray water from human wastewater do not exactly fit under ecological engineering; composting systems may be just on the border between the two fields.

Industrial Ecology

Industrial ecology is a term that has been used since the early 1990s to describe a general approach that attempts to reduce or even eliminate the release of chemicals from industry through techniques such as recycling and material replacement. It is described as a way of finding innovative solutions to complicated environmental problems, particularly in an industrial setting. The field has now been described in several texts (e.g., Graedel and Allenby, 1995), has its own scholarly *Journal of Industrial Ecology,* and is supported by the International Society of Industrial Ecology. Examples of industrial ecology studies are examining the flow of a certain material or element through an industrial system, material and energy flows studies ("industrial metabolism"), extended producer responsibility ("product stewardship"); eco-industrial parks ("industrial symbiosis"); and ecoefficiency. Other examples of technology-oriented approaches are *life-cycle assessments,* whereby any product is managed through its life cycle and all material and energy flows from raw material extraction to product completion and disposal are considered. It is our opinion that ecological engineering and industrial ecology are not the same fields but are certainly compatible fields designed to solve some of our most difficult environmental problems.

3

CLASSIFICATION OF ECOLOGICAL ENGINEERING

Part II of this book is devoted to specific applications of ecological engineering and ecosystem restoration. To simplify the decision as to whether or not an approach fits in these fields, we attempt here to illustrate both the wide breadth of fields now existing that could be assumed to be under the umbrella of ecological engineering/ecosystem restoration and a practical, yet simple classification system of applications. It may be a first attempt, but it may prove to be useful as we try to define the breadth of this new field. The two important messages from Chapter 2 are:

1. The practices need to involve the creation and restoration of ecosystems, not simply the enhancement of one or two ecological or biological processes (otherwise, yogurt making would be ecological engineering).
2. There have to be benefits to both humans and to nature.

The classifications we propose are comprehensive without being complete, because it is not our goal to present all methods. Rather, we propose to give readers a good understanding of how, where, and when ecotechnology can be applied and to illustrate the approaches and their ecological soundness.

3.1 SPECTRUM OF ECOLOGICAL ENGINEERING

One way to view ecological engineering and ecosystem restoration, since it covers such a wide varieties of fields, is to consider the various fields on a spectrum of how much "conventional" engineering and hence reliance on human structures and our fossil-fuel economy are involved (Figure 3.1). On

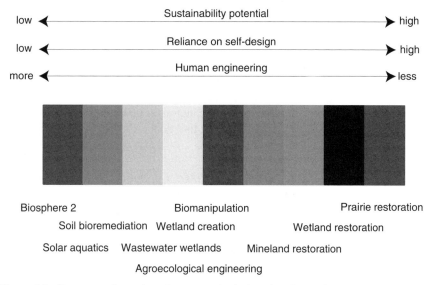

Figure 3.1 Spectrum of practices that are ecological engineering and ecosystem restoration. (From Mitsch, 1998; copyright 1998; reprinted with permission from Elsevier Science.)

the right side of this spectrum are ecological approaches such as prairie restoration or wetland restoration, where quite often only a small effort of conventional engineering is needed. For example, a prairie can be restored in theory with only a match to start a fire. In some cases, wetlands (at least the hydrology) can almost be restored in a few hours or days if a drainage tile is broken or blocked, allowing the return of the original hydrology. In the middle of the spectrum are approaches such as biomanipulation or wetland creation, where clearly more management and engineering are required. This is probably also true for wetlands built for wastewater or nonpoint-source pollution control or for developing multispecies agroecosystems. At the far left of our theoretical spectrum are examples that involve significant amounts of energy. The solar aquatic systems as developed by John Todd (Todd and Josephson, 1996) and the dynamic aquaria and coral reefs of Walter Adey and colleagues at the Smithsonian Institution (Adey and Loveland, 1991; Luckett et al., 1996) are examples of systems where structures such as greenhouses and even pumps are needed. An intensely engineered system such as Biosphere 2 (see Marino and Odum, 1999) is ecological engineering in the sense that ecosystems are being created, albeit enormously subsidized. Are there other types of ecological engineering projects that fit within this spectrum? When do systems become so subsidized that they cease to be ecological engineering? Although the spectrum given in Figure 3.1 shows a general sustainability potential increasing from left to right (remember the word *sustainable* is defined in Chapter 2), it is unclear "how" sustainable a system needs to be to be considered an example of ecological engineering. It is generally understood that in true restoration, the sustainability needs to be complete.

3.2 CLASSIFICATION ACCORDING TO FUNCTION

We suggest that there are five general categories of ecological engineering and ecosystem restoration according to the general type of application or function:

1. Ecosystems are used to reduce or solve a pollution problem that would otherwise be harmful to other ecosystems.
2. Ecosystems are imitated or "copied" to reduce or solve a resource problem.
3. The recovery of ecosystems is supported after significant disturbances.
4. Existing ecosystems are modified in an ecologically sound way to solve an environmental problem.
5. Ecosystems are used for the benefit of humankind without destroying the ecological balance.

Examples of ecological engineering of each of these approaches are presented in Table 3.1. The diversity of approaches, for both aquatic and terrestrial ecosystems, shows the breadth of ecological engineering. It should also

Table 3.1   Examples of ecological engineering approaches for terrestrial and aquatic systems according to types of applications

Ecological Engineering Approaches	Terrestrial Examples	Aquatic Examples
1. Ecosystems are used to solve a pollution problem.	Phytoremediation	Wastewater wetland
2. Ecosystems are imitated or copied to reduce or solve a resource problem.	Forest restoration	Replacement wetlands
3. The recovery of an ecosystem is supported after disturbance.	Mine land restoration	Lake restoration
4. Existing ecosystems are modified in an ecologically sound way.	Selective timber harvest	Biomanipulation
5. Ecosystems are used for benefit without destroying ecological balance.	Sustainable agro-ecosystems	Multispecies aquaculture

be noted that ecological engineering encompasses parts of some applied fields, such as forestry and fisheries, when a systems approach is taken toward these activities. Applications range from constructing new ecosystems for solving environmental problems to ecologically sound harvesting of existing ecosystems. More information on some of these examples is presented below, although more detailed approaches are presented in Part II.

Solving or Reducing a Pollution Problem

Figure 3.2 illustrates the difference between traditional environmental technology and ecotechnology for the control of eutrophication. Eutrophication of a lake is considered to be a pollution problem, and environmental technology can be used to reduce the input of nutrients (e.g., phosphorus, by chemical precipitation, or nitrogen, by ion exchange or denitrification). However, the reduction in nutrient input from wastewater is often not sufficient to control eutrophication due to the contributions from nonpoint sources. Furthermore, the retention time of a lake is long, so more rapid improvements would be advantageous. Here ecotechnology comes into the picture. A littoral wetland is restored to trap the nutrients in the inflow, and nutrient-rich hypolimnetic water is siphoned off downstream of the lake. The wetland then also provides a different habitat for wildlife. This example shows that environmental technology alone is not sufficient and that ecotechnological methods can be used to supplement them.

Figure 3.3 illustrates even more clearly the difference between the environmental technological and ecotechnological approaches. In an environmental technology approach, sludge from a wastewater treatment plant is incinerated, causing air pollution problems, and the slags and ash still have

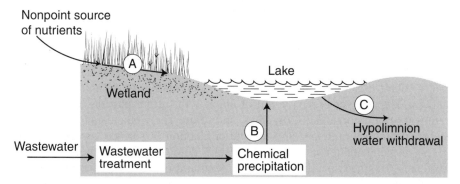

Figure 3.2 Control of lake eutrophication with a combination of a wetland to remove phosphorus from nonpoint sources of nutrients (pathway A, ecological engineering), chemical precipitation for phosphorus removal from wastewater (pathway B, environmental technology), and a siphoning of hypolimnetic, nutrient-rich water (pathway C, an ecotechnique).

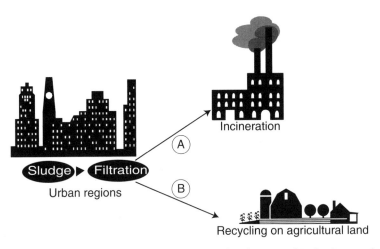

Figure 3.3 Sludge can be treated through the use of environmental technology such as incineration (pathway A) or ecotechnology by applying the material to farmland and growing crops (pathway B).

to be deposited. The ecotechnological solution, on the other hand, recognizes the sludge as a resource of nutrients and organic matter, and deposits it on agricultural land, where the resources can be utilized. This practice has frequently been thwarted because the sludge might contain metals that appropriately discourage its use. But if there are few industrial sources feeding the wastewater treatment plant, this ecotechnological solution makes ecological sense.

Figure 3.4 gives a third example, concerned with nitrate removal from drinking water or wastewater. Environmental technology would use either ion exchange or denitrification. The first method has high operating costs and produces a regeneration solution, which must be deposited. The second method is expensive in installation (4 to 8 hours of retention time is required), and the water requires disinfection. The ecotechnological method uses a treatment wetland. To optimize the denitrification potential, cellulose can be placed in a layer about 1 m under the wetland surface. This causes a high rate of denitrification, owing to the high accumulation of organic matter. Finally, a deeper layer of sandy soil of about 1 m further purifies the water by removal of microorganisms, organics, and so on. The wetland can be designed to be a self-sustaining system with the production of surface vegetation supplying organic matter necessary to sustain denitrification in the long term. Wetlands designed specifically for the treatment of wastewater are described in Chapter 10.

Imitating or Copying Ecosystems

In many cases we are interested in imitating an existing ecosystem by creating or restoring one that mimics a natural ecosystem to solve a pollution or re-

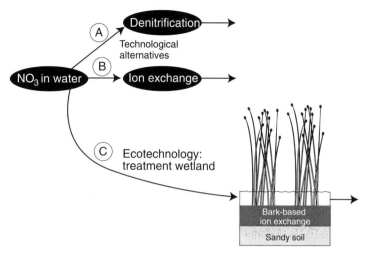

Figure 3.4 Nitrates can be removed from drinking water by environmental technologies such as denitrification and ion exchange (pathways A and B) and ecological engineering such as a treatment wetlands (pathway C).

sources problem. For example, artificial fishponds can be designed to imitate natural aquatic ecosystems with resulting optimization of food production and pollution minimization. Aquatic ecosystems are complex systems with many pathways of nutrient recycling and food transfer, whereas modern aquaculture systems in the West are typified as having a simple system of one fish being fed by fish food and some algal production (Figure 3.5a). The result is some production and a polluted water column. Furthermore, electricity-based fountains are often used to enhance oxygen transfer into low-oxygen water columns to allow even more fish production per unit area. In integrated fish production, the cycling of mass and energy is imitated (Figure 3.5b). Herbivorous fish are used to take care of excess algal production, and detritus feeders utilize the detritus and serve as feed for other fish. Pollution is thereby eliminated, or at least almost so, and two species of fish are harvested from the system. A model that simulated the multispecies aspects of a Chinese fishpond and illustrated that taking just one component out made the system suboptimal is described in Chapter 13.

Creating and restoring wetlands is another example of imitating or copying ecosystems to compensate for a resource loss. Wetlands are often restored or created in the United States as "mitigation" for the destruction of wetlands that occur during development projects such as shopping malls or highways. It is a controversial practice, as many believe that it is a license to destroy wetlands and will result in less functional wetlands (National Resource Council, 2001). Yet if a wetland can be designed and created to fit the existing hydrology, a successful compensation can be made (Figure 3.6) and the economic development project in question can proceed. As shown in Figure 3.6, it is important to distinguish between legal success and ecological suc-

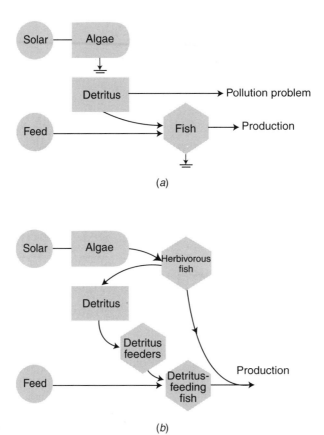

Figure 3.5 Comparison of (a) conventional aquaculture and (b) ecologically engineered integrated fishponds that use more species that better imitate natural ecosystems.

cess. Sometimes, too much emphasis is given to the former. The creation and restoration of wetlands for habitat development is described in more detail in Chapter 8.

Supporting Ecosystem Recovery

Examples of support for ecosystem recovery are coal mine and disturbed land recovery and the restoration of lakes, estuaries, and rivers. In all cases, a damaged part of the landscape is accelerated to its recovery. Techniques to accelerate the restoration of aquatic systems are described in Chapters 6 to 9. One of the issues of this general classification is the time that is required for that restoration to take place. The pattern of ecosystem recovery, sometimes referred to as its *ecological trajectory,* has been attempted in a number of cases, but exactly what is being recovered or restored needs to be identified first.

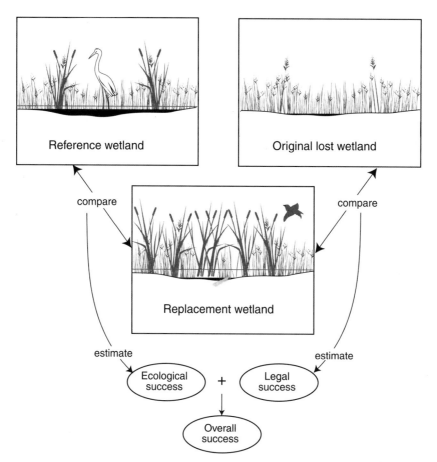

Figure 3.6   Wetlands are often created to copy or replace wetlands lost elsewhere. These are sometimes referred to as *mitigation wetlands* or *replacement wetlands*. Overall, the success can be legal where the wetland does what the permits require. Better is ecological success where the wetland does what natural wetlands in similar ecological settings would do. (From Mitsch and Gosselink, 2000; copyright 2000; reprinted with permission from John Wiley & Sons, Inc.)

Some examples of predicted trajectory of ecosystem development or recovery are illustrated in Figure 3.7. In Figure 3.7a, possible patterns of wetland successional development with different initial conditions (e.g., planting versus not planting) are illustrated. A wetland could be "planted" or otherwise influenced to develop adequate ecosystem structure to pass an ecological success milestone in five years' time. In essence, the results are obtained by influencing the initial conditions significantly. This pattern of initial conditions affecting the initial trajectory of a system is known by anyone who is familiar with dynamic ecological modeling. The transient response due to the initial conditions can have a major effect at the beginning of a model simulation.

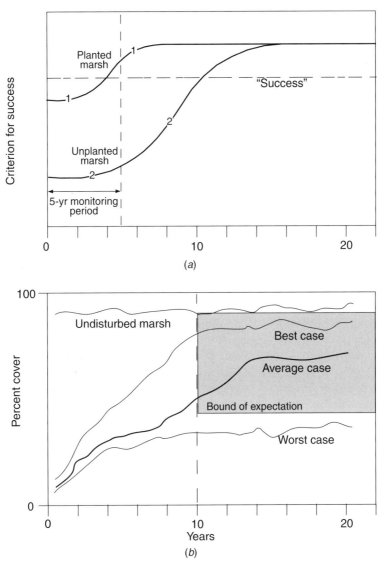

Figure 3.7 Patterns of projected ecosystem trajectory subsequent to wetland creation/restoration. (a) Suggested pattern of a criterion of ecosystem function with time for (1) a planted marsh and (2) an unplanted marsh. Planted marsh is shown to be "successful" after a five-year monitoring period, whereas unplanted marsh is not. By the fifteenth year, both marshes are identical, as planting was effective only on early ecosystem function, and ultimately both marshes converged. (From Mitsch and Wilson, 1996.) (b) Hypothetical curves for salt marsh vegetation cover for 20 years after restoration, with comparison to a natural marsh and the suggestion of expected bounds. (From Weinstein et al., 1997.)

What is also hypothesized from the graph is that, ultimately, the two systems with different initial conditions end up at the same general level of ecosystem "success." It just takes the system with lower initial development longer to get there. In the United States, the initial condition of created and restored wetlands is often either the seed bank (restored wetland) or planting (created or restored wetland). If such a seed bank exists or planting takes place, it is probable that the system will develop criteria quicker that are viewed as successful. In the long term, the effects of these introductions are not as well known, although the long-term experiments with freshwater marshes described in Chapter 10 have been enlightening.

In Figure 3.7*b*, a salt marsh restoration was projected with a specific number of years given for "success" because of legal obligations. This coastal restoration project, described in Chapter 9, required that sites have more than 40 percent desirable vegetation after 10 years. Three cases are suggested, including best, average, and worst cases. In all but the worst cases, the percent cover by desirable vegetation meets the "bounds of expectation." The patterns of restoration are also compared to vegetation cover in undisturbed marshes, with the expectation that the restored marsh will eventually reach that level.

Modifying Existing Ecosystems in an Ecologically Sound Way

Biomanipulation of lakes and reservoirs (see Hosper and Jagtman, 1990) is an accepted way to manipulate water quality. When predatory fish such as pike (*Esox lucius*) are added to bodies of water, zooplanktivorous fish are consumed. This, in turn, leads to an abundance of zooplankton that, in turn, removes greater amounts of algae, thus helping to cure the turbid symptoms of eutrophication. It has been argued that even with the removal of nutrients, there are alternative steady states to the tropic status of lakes that can be influenced by the design of the aquatic food chain (Figure 3.8). Hosper and Meijer (1993) describe biomanipulation of shallow lakes in the Netherlands as the "drastic reduction of the fish-stock in order to trigger a shift from a stable, turbid and algae-dominated system to a stable, clear, and macrophyte-dominated system." In the process of biomanipulation, predatory fish are often added to get the lake recovery started and to reduce planktivorous and benthivorous fish. A 75 percent reduction is considered minimum to see improvement in water quality. Since the planktivorous fish consume zooplankton and the benthivorous fish cause sediment resuspension, their removal by predation and netting will lead to improved water quality. The result is a clear-water, middle-nutrient steady-state condition in the lake, analogous to the left sides of the middle graphics in Figure 3.8.

Using Ecosystems without Destroying the Ecological Balance

Utilization of the ecosystem on an ecologically sound basis is the most general category. Typical examples are the use of agroecosystems and a sound

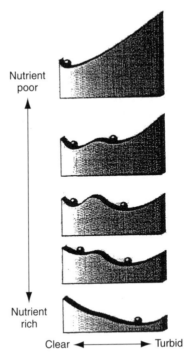

Figure 3.8 Steady states of lakes that could occur, depending on biomanipulation. Nutrient-poor (oligotrophic) and nutrient-rich (highly eutrophic) lakes have only one steady state each—clear and turbid conditions, respectively. Alternative steady-state conditions can exist for intermediate conditions of eutrophication if biomanipulation is practiced through the introduction of predatory fish and conditions can shift to less-turbid conditions. (From Hosper and Meijer, 1993; copyright 1993; reprinted with permission from Elsevier Science.)

ecological basis for the harvest of renewable resources (fish, timber, etc.). It is sometimes called *sustainable resource management* and is now discussed and practiced in forestry and fisheries management in some cases. Often, "yields" in the old sense of tons of fish or board-feet of wood are lower than in intensive management schemes practiced on newly found resources, but the concept is that the resource will recover on its own and be available for sound harvesting on another day or year.

3.3 CLASSIFICATION ACCORDING TO SCALE

Early development of ecological engineering in the West has stressed a partnership with nature and has been investigated primarily in experimental ecosystems rather than in full-scale applications. Some of the more significant experiments that have been conducted or are currently under way in ecological engineering relate to aquatic systems, particularly shallow ponds and wetlands. Ecological engineering as practiced in China and the East has been used in a wide variety of natural resource and environmental problems, ranging from fisheries and agriculture to wastewater control and coastline protection. The emphasis in the Chinese systems has been on applications rather than on experimentation and on the production of food and fiber more than on environmental protection (see Chapter 13).

To simplify the variety of approaches and systems used in ecological engineering, Mitsch (1993) divided ecological engineering case studies into three spatial scales: (1) mesocosms, (2) ecosystems, and (3) regional systems. Examples of systems in each of these categories are given in Table 3.2. Mesocosms are generally artificially enclosed systems and can range in size from laboratory bench-sized systems to Biosphere 2 in Arizona. Understanding of ecosystem function can come from the construction of scale-model ecosystems. Scale models of ecosystems (microcosms and mesocosms) have been built around the world, including scale models of both the Everglades and Chesapeake Bay and experimental mesocosms to investigate the role of hydroperiods in nutrient and metal retention in marshes, the competition of macrophytes in high- and low-nutrient conditions, and the use of alternative liner materials in treatment wetlands.

Ecosystem applications have been dominated by wetlands and water pollution control ecosystems and is probably the scale for which we have the most examples of ecological engineering today. Early full-scale experiments and demonstration projects in Gainesville Florida and Houghton Lake, Michigan, in the early 1970s investigated the use of natural wetlands for wastewater recycling. These projects are described in more detail in Chapter 10. Other ecosystem-scale projects have been developed to investigate the role of planting on wetland function and on the use of Christmas tree recycling to enhance land building and marsh restoration in Louisiana.

Regional systems involve the construction or restoration of a multiplicity of ecosystems that are all interconnected in reinforcing patterns and pathways. In some cases, the economic benefits of these multiple-ecosystem networks are illustrated and humans are part of the system design. Many examples of this type of system are found in China, where human nutrition is tied into a functional ecosystem or set of ecosystems. Large-scale riparian and river restoration projects have also been carried out over several years in northeastern Illinois and central France.

3.4 WHEN TO USE ECOTECHNOLOGY

If a proper use of ecological engineering is achieved during the next decade or two, it will most probably imply that all planning of projects with related pollution control problems will have been subjected to the following examinations and considerations before they are launched:

Table 3.2 Examples of ecological engineering studies at mesocosm, ecosystem, and regional scales

Ecological Engineering Project	Location	Purpose	References
Mesocosm Scale			
Treatment of septage wastes	Harwich, Massachusetts	To produce clean water (drinking water standards) from septage in an un-lined landfill lagoon	Guterstam and Todd, 1990; Teal and Peterson, 1991
Scale models of Everglades and Chesapeake Bay	Washington, DC	To simulate physical and biological function of large-scale ecosystems	Adey and Loveland, 1991
Biosphere 2	Catalina Mountains, Arizona	1.5-ha glass-enclosed system to investigate ecology and humans in en-closed systems	Marino and Odum, 1999
Wetland mesocosms	San Diego, California	To investigate the role of hydroperiods on marsh retention of nutrients and metals	Busnardo et al., 1992; Sinicrope et al., 1992
Wetland mesocosms	Columbus, Ohio	To compare competitive growth of two wetland macrophyte species in low and high nutrients	Svengsouk and Mitsch, 2001
Wetland mesocosms	Columbus, Ohio	To investigate the use of sulfur scrub-ber material as liners for treatment wetlands	Ahn et al., 2001; Ahn and Mitsch, 2002b
Peatland restoration plots	Waikato region, NorthIsland, New Zealand	To determine the role of fertilizer, seed additions, and cultivation techniques for peatland restoration	Schipper et al., 2002
Ecosystem Scale			
Experimental estuarine ponds	Morehead City, North Caro-lina	To investigate estuarine ponds receiv-ing a mixture of wastewater and salt water	Odum, 1985, 1989b

Forested wetlands for recycling	Gainesville, Florida	To experimentally investigate forested cypressdomes for wastewater recycling and conservation	Odum et al., 1974; Ewel and Odum, 1984; Dierberg and Brezonik, 1985
Created flow-through riverine wetlands	Columbus, Ohio	To experimentally determine the long-term effects of planting on ecosystem function	Mitsch et al., 1998; work in progress
Root-zone wetlands for wastewater treatment	Snogeröd, Sweden	To investigate use of root-zone wetlands to provide tertiary treatment of wastewater from slall town	Gumbricht, 1992
Nonpoint-source control wetlands	Central and southern Norway	To estimate the interaction between wetland retention efficiency and nutrient loss from agricultural watersheds	Braskerud, 2002a,b
Surface-water wetlands for wastewater treatment	Houghton Lake, Michigan	To use natural peatlands to treat wastewater from municipality to prevent lake pollution	Kadlec and Knight, 1996
Renovation of coal mine drainage	Athens County, Ohio	To study iron retention from coal mine drainage with *Typha* wetland	Mitsch and Wise, 1998
River pollution control	Suzhou, China	To use water hyacinths (*Eichhornia crassipes*) systems for water pollution control and production of fodder	Ma and Yan, 1989
Nonpoint-source pollution control	Central Illinois	To create wetlands to remove nutrients from midwestern agricultural runoff	Kovacic et al., 2000; Larson et al., 2000

(continues)

Table 3.2 (Continued)

Ecological Engineering Project	Location	Purpose	References
Intertidal sediment fences	Southern Louisiana	To construct intertidal fences made from recycled Christmas trees to revegetation on mudflats and increased sediment trapping	Boumans et al., 1997
Regional Scale			
Restoration of riparian landscape	Lake County, Illinois	To restore midwestern U.S. river floodplain and determine design procedures for restored wetlands	Hey et al., 1989; Mitsch, 1992; Sanville and Mitsch, 1994
Regional landscape restoration	Central Florida	To reconstruct wetland/upland landscape at phosphate mine	Brown et al., 1992
Agroecological engineering	Several thousand sites in China	To have multiple-product farming with extensive recycling	R. Zhang et al., 1998
Fish production/wetland systems	Yixing County, Jiangsuu Province, China	To produce fisheries synchronized to *Phragmites* wetland production and harvesting	Mitsch, 1991
Salt marsh creation	China's east coast, especially Wenling, Zhejiang Province	To develop *Spartina* marshes on former barren coastline for shoreline protection and food and fuel production	Chung, 1985, 1989; Qin et al., 1997
Salt marsh restoration	Delaware Bay, New Jersey	To restore salt marshes from salt hay farms and *Phragmites*-dominated marshes	Weinstein et al., 1997, 2001; Teal and Weinstein, 2002
River backwater restoration	Rhône river, central France	To restore and enhance river and river backwater connectivity	Henry and Amoros, 1995; Henry et al., 2002

1. The parts of nature that are touched by the project, directly or indirectly, with various pollution control alternatives must be determined.
2. Quantitative assessment (by use of models) of the environmental impact for all alternatives must be carried out.
3. The project is optimized by taking into consideration the entire system, that is, human society and the affected ecosystem. Ecotechnological solutions to existing problems should be included in this step.
4. The optimization must include short- as well as long-term effects, and ecology as well as economy. In this context, the application of ecological economic models becomes a very strong tool.
5. The renewable and nonrenewable resources involved in the various alternatives should be quantified.
6. All such examinations carry an uncertainty, which should be stated, and the uncertainty should be accounted for at least equally in ecological and economic considerations.

We have introduced the application of points 1 and 2 during the last two decades, but points 3 to 6 are open for introduction by ecological engineering and ecosystem restoration. However, before we can apply any ecological engineering approach to project planning, we need more experience in the application of ecotechnology, and this again requires that we reinforce ecological research in its broadest sense, including systems ecology, applied ecology, and ecological modeling. More resources must be allocated to this research in the coming years to assure a proper application of all the considerations mentioned above. Ecological engineering will not flourish without a strong base in ecology.

Management of ecosystems is not an easy task; building and restoring ecosystems and predicting their behavior are even more difficult. But complex problems of complex systems should not be expected to have easy solutions. Most environmental problems require the application of environmental technology as well as ecotechnology to find an optimal solution. This will again require deep ecological knowledge to understand the processes and reactions of ecosystems to possible management strategies. Recognition of nature's ability to self-design to its forcing function is also necessary to find the right ecotechnological methods.

4

ECOSYSTEMS

It is clear from the definitions of ecological engineering and ecosystem restoration that these are approaches operating in ecosystems. It is therefore necessary before we go into the applications of ecotechnology that we understand some characteristic properties of ecosystems. *Ecology* is the scientific study of the relationship between organisms and their environment. In accordance with the definition of the word ecology, it covers the household of nature (i.e., how mass and energy are used in natural systems). *Systems ecology,* a subdiscipline of ecology, focuses on the system properties of ecosystems. Sometimes it is called *ecosystem ecology,* although that field sometimes shies away from models and quantification. As described in Chapter 2, ecology, including systems ecology, is the scientific basis of ecological engineering and ecosystem restoration. This chapter is about how ecosystems work and some of the emergent properties that all systems may have that are pertinent to the way we could and do evaluate created and restored ecosystems.

4.1 TWO ECOLOGIES

Ecology may be approached from two sides: a *reductionistic approach,* where the relationships are found one by one and put together afterward; and a *holistic approach,* where the entire system is considered and it is attempted to reveal properties on the system level. Reductionism is a watchmaker apprentice's view of nature. A watch can be disassembled into its components, and it can be assembled again from these parts. The apprentice provides the watchmaker's parts, assembling them in nice boxes and drawers, watching over the shoulders of the watchmaker as he or she assembles and repairs watches. Reductionists think of the most complex systems as being made up of components that have been combined by nature in countless ingenious ways and which can be assembled and disassembled. It is far easier to take a complex system apart than it is to reassemble the parts and restore the important functions! The apprentice is given more responsibility for repairing and building whole watches when ready for such a responsibility.

Reductionistic ecology examines the relationships between organisms and their environment one by one. The ecological scientific journals are full of papers that reveal such relationships: carrying capacity versus nitrate concentration, relationship between the abundance of two populations in competition for space or resources, light intensity and photosynthesis, primary production versus diversity, and so on. We indeed need reductionistic ecology in ecological theory and practice. Many of the relationships found in this ecology have served as ideas, perceptions, or inspirations for holistic ecology, but because of the many pressing global problems, we need urgently to think and work much more holistically.

We face complex problems that sometimes need such reductionism for understanding some causes and effects. But understanding and predicting the behavior of entire ecosystems cannot be analyzed, explained, or predicted without a holistic approach that is able to deal with complex and multivariate phenomena. We need a science that can with irreducible systems as ecosystems or the entire ecosphere—systems that cannot be reduced to simple relationships as we have been used to in mechanical physics.

Ecology deals with irreducible systems. We cannot completely design simple experiments that reveal a relationship that can in all detail be transferred from one ecological situation and one ecosystem to another situation in another ecosystem (Jørgensen, 2002). That is possible, for instance, with Newton's laws on gravity, because the relationship between forces and acceleration is reducible. The relationship between force and acceleration is linear, but growth of living organisms depends on many interacting factors, which again are functions of time.

Reductionistic ecology has also been questioned as to whether it provides the proper scale for understanding how ecosystems work. Both Carpenter (1998) and Schlindler (1998) have argued that reductionistic experiments, when run at small scales, can lead to erroneous management decisions if the results are extrapolated to full-scale ecosystems—the very level at which ecological engineering and restoration often work. Small scales represent convenient science to obtain replication of results, but full-scale ecosystems have ecological feedbacks, particularly from larger organisms that cause the ecosystems to behave in exactly the opposite directions (Petersen et al., 1997; Kemp et al., 2001). Ahn and Mitsch (2002a) compared a wetland ecosystem ($10,000 \, m^2$) and wetland mesocosms ($1 \, m^2$) and showed that the mesocosms, with their simple structure compared to the full-scale ecosystem, was able to replicate some functions fairly accurately but that the rates of biogeochemical

processes were quite different at the different scales (Figure 4.1). They concluded that the scale of experiments or experiment artifacts must be considered before their results are generalized to large field-scale wetlands. This
study also pointed out the importance of ecosystem complexity on ecosystem
function at the various scales.

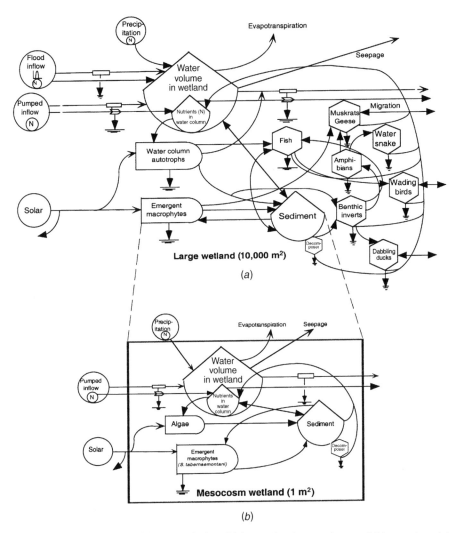

Figure 4.1 Comparison of the complexity of (a) a wetland ecosystem and (b) experimental
mesocosms meant to simulate behavior of wetland. Note the much simpler structure in the
experimental mesocosms, which suggests that the scale of experiment could have some relation to experimental results compared to a full-scale system. Based on created wetlands and
mesocosms at the Olentangy River Wetland Research Park in Ohio. (From Ahn and Mitsch,
2002a; copyright 2002; reprinted with permission from Elsevier Science.)

Feedback mechanisms regulate all the factors and rates simultaneously, and they also interact and are functions of time (Straskraba, 1980). In the experimental wetlands described by Ahn and Mitsch (2002a), large animals (e.g. muskrats, geese) had a profound effect on ecosystem function in wetlands in much the same way that fish can affect the trophic state of lakes. Small-scale experiments would not show these elements of the ecosystem. Table 4.1 shows the hierarchy of regulation mechanisms in ecosystems that are operating at the same time. From this example the complexity alone clearly prohibits the reduction to simple relationships that can be used repeatedly.

An ecosystem consists of so many interacting components that it is impossible ever to be able to examine all these relationships, and even if we could, it would not be possible to separate one relationship and examine it carefully to reveal its details, because the relationship is different when it works in nature with interactions from many other processes and when we examine it in a laboratory with the relationship separated from the other ecosystem components. The observation that it is impossible to separate and examine processes in real ecosystems corresponds to that of the examinations of organs that are separated from the organisms in which they are working. Their functions are completely different when separated from their organisms and examined in, for instance, a laboratory and when they are placed in their right context and in "working" condition. These observations are indeed expressed in *systems ecology,* focusing on ecosystems as systems. This is the ecology that provides a more useful basis for ecotechnology.

Table 4.1 Hierarchy of regulating feedback mechanisms

Level	Explanation of Regulation Process	Exemplified by Phytoplankton Growth
1	Rate by concentration in medium	Uptake of phosphorus in accordance with phosphorus concentration
2	Rate by needs	Uptake of phosphorus in accordance with intracellular concentration
3	Rate by other external factors	Chlorophyll concentration in accordance with previous solar radiation
4	Adaptation of properties	Change of optimal temperature for growth
5	Selection of other species	Shift to better-fitted species
6	Selection of another food web	Shift to better-fitted food web
7	Mutations, new sexual recombinations, and other shifts of genes	Emergence of new species or shifts of species properties

Source: After Jørgensen (1988).

4.2 SYSTEMS ECOLOGY

The following two phrases give the ideas behind the characteristics of eco-systems as systems: "Everything is linked to everything" and "The whole is greater than the sum of the parts" (Allen 1988). Both statements imply that it may be possible to examine the parts by reduction to simple relationships, but when the parts are put together, they will form a whole that behaves differently from the sum of the parts. This statement requires a more detailed discussion of how an ecosystem works. Allen (1988) claims that the latter statement is correct because of the evolutionary potential that is hidden within living systems. The ecosystem itself contains within itself the possibilities of becoming something different (i.e., of adapting and evolving). The evolution-ary potential is linked to the existence of microscopic freedom, represented by stochastic and nonaverage behavior, resulting from the diversity, complex-ity, and variability of its elements.

Microscopic and macroscopic diversity add to the complexity to such an extent that it is impossible to cover all the possibilities and details of phe-nomena observed. We attempt to capture at least a part of the reality by the use of models. It is not possible to use one or a few simple relationships, but models that include several of the relationships may be the only useful tools when we are dealing with irreducible systems. However, one model alone is so far from reality that we need many complementary models, which are used simultaneously to capture reality. It seems our only possibility to deal with the very complex systems, especially ecosystems.

So, we need system ecology, because:

1. The basic conditions determined by the external factors for our analysis are changing constantly (one factor is typically varied by an analysis, while all the others are assumed constant) in the real world and the analytical results are therefore not necessarily valid in the system con-text.

2. The interaction from all the other processes and components may sig-nificantly change the processes and properties of all biological com-ponents in the real ecosystem, and the analytical results are therefore not valid at all.

3. A direct overview of the many processes working simultaneously is not possible, and wrong conclusions may be the result if it is attempted anyhow.

The conclusion is therefore that we need a tool to overview and synthesize the many interacting processes. Systems ecology tries to synthesize what we know about the ecology of an ecosystem (1) to obtain an image of the eco-system, (2) to capture its properties on the system level, and (3) to predict ecosystem behavior sometimes in a set of new or different conditions. The first step is made by "putting together" various analytical results, but afterward we most often need to make changes to account for additional effects, resulting from the fact that the processes are working together and thereby become more than the sum of the parts. The interacting parts show, in other words, a synergistic effect—a symbiosis. Modeling (which is presented in detail in Chapter 14 as a tool of prediction for ecosystem restoration) is able to meet the needs for a synthesizing tool. It is our only hope that further synthesis of our knowledge to attain a system-level understanding of ecosystems will enable us to cope with environmental problems and ecological restoration. The basic environmental problem is that humankind has made immense progress, unique in our history, but we have not understood and controlled the full consequences on all levels of this progress.

Ulanowicz (1986) called for holistic descriptions of ecosystems. Holism is taken to mean a description of the system-level properties of an ensemble rather than simply an exhaustive description of all the components. It is thought that by adopting a holistic viewpoint, certain properties become apparent, and other behaviors are made visible that otherwise would be undetected. It means that we can only try to reveal the basic properties behind the complexity. Systems ecology should be the basic science supporting the application and development of ecological engineering and ecosystem restoration.

4.3 HIERARCHY THEORY

The components of ecosystems and their related processes are organized hierarchically. *Ecological hierarchy* is the well-known hierarchy of genes, cells, organs, organisms, populations, communities, and ecosystems. On each level in this hierarchy, processes and regulations will take place. Each level works as a unit, which can be influenced (controlled), however, from a level higher and lower in the hierarchy. Figure 4.2 shows various levels of the ecological hierarchy and the time and space scale some major environmental issues. Notice that there are distinct levels in both time and space.

Hierarchy theory (Allen and Starr, 1982) contends that the higher-level systems have emergent properties that are independent of the properties of their lower-level components. This compromise between the two other concepts seems to be consistent with our observations in nature. The hierarchical theory is a very useful tool to understand and describe such complex "medium number" systems as ecosystems (O' Neill et al., 1986).

During the 1990s, a debate arose on whether bottom-up (limitation by resources) or top-down (control by predators) affects primarily control the system dynamics. Sometimes the effect of the resources may be most dominant, sometimes the higher levels control the dynamics of the system, and sometimes both effects determine the dynamics of the system. When a system's view is applied, it can easily be concluded what the answer is to this

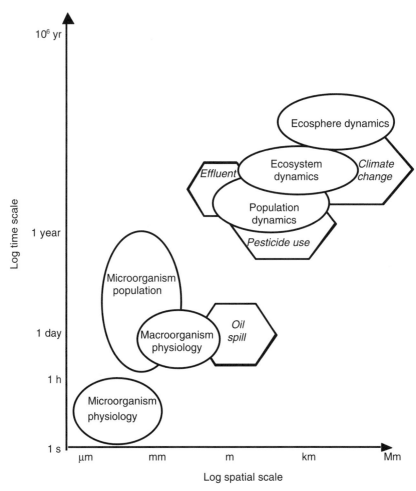

Figure 4.2 Spatial versus time scale for various environmental problems (hexagons, italic) and for the various levels of the ecological hierarchy (ovals, nonitalic). (From Jørgensen and Bendoricchio, 2001; copyright 2001; reprinted with permission from Elsevier Science.)

control question. Both can be important, depending on the resource availability and the complexity of the ecosystem structure.

4.4   CONSERVATION AND CYCLING OF MATTER AND ENERGY

Energy and matter are conserved according to basic physical concepts that are also valid for ecosystems. This means that energy and matter are neither created nor destroyed. The expression "energy *and* matter" is used, as energy can be transformed into matter, and matter into energy. The unification of the two concepts is possible by the use of *Einstein's law:*

$$E = mc^2 \quad (ML^2T^{-2}) \tag{4.1}$$

where E is the energy, m the mass, and c the velocity of electromagnetic radiation in vacuum ($= 3 \times 10^8$ m s^{-1}).

The transformation from matter to energy, and vice versa, is of interest only for nuclear processes and does not need to be applied on ecosystems on Earth. We might therefore break the proposition down to two more useful propositions when applied in ecology:

1. Ecosystems conserve matter.
2. Ecosystems conserve energy (the first law of thermodynamics).

Mass Balances

The conservation of matter may be expressed mathematically as follows:

$$\frac{dm}{dt} = \text{input} - \text{output} \quad (MT^{-1}) \tag{4.2}$$

where m is the total mass of a given system. The increase in mass is equal to the input minus the output. The practical application of the statement requires that a system is defined, which implies that the boundaries of the system must be indicated.

If the law of mass conservation is used for chemical compounds that can be transformed to other chemical compounds, equation (4.2) must be changed to

$$\frac{dm}{dt} = \text{input} - \text{output} + \text{formation} - \text{transformation} \quad (MT^{-1}) \tag{4.3}$$

The principle of mass conservation is used widely in the class of ecological models called *biogeochemical models*. Figure 4.3 shows nitrogen and phosphorus cycles in a lake as a typical example of cycling of mass in ecosystems. Matter forms local and global cycles that are crucial for the maintenance of life on Earth. Equations can be set up for any relevant elements or organic matter in ecosystems such as dry weight, organic matter, macronutrients such as C, P, N, O_2, and Si, or minor elements or even toxic chemicals such as Hg and Cd.

For plants, the conservation of mass can be described as

$$\begin{aligned} &\text{net retention (mass)} \\ &\quad = \text{uptake} - \text{releases (respiration, grazing, detrital loss, etc.)} \end{aligned} \tag{4.4}$$

Similarly, the mass flow through secondary consumers is mapped by use of the mass conservation principle. The food taken in by one level in the food chain is used in respiration, waste food, undigested food, excretion, growth, and reproduction. If growth and reproduction are considered as the net production, it can be stated that

net secondary production
$$= \text{intake of food} - \text{respiration} - \text{excretion} - \text{waste food} \qquad (4.5)$$

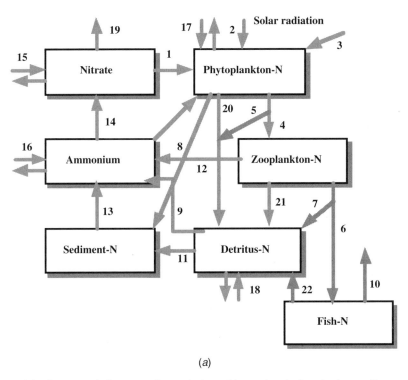

(a)

Figure 4.3 Conceptual diagrams of mass balance biogeochemical cycles in aquatic ecosystems: (a) nitrogen; (b) phosphorus. For the nitrogen cycle (a), the processes are (1) uptake of nitrate and ammonium by algae; (2) photosynthesis; (3) nitrogen fixation; (4) grazing with loss of undigested matter; (5), (6), and (7) predation and loss of undigested matter; (8) settling of algae; (9) mineralization; (10) fishery; (11) settling of detritus; (12) excretion of ammonium from zooplankton; (13) release of nitrogen from the sediment; (14) nitrification; (15), (16), (17), and (18) inputs/outputs; (19) denitrification; (20), (21), and (22) mortality of phytoplankton, zooplankton, and fish. For the phosphorus cycle (b), the processes are (1) uptake of phosphorus by algae; (2) photosynthesis; (3) grazing with loss of undigested matter; (4) and (5) predation with loss of undigested material; (6), (7), and (9) settling of phytoplankton; (8) mineralization; (10) fishery; (11) mineralization of phosphorous organic compounds in the sediment; (12) diffusion of pore water P; (13), (14), and (15) inputs/outputs; (16), (17), and (18) mortalities; (19) settling of detritus. (From Jørgensen and Bendoricchio, 2001; copyright 2001; reprinted with permission from Elsevier Science.)

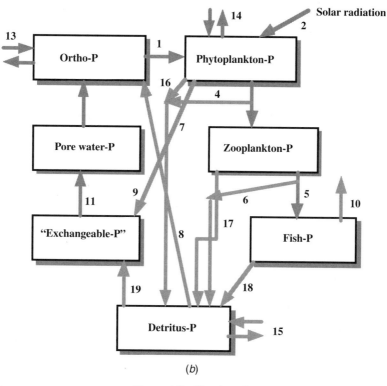

Figure 4.3 (*Continued*)

The ratio of the net production to the intake of food is termed the *net efficiency*. The net efficiency depends on several factors, but is often as low as 10 to 20 percent. Any toxic matter in the food is unlikely to be lost through respiration and excretions, because it is much less biodegradable than the normal components in the food. This being so, the net efficiency of toxic matter is often higher than for normal food components, and as a result, some chemicals, such as chlorinated hydrocarbons, including DDT and polychlorinated biphenyls, will be magnified in the food chain. This phenomenon is called *biological magnification* and is illustrated for DDT in Table 4.2. DDT and other chlorinated hydrocarbons have an especially high biological magnification, because they have a very low biodegradability and are excreted from the body only very slowly, due to dissolution in fatty tissue. As humans are the last links of the food chain, relatively high DDT concentrations have been observed in human body fat.

Energy Flow

Like all living systems, ecosystems are nonisolated systems, meaning that they must be open to an inflow of energy that is needed to maintain the system

Table 4.2   Concentration of DDT in materials and living matter (mg kg dry matter^{-1}) as indication of biological magnification

Atmosphere	0.000004
Rainwater	0.0002
Atmospheric dust	0.04
Cultivated soil	2.0
Fresh water	0.00001
Seawater	0.000001
Grass	0.05
Aquatic macrophytes	0.01
Phytoplankton	0.0003
Invertebrates on land	4.1
Invertebrates in sea	0.001
Freshwater fish	2.0
Sea fish	0.5
Eagles, falcons	10.0
Swallows	2.0
Herbivorous mammals	0.5
Carnivorous mammals	1.0
Human food, plants	0.02
Human food, meat	0.2
Humans	6.0

far from thermodynamic equilibrium. Without the inflow of energy, the system would move toward thermodynamic equilibrium according to the second law of thermodynamics, which it would reach sooner or later. At thermodynamic equilibrium no life can exist, and all structures, differences, and gradients are eliminated. Morowitz (1968) has shown that an inflow of energy is sufficient to create at least one cycling of matter/energy that is a prerequisite for life because matter is limited and must therefore be recycled to maintain life.

The understanding of the principle of conservation of energy is called the *first law of thermodynamics:* Energy can be neither created nor destroyed. If the concept internal energy, U, is introduced,

$$dQ = dU + dW \qquad (ML^2T^{-2}) \tag{4.6}$$

where dQ is the thermal energy added to the system, dU the increase in internal energy of the system, and dW the mechanical work done by the system on its environment.

The internal energy in the general sense can include several forms of energy: mechanical, electrical, chemical, magnetic energy, and so on. For ecological systems, it is represented by the energy equivalent of organic matter or the biomass of organisms (Table 4.3). The energy content per gram of ash-free organic material is surprisingly uniform. Biomass can be translated into

Table 4.3 Energy content of biological material (ash-free dry weight)

Component	kcal g^{-1}	kJ g^{-1}
Plant material		
Green grass	4.37	18.3
Standing dead vegetation	4.29	17.9
Litter	4.14	17.3
Roots	4.17	17.4
Green herbs	4.29	17.9
Animals		
Terrestrial flatworm	5.68	23.8
Aquatic snail	5.41	22.6
Brine shrimp	6.74	28.2
Cladocera	5.60	23.4
Spit bug	6.96	29.1
Mite	5.81	24.3
Beetle	6.31	26.4
Guppie	5.82	24.3
Spring birds	7.04	29.4
Food		
Milk	5.65	23.5
Fruits	5.20	21.8
Grain	5.80	24.3
Yeast	5.00	20.9

energy, and this is also true of transformations through food chains. The transformation of solar energy to chemical energy by plants conforms to the first law of thermodynamics (see Figure 4.4a):

$$GPP = NPP + R_p \qquad (4.7)$$

where GPP is the gross primary productivity (gross rate of radiant energy converted to chemical energy in photosynthesis or solar energy assimilated by vegetation), NPP the net primary productivity (net amount of energy converted to biomass after subtracting vegetation maintenance), and R_p is plant respiration.

For consumers in food chains, the energy balance can also be set up as shown in Figure 4.4b:

$$F = A + UD = G + R_c + UD \qquad (ML^2T^{-2}) \qquad (4.8)$$

where F is the food intake converted to energy, A the energy assimilated by consumers, UD the undigested food or the chemical energy of feces/urine, G the chemical energy of animal growth, and R_c the heat energy of consumer respiration.

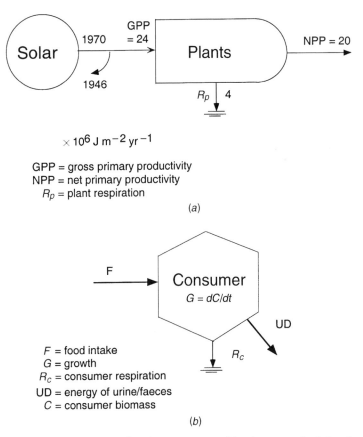

Figure 4.4 Examples of energy flow in ecosystems: (a) primary productivity of perennial grass–herb vegetation of an old field community; (b) secondary productivity consumer. Symbols defined in Figure 4.5.

As the energy of feces/urine is generally 20 percent of the respiration rate for freshwater fish (Mann, 1975), the growth equation for those organisms can be simplified as

$$G = F - 1.20R_c \tag{4.9}$$

Energy flow in ecosystems, of course, cannot commonly be measured directly, as the amount of energy is quite small relative to other energy fluxes that can be measured directly, through, for example, temperature changes. We usually estimate the flow of energy indirectly by measuring the flux of carbon, oxygen, or organic matter according to the reaction

$$6CO_2 + 12H_2O + 2960 \text{ kJ} \leftrightarrow C_6H_{12}O_6 + 6O_2 + 6H_2O \tag{4.10}$$

When the equation goes to the right, it reflects photosynthesis, and energy is required and carbon is reduced; when the equation goes to the left, it illustrates respiration and energy is released and organic carbon is oxidized. Table 4.4 summarizes both useful conversion factors among energy and power (energy per unit time) units and conversions between material flows and energy as used in ecological studies. For example, 1 g of organic carbon is generally assumed to be equivalent to about 40 kJ in equivalent energy [(2960 kJ)/(12 × 6 g C) from equation (4.10)].

Table 4.5 expresses the energy available for the above and other chemical oxidation processes to build adenosine-5'-triphosphate (ATP). The first equation is a simplification of respiration illustrated in equation (4.10). The other equations, for which less energy per equivalent is gained, represent chemicals other than oxygen being used as the terminal electron acceptor (e.g., NO_3^-, Mn^{4+}, Fe^{3+}, SO_4^{2-}, and CO_2).

Ecological energy flows are of considerable environmental interest, as calculations of biological magnifications are based on energy flows. There is a close relationship between energy flow rates and organism size, such as this one applicable for fish:

$$R = \sum [N(t)AW(t)^{0.8}] \tag{4.11}$$

where R is the respiration rate, $N(t)$ the number of fish at a given time, A is a constant $[= f(\text{temperature})]$, and $W(t)$ is the average weight per fish at a given time.

As many ecological rates are closely related to the rate of energy exchange, it is possible to find unknown parameters for various organisms on the basis of knowledge of the same parameters for other organisms, provided that the sizes of the organisms are known. Any self-sustaining ecosystem will contain a wide spectrum of organisms, ranging in size from tiny microbes to large animals and plants. Small organisms account in most cases for most of the respiration (energy turnover), whereas the larger organisms comprise most of the biomass. It is therefore important for the ecosystem to maintain both small and large organisms, as that means that both the energy turnover rates and the energy storage in the form of biomass are maintained.

Ecosystem Stoichiometry

Just as energy and measures of carbon, oxygen, and organic matter are convertible in describing ecosystem fluxes, so, too, can the general stoichiometry of mineral flows be estimated from measures of biomass change (see Table 4.4). The well-known *Redfield ratio* calculated from the average chemical concentration of plankton in the ocean by Redfield (1958), $(CH_2O)_{106}(NH_3)_{16}(H_3PO_4)$, illustrates that the ratio of carbon to nitrogen to phosphorus in organisms approximates

Table 4.4 Conversion factors for energy units; power units; carbon, oxygen, organic matter, and energy in ecosystems; and ratios among carbon, nitrogen, and phosphorus in organic matter

Original Units	Multiply by:	To Obtain:
Energy		
British thermal unit (Btu)	0.2530	kilocalorie (kcal)
	1054	joule (J)
calorie (cal)	4.1869	joule (J)
	0.001	kilocalorie (kcal)
joule (J)	0.239	calorie (cal)
	2.390×10^{-4}	kilocalorie (kcal)
kilocalorie (kcal)	1000	calorie
	3.968	British thermal unit (Btu)
	4183	joule (J)
	4.183	kilojoule (kJ)
	0.001162	kilowatthour (kWh)
kilojoule (kJ)	0.239	kilocalorie (kcal)
kilowatthour (kWh)	860.5	kilocalorie (kcal)
	3.6×10^6	joule (J)
langley[a] (ly)	1	calorie per square centimeter (cal cm^{-2})
	10	kilocalorie per square meter (kcal m^{-2})
Power		
horsepower	0.7457	kilowatt (kW)
	10.70	kilocalorie per minute (kcal min^{-1})
kilocalorie/day (kcal day^{-1})	6.4937×10^{-5}	horsepower (hp)
	4.8417×10^{-5}	kilowatt (kW)
kilowatt (kW)	1.341	horsepower (hp)
	14.34	kilocalorie per minute (kcal min^{-1})
	1000	watt (W)
watt (W)	1	joule per second (J s^{-1})
Primary productivity/energy flow[b]		
gram dry weight (g-dw)	4.5	kilocalorie (kcal)
	0.45	gram C (g-C)
grams O$_2$ (g-O$_2$)	3.7	kilocalorie (kcal)
	0.375	gram C (g-C)
gram C (g-C)	10	kilocalorie (kcal)
	2.67	gram O$_2$ (g-O$_2$)
kilocalorie (kcal)	0.1	gram C (g-C)
Stoichiometry of organic matter[c]		
molar ratio	106C : 16N : 1P	
weight ratio	41C : 7.2N : 1P	

Source: After Mitsch and Gosselink (2000).

[a]Solar constant = radiant energy at outer limit of Earth's atmosphere ~ 2.00 langleys per minute (ly min^{-1}).

[b]Based generally on the production of glucose: $6CO_2 + 12H_2O + (118 \times 6)$ kcal $\rightarrow C_6H_{12}O_6 + 6O_2 + 6H_2O$

[c]Based on Redfield (1958), molecule of plankton organic matter: $(CH_2O)_{106}(NH_3)_{16}(H_3PO_4)$. Useful atomic weights: H, 1; C, 12; N, 14; O, 16; P, 31

Table 4.5 Energy available per equivalent to build ATP for various oxidation processes of organic matter at pH 7.0 and 25°C

Reaction	Available kJ/Eq
$CH_2O + O_2 \rightarrow CO_2 + H_2O$	125
$CH_2O + 0.8NO_3^- + 0.8H^+ \rightarrow CO_2 + 0.4N_2 + 1.4H_2O$	119
$CH_2O + 2MnO_2 + H^+ \rightarrow CO_2 + 2Mn^{2+} + 3H_2O$	85
$CH_2O + 4FeOOH + 8H^+ \rightarrow CO_2 + 7H_2O + Fe^{2+}$	27
$CH_2O + 0.5SO_4^{2-} + 0.5H^+ \rightarrow CO_2 + 0.5HS^- + H_2O$	26
$CH_2O + 0.5CO_2 \rightarrow CO_2 + 0.5CH_4$	23

$$C:N:P = 106:16:1 \qquad \text{(by moles)} \qquad (4.12)$$

When converted to weight, the ratio is

$$C:N:P = 41:7:1 \qquad \text{(by weight)} \qquad (4.13)$$

This ratio suggests, for example, that for every 100 g of productivity or uptake of carbon, approximately 17 g of nitrogen and 2.4 g of phosphorus are involved as parallel transformations (e.g., incorporation into biomass or uptake). This ratio is, of course, approximate and does not account for luxury uptake of nutrients, for example, a common phenomenon in eutrophic conditions.

Ecological Language

Ecological modeling often focuses on the energy or mass flows in an ecosystem, because these flows determine further development of the system and characterize the present conditions of the system. H. T. Odum (1971, 1973, 1983) developed an energy language (sometimes called *energese*), which is a useful tool for incorporating much information into energy, mass, and information flow diagrams. The symbols in Figure 4.5 allow an ecologist to consider not only the flows and storages, but also the feedback mechanisms and the nonlinear rate regulators. For example, see how the consumer module in Figure 4.5 includes autocatalytic feedback on the inflow as a requirement for living organisms. Examples of model diagrams are shown in Figures 4.1 and 4.4. The strength of the Odum symbols is that they implicitly include the importance of self-design, feedback design, and autocatalytic reactions into energy diagrams to be able to consider the role of these mechanisms on the utilization of energy. No transfer of energy is possible without matter and information, and the higher the levels of information, the higher the utilization of matter and energy for further development of ecosystems away from thermodynamic equilibrium. These diagrams make it possible to illustrate energy, matter, and information flows.

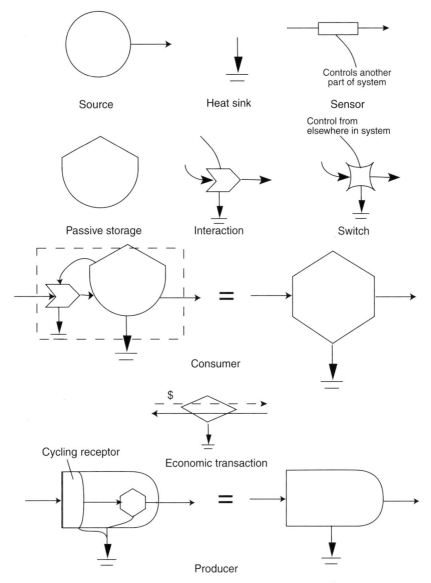

Figure 4.5 Energy circuit language (sometimes called *energese*) as developed by H. T. Odum (1971, 1973, 1983).

4.5 POLLUTION EFFECTS: THRESHOLD AND NONTHRESHOLD AGENTS

A pollutant can be defined as any material or set of conditions that creates a stress or unfavorable alternation of an individual organism, population, community, or ecosystem beyond the point that is found in normal environmental conditions (Cloud, 1971). The range of tolerance to stress varies considerably with the type of organism and the type of pollutant. To determine whether an effect is unfavorable may be a very difficult and often highly subjective process.

It is important to recognize that there are both natural and human-generated pollutants. Of course, the fact that nature sometimes concentrates chemicals in "pollution" concentrations (e.g., volcanoes, hot springs, upwelling, etc.) does not justify the extra addition of such pollutants, as this might result in the threshold level being reached.

In general, we can classify pollutants into two groups: (1) threshold agents, which have a harmful effect only above or below some concentration or threshold level (lines A in Figure 4.6), and (2) nonthreshold or gradual agents, which are potentially harmful in almost any amount (lines B in Figure 4.6).

For the former class we come closer to the limit of tolerance for each increase or decrease in concentration, until finally, like the straw that broke the camel's back, the threshold is crossed. For nonthreshold agents, which include several types of radiation, many human-made organic chemicals (which do not exist in nature), and some heavy metals, such as mercury, lead, and cadmium, there is theoretically no safe level. In practice, however, the degree of damage at very low trace levels is considered negligible or worth the risk relative to the benefits obtained from using the products or processes.

Threshold agents can include various nutrients, such as phosphorus, nitrogen, silica, carbon, vitamins, and minerals (e.g., calcium, iron, and zinc). When they are added or taken in excess, the organism or ecosystem can be

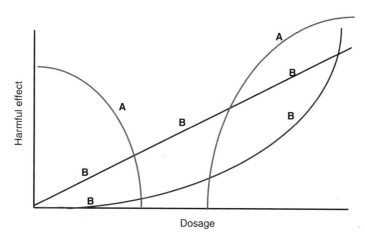

Figure 4.6 Models of pollution dosage effects on ecosystems. Lines A indicated threshold agent effects; lines B indicate nonthreshold or gradual agent effects. To have a threshold agent, it is sufficient that one of the two A-plots is valid. The two B-plots represent two different dose–response curves.

overstimulated, and the ecological balance is damaged. Examples are the eutrophication of lakes, streams, and estuaries from fertilizer runoff or municipal wastewater. The threshold level and type and extent of damage vary widely with different organisms and stresses. The thresholds for some pollutants may be quite high, whereas for others they may be as low as 1 part per billion (ppb).

Threshold levels are closely related to concentrations found in nature under normal environmental conditions. Even uncontaminated parts of the ecosphere contain almost all elements, although some only in very low concentrations. In addition, an organism's sensitivity to a particular pollutant varies at different times of its life cycle [e.g., threshold limits are often lower in the juvenile stage (where body defense mechanisms may not be fully developed) than in the adult stage]. This is especially true for chlorinated hydrocarbons such as DDT and heavy metals, both of which represent some of the most harmful pollutants.

Pollutants can also be characterized by their longevity in process organisms and ecosystems. Degradable pollutants will naturally be broken down into more harmless components if the system is not overloaded. Nondegradable or persistent pollutants will, however, not be broken down or be broken down very slowly, and the intermediate components are often as toxic as the pollutants. Knowledge of the degradability of the different components is naturally of great importance to environmental management, since an undesired concentration of pollutants in the environment is a function not only of the input and output, but also of the processes that take place in the environment. These processes must, of course, be considered whenever the concentration of pollutants is being computed. Ecological engineering requires not only knowledge of the concentration and the form of a pollutant, but also of the processes that the pollutant might undergo in the environment.

4.6 LIMITING FACTORS AND GROWTH

The conservation laws of energy and matter set limits to the further development of "pure" energy and matter, while information may be amplified (almost) without limit. These limitations lead to the concept of limiting factors that plays a significant role in ecosystems. Biological growth is limited by the component that is less abundant in the environment relatively to the amount required for growth. Phosphorus is for instance the limiting factor for algae growth in many lakes. A typical composition of phytoplankton of $C:N:P = 41:7:1$ on weight basis (see Redfield ratio, above) implies that phosphorus is limiting when the total nitrogen concentration is more than 7 times the phosphorus concentration and the carbon concentration is more than 41 times the phosphorus concentration. The relationship between the growth and the concentration of the limiting factor (which could also be light for the photosynthesis) is often expressed by the so-called *Michaelis-Menten equation:*

$$\mu = \mu_{max}\frac{c}{k_m + c} \tag{4.14}$$

where μ is the growth rate, μ_{max} the maximum growth rate, c the concentration of the limiting factor, and k_m is a constant called the Michaelis–Menten constant, which is the concentration where the growth rate is 0.5 times the maximum growth rate.

A graph of the relationship is shown in Figure 4.7. At a certain concentration the maximum growth rate will be attained, which indicates that another factor will now become limiting for further increase in the growth rate. It is possible by plotting the inverse of μ, $1/\mu$, versus the inverse of c, $1/c$, to assess whether the relationship between μ and c corresponds to a Michaelis–Menten equation. If the graph, denoted *Lineweaver–Burke's plot,* of $1/\mu$ versus $1/c$ is a straight line, the Michaelis–Menten equation is valid, as can easily be shown:

$$\frac{1}{\mu} = \frac{k_m}{\mu_{max}}\frac{1}{c} + \frac{1}{\mu_{max}} \tag{4.15}$$

Life-building processes require the presence of the elements characteristic to the biosphere. The composition of the biosphere is closely related to the function of the elements. High concentrations of C, H, and O are due to the

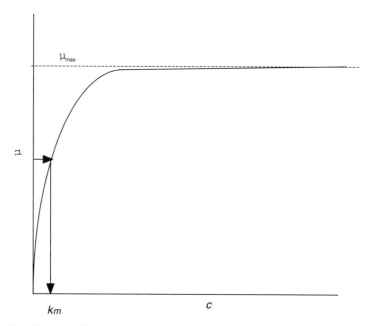

Figure 4.7 Michaelis–Menten limiting factor curve as described in equation (4.14). The rate is half of the maximum rate, μ_{max}, at $c = k_m$.

composition of organic compounds. Nitrogen concentrations result from the presence of proteins, including enzymes and polypeptides, and nucleotides. Phosphorus is used as a matrix material in the form of calcium compounds, in phosphate esters, and in ATP, which is involved in all energetically coupled reactions.

The pattern of the biochemistry determines the relative need for a number of elements (on the order of 20 to 25), whereas other elements (the remaining 65 to 70 elements) are more or less toxic. Some elements needed by some species are toxic to others in all concentrations. This does not imply that a particular species has a fixed composition of the required elements. The composition might vary within certain ranges.

In arid climates primary production (the amount of solar energy stored by green plants) is strongly correlated with precipitation, as water is obviously the "limiting factor." Among many limiting factors, frequently the most important are various nutrients, water, and temperature. The relation between nutrient concentration and growth has long been known for crop yields. This is expressed in Michaelis–Menten equation (4.14) but is also sometimes expressed as *Liebig's law of the minimum:* If a nutrient is at a minimum relative to its use for growth, there is a linear relation between growth and the concentration of the nutrient. If the supply of other factors is at a minimum, further addition of the nutrient will not influence growth.

Multiple Limiting Nutrients

The situation in nature is, however, often not as simple, as two or more nutrients (resources) may be limiting simultaneously. This can be described by

$$\mu = \mu_{max} \left(\frac{N_1}{k_{m1} + N_1} \times \frac{N_2}{k_{m2} + N_2} \right) \qquad (4.16)$$

where N_1 and N_2 are nutrient concentrations and k_{m1} and k_{m2} are half-saturation constants related to N_1 and N_2. This equation will, however, often limit growth too much and is in disagreement with many observations. The following equation seems to overcome these difficulties:

$$\mu = \mu_{max} \min \left(\frac{N_1}{k_{m1} + N_1}, \frac{N_2}{k_{m2} + N_2} \right) \qquad (4.17)$$

Although nutrients are necessary for plant growth, they may produce deterioration in life conditions for other forms of life. Ammonia is extremely toxic to fish, whereas ammonium, the ionized form, is harmless. Fish growth may be reduced at even relatively low concentrations of ammonia. The relation between ammonium and ammonia depends on pH:

$$NH_4^+ \leftrightarrow NH_3 + H^+ \tag{4.18}$$

$$pH = pK + \log \frac{[NH_3]}{[NH_4^+]} \tag{4.19}$$

where $pK = -\log K$ and K is the equilibrium for process (4.18). The pH value as well as the total concentrations of ammonium and ammonia are important according to these equations. This implies that the situation is very critical in many hypereutrophic lakes and wetlands during the summer, when photosynthesis is most pronounced, causing the pH to increase when the acidic component CO_2 is removed or reduced by this process. The pK value is about 9.3 in distilled water at 25°C, but increases with increasing salinity. At chlorinities of 2 percent, pK is increased to about 9.7 at 25°C.

4.7 BIOACCUMULATION

Mass flow through consumers was discussed earlier in this chapter. The food taken in by one level in the food chain (F), is used in respiration (R), non-utilized (wasted) food (NUF), undigested food (feces, Fe), excretion (urine, E), and growth and reproduction (G). If growth and reproduction are considered as the net secondary productivity (NSP), we can state that

$$NSP = G = F - (R + NUF + Fe + E) \tag{4.20}$$

and

$$\text{net efficiency} = \frac{NSP}{F} \times 100\% \tag{4.21}$$

The efficiency depends on several factors, but may be as low as 10 percent or less. Toxic matter in the food is unlikely to be lost through respiration and excretion, because it is usually much less biodegradable than the normal components in the food. As a result, several chemicals, such as chlorinated hydrocarbons, including DDT and some heavy metals, can be magnified in the food chain. This phenomenon is denoted as *biomagnification*.

The assimilated food, A, is the food used for (growth + reproduction) + respiration + excretion):

$$A = F - (NUF + Fe) \tag{4.22}$$

Many organic toxic compounds are taken up (assimilated) by a high efficiency value (greater than 90 percent); that is, the loss by feces is low. Fortunately, heavy metals have a low assimilation efficiency. Approximately only 5 to 10 percent of their content in food is assimilated, but as they are excreted slowly and not removed by respiration, they still have a relatively high biomagnification.

Many organic compounds, including chlorinated hydrocarbons, have a particularly high biomagnification because they have (1) high assimilation efficiency, (2) very low biodegradability, and (3) are excreted from the body only very slowly, because they are dissolved in fatty tissue. This is illustrated for DDT in Table 4.2 by a comparison of concentrations of DDT in various abiotic and biotic components. As the human being is on the last level of the food chain, relatively high DDT concentrations have been observed in human body fat. The concentration of a toxic component in an organism (Tox) can be followed approximately by use of a simplified differential equation:

$$\frac{d\text{Tox}}{dt} = \text{daily uptake via respiration and food} - k\text{Tox} \qquad (4.23)$$

where k is the first-order excretion coefficient.

The total daily uptake is found from the concentration in the ambient air of the toxic component times the efficiency of uptake via respiration plus the concentration in the food times the assimilation efficiency. It is assumed that excretion follows a first-order reaction, which is usually correct. Concentrations of lead in food items are shown in Table 4.6 to illustrate the presence of toxic substances in our food. The concentrations in the table are taken from the mid-1980s (i.e., before the introduction of lead-free gasoline had shown any significant effect).

Equation (4.23) explains why the concentration of a toxic substance increases with increasing weight and age of the organism. This is illustrated in

Table 4.6 Typical lead concentrations in food in selected European countries (mg kg fresh weight^{-1}) in the mid-1980s

Food Items	England	The Netherlands	Denmark
Milk	0.03	0.02	0.005
Cheese	0.10	0.12	0.05
Meat	0.05	<0.10	<0.10
Fish	0.27	0.18	0.10
Eggs	0.11	0.12	0.06
Butter	0.06	0.02	0.02
Oil	0.10	—	—
Corn	0.16	0.045	0.05
Potatoes	0.03	0.1	0.05
Vegetables	0.24	0.065	0.15
Fruits	0.12	0.085	0.05
Sugar	—	0.01	0.01
Soft drinks	0.12	0.13	—

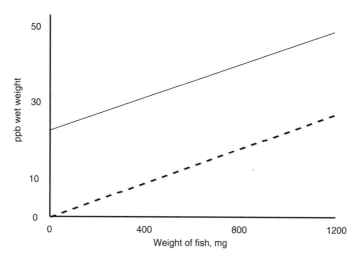

Figure 4.8 Increase in pesticide residues in fish as weight of the fish increases. Top line, total residues; bottom (dashed) line, DDE only. (After Cox, 1970.)

Figure 4.8 for fish with increasing weight. A steady-state concentration, $d\text{Tox}/dt = 0$, can be found from equation (4.23) as Tox = daily uptake via respiration and food$/k$. The coefficient k is low for many toxic compounds, including most toxic heavy metals, such as mercury, lead, and cadmium. This implies that the concentration of Tox becomes high and that it takes many years to reach a concentration close to the steady-state situation.

It is important to quantify any imbalance in the concentrations of life-essential components, as well as toxic components, for assessment of the life conditions in an ecosystem. This is, of course, valid on the global level as well as on the regional and local levels. Many toxic substances are widely dispersed, and a global increase in the concentration of heavy metals and pesticides has been recorded. The relationship between a global and regional pollution problem and the role of dilution for this relationship are illustrated in Table 4.7, where the ratios of heavy metal concentrations in the River Rhine and in the North Sea are shown. Lead concentrations in the Rhine are as high as 700 times that in the receiving North Sea.

4.8 COMPLEXITY, DIVERSITY, AND STABILITY

Complexity

There are many millions of species on Earth (on the order of 10^7), and there are on the order of 10^{21} organisms (the number is, of course, very uncertain). Organisms belonging to the same species have a high extent of similarity, but every organism is nevertheless different from all other organisms, as each

Table 4.7   Metal flux in the River Rhine and ratio of river concentrations to those in the North Sea

Metal	River Rhine (metric Tons yr^{-1})	$\dfrac{\text{Conc. in the Rhine}}{\text{Conc. in the North Sea}}$
Cr	1,000	20
Ni	2,000	10
Zn	20,000	40
Cu	200	40
Hg	100	20
Pb	2,000	700

Homo sapiens is different from his or her neighbor. Complexity certainly increases as the number of components increases, but the number of components is not the only measure of complexity. One mole consists of 6.02×10^{23} molecules. Yet physicists and chemists are able to make predictions related to pressure, temperature, and volume not despite, but because of, the large numbers of molecules. The reason is that all the organisms are different, whereas the molecules are essentially identical (there may be a few, but only a few, different types of molecules: oxygen, nitrogen, carbon dioxide, etc.). Interactions of molecules are random, and overall system averages are easily calculated. We are therefore able to apply statistical methods on the molecules but not to the much lower number of very different organisms. Ecosystems or the entire ecosphere are "medium number" systems in terms of complexity. They include most systems and are characterized by an intermediate number of different components and structured interrelationships among these components.

Diversity

Species diversity indices in ecological systems attempt to demonstrate the breadth of the number of species. Among the early measures of species diversity that have been used over the years are

$$D = \frac{s}{1000 \text{ individuals}} \tag{4.24}$$

$$D = \sum \frac{n_1}{N} \quad \text{(Simpson's index)} \tag{4.25}$$

where s is the number of species observed, n_1 the number of individuals of species i, and N the total number of individuals. The problem with simple indices such as these is that they are only a measure of *richness*. To incorporate *evenness* or *equitability* of the system, the Shannon–Weaver (more properly, the Shannon–Wiener) diversity index (*H*) should be used:

$$H = -\sum \frac{n_1}{N} \ln \frac{n_1}{N} \tag{4.26}$$

A natural log (ln) is illustrated here, but logs to the base 2 and logs to the base 10 have also been used for this formula. The Shannon–Weiner index is not without problems as well, and there is some support for the idea of simply using the species richness count (*s*), as long as it is always done with the same sampling effort (Colinvaux, 1993).

Diversity and Stability

The high number of species gives an extremely high number of possible connections and different relations in ecosystems. However, a model with many components and a high number of connections is not necessarily more stable than a simple one (May, 1981). There was intense discussion of stability and diversity of ecosystems in the 1960s and 1970s, and it has been renewed with the more recent discussion on the value of biodiversity to humans. *Ecological stability* is the ability of an ecosystem to resist changes in the presence of perturbations. It has been the governing theory that there is no (simple) relation between stability and diversity. It is possible in nature to find very stable and simple ecosystems, and it is possible to find rather unstable, very diverse ecosystems. May (1981) claims that *r-selection* is associated with a relatively unpredictable environment and simple ecosystems, while *K-selection* is associated with a relatively predictable environment and a complex biologically crowded community.

Reactions of ecosystems to perturbations have been discussed widely in relation to the stability concepts. However, this discussion has in most cases not considered the enormous complexity of regulation and feedback mechanisms. The stability concept *resilience* is understood as the ability of the ecosystem to return "to normal" after perturbations. This concept has more interest in a mathematical discussion of whether equations may be able to return to steady state, but the shortcomings of this concept in a real ecosystem context are clear. An ecosystem is a soft system that will never return again to the same point. It will be able to maintain its functions on the highest possible level, but never with exactly the same biological and chemical components in the same concentrations again. The species composition or food web may or may not have changed, but at least it will not be the same organisms with the same properties. In addition, it is unrealistic to consider that the same conditions will occur again.

We can observe that an ecosystem has the property of resilience in the sense that ecosystems have a tendency to recover after stress, but a complete recovery, understood as meaning that exactly the same situation will appear again, will never be realized. The combination of external factors—the impact of the environment on the ecosystem—will never appear again, and even if they would, the internal factors—the components of the ecosystem—have meanwhile changed and can therefore not react in the same way as the previous internal factors did. The concept of resilience is therefore not a realistic quantitative concept. If it is used realistically, it is not quantitative, and if it is used quantitatively, for instance in mathematics, it is not realistic. Resilience covers the ecosystem property of elasticity to a certain extent, but in fact, the ecosystem is more flexible than elastic. It will change to meet the challenge of changing external factors, not try to struggle to return to exactly the same situation.

Resistance is another widely applied stability concept. It covers the ability of the ecosystem to resist changes when external factors are changed. This concept needs a more rigorous definition, however, and needs to be considered multidimensionally to be able to cope with real ecosystem reactions. An ecosystem will always be changed when the conditions are changed; the question is: What is changed, and by how much?

Webster (1979) examined by models the ecosystem reactions to the rate of nutrient recycling. He found that an increase in the amount of recycling relative to input resulted in a decreased margin of stability, faster mean response time, greater resistance [i.e., greater buffer capacity (see the definition below)], and less resilience. Increased storage and turnover rates resulted in exactly the same relationships. Increases in both recycling and turnover rates produced opposite results, leading, however, to a larger stability margin, faster response time, smaller resistance, and greater resilience. O'Neill (1976) examined the role of heterotrophs on resistance and resilience and found that only small changes in heterotroph biomass could reestablish system equilibrium and counteract perturbations. He suggested that many regulation mechanisms and spatial heterogeneity should be accounted for when the stability concepts are applied to explain ecosystem responses.

These observations explain why it has been very difficult to find a relationship between ecosystem stability in its broadest sense and species diversity. We do know that diversity is changed by an ecosystem's trophic status. In lakes, it has been observed that increased phosphorus loading gives decreased diversity (Ahl and Weiderholm, 1977; Weiderholm, 1980), yet very eutrophic lakes are very stable. Figure 4.9a gives the result of a statistical analysis from a number of Swedish lakes. The relationship shows a correlation between number of species and the system's eutrophication, measured as chlorophyll *a* in μg L^{-1}. A similar relationship was obtained by Moore et al. (1989) for fertile and infertile wetlands (as measured by wetland plant standing crop) in Ontario, Canada. They found the greatest species richness where biomass and nutrient availability were low and the lowest species richness where productivity was high (Figure 4.9b). Wisheu and Keddy (1992) described the low-diversity *Typha* marshes as the core wetland habitat of this

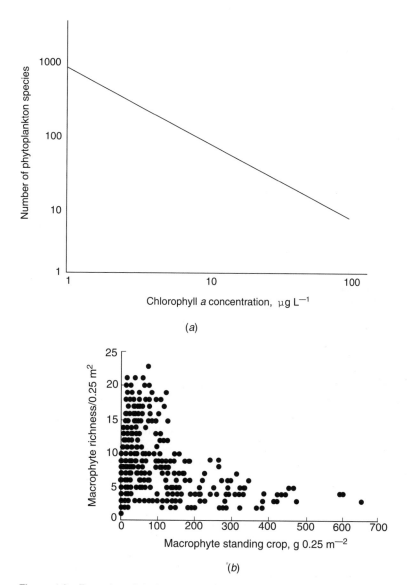

Figure 4.9   Examples of the inverse relationship between diversity or richness and measures of productivity. (a) Relationship for a number of Swedish lakes showing the number of phytoplankton species (richness as a measure of diversity) and eutrophication, expressed as algal biomass (chlorophyll a). (From Weiderholm, 1980.) (b) Relationship between macrophyte species richness and vegetation biomass (as a measure of macrophyte productivity) for several wetland sites in Canada. (From Moore et al., 1989.)

region, with variation from that stable system only when there were resource or stress conditions. High nutrient conditions were most stable, yet lowest in diversity.

Buffer Capacity

Buffer capacity (β) is defined by Jørgensen (1982, 1988, 1994, 2002) as

$$\beta = \frac{\Delta(\text{forcing function})}{\Delta(\text{state variable})} \tag{4.27}$$

Forcing functions are the external variables that are driving a system, such as discharge of wastewater, precipitation, wind, and so on, while state variables are the internal variables that determine a system: for instance, the concentration of phosphorus in the soil, the concentration of zooplankton in a cubic meter of lake water, and so on. As seen, the concept of buffer capacity has a definition that allows us to quantify, for instance in modeling, and it is also applicable to real ecosystems, as it acknowledges that *some* changes will always take place in the ecosystem in response to changed forcing functions. The question is: How large are these changes relative to changes in the conditions (the external variables or forcing functions)?

The concept should be considered multidimensionally, as we may consider all combinations of state variables and forcing functions. It implies that even for one type of change there are many buffer capacities, corresponding to each of the state variables. Ecological stability is very close to buffer capacity, but it is lacking the multidimensionality of ecological buffer capacity.

The relation between forcing functions (impacts on the system) and state variables (Figure 4.10) indicates that the conditions of the system are rarely linear and buffer capacities are therefore not constant. It may therefore be important in restoration and creation projects to reveal the relationships between forcing functions and state variables to observe under which conditions buffer capacities are small or large.

4.9   DEVELOPMENT AND EVOLUTION

Ecosystems are forever changing. In the short term, the changes are called *development* or *succession*. In the long term, ecosystems change as a result of evolution.

Development and Growth

Ecological development is the changes over time in nature caused by the dynamics of the external factors, giving the system sufficient time for the reactions. It encompasses a number of processes, including adaptation and shifts in species composition. The word *succession* is used to describe that

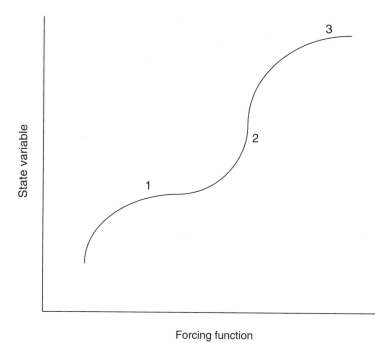

Figure 4.10 The relation between state variables and forcing functions illustrates the buffering capacity. At points 1 and 3 the buffering capacity is high; at point 2 it is low.

the species composition of an ecosystem is changing with the change in external factors. Characteristics of mature ecosystems consistent with E. P. Odum's (1969, 1971) attributes are shown in Table 4.8. Successional characteristics are distinguished between a poorly developed ecosystem and a mature, well-developed ecosystem. The properties listed in Table 4.8 have been used for decades to describe ecosystem development, and most are still valid indicators, mostly for terrestrial systems.

Growth forms are another way of describing ecosystem change. Three growth forms have been described by Holling (1986) as:

1. *Growth of physical structure* (biomass), which is able to capture more of the incoming energy in the form of solar radiation, but also requires more energy for maintenance (respiration and evaporation). This growth form leads to a larger biomass, intrabiotic inorganic nutrients, and a low P/B ratio.
2. *Growth of network,* which means more cycling of energy and matter. This growth form corresponds to complex cycles, well-organized patterns, and feedback-controlled growth.
3. *Growth of information* (more plants and animals with more genes) from *r*-strategists to *K*-strategists, which waste less energy but also usually

Table 4.8 Differences between initial stages and mature stages in ecological succession

Property	Early Stage	Late or Mature Stage
Energetics		
P/R	$>>1$ or $<<1$	Close to 1
P/B	High	Low
Yield	High	Low
Specific entropy	High	Low
Entropy production per unit of time	Low	High
Exergy	Low	High
Information	Low	High
Structure		
Total biomass	Small	Large
Inorganic nutrients	Extrabiotic	Intrabiotic
Diversity, ecological	Low	High
Diversity, biological	Low	High
Patterns	Poorly organized	Well organized
Niche specialization	Broad	Narrow
Size of organisms	Small	Large
Life cycles	Simple	Complex
Mineral cycles	Open	Closed
Nutrient exchange rate	Rapid	Slow
Selection and homeostatis		
Internal symbiosis	Undeveloped	Developed
Stability (resistance to external perturbations)	Poor	Good
Ecological buffer capacity	Low	High
Feedback control	Poor	Good
Growth form	Rapid	Feedback-controlled

Source: After E. P. Odum (1969).

are larger and carry more information. This growth form is covered by a high level of information, a narrow niche specialization, large organisms, long life span, and the shift from *r*-strategists to *K*-strategists.

Holling (1986) has suggested that ecosystem progress through sequential phases of renewal (primarily, growth form 1), exploitation (primarily, mainly growth form 2), conservation (dominant growth form 3), and *creative destruction*. The last phase also fits into the three growth forms but will require further explanation. The creative destruction phase is either a result of external or internal factors. In the first case (e.g., hurricanes and volcanic activity), further explanation is not needed, as an ecosystem has to use the growth forms under the prevailing conditions, which are determined by the external factors. If the destructive phase is a result of internal factors, the question is:

Why would a system be self-destructive? A possible explanation is that as a result of the conservation phase, almost all nutrients are contained in organisms, which implies that no nutrients are available to test new and possibly better solutions to move further away from thermodynamic equilibrium or, expressed in Darwinian terms, to increase the probability of survival. This is also indicated implicitly by Holling as he talks about creative destruction. Therefore, when new solutions are available, it would in the long run be beneficial for the ecosystem to decompose the organic nutrients into inorganic components which can be utilized to test the new solutions. The creative destruction phase can be considered a method to utilize the three other phases and the three growth forms more effectively in the long run.

Evolution

Evolution, on the other hand, is related to the genetic pool. It is the result of the relation between the dynamics of the external factors and the dynamics of the genetic pool. The external factors steadily change the conditions for survival, and the genetic pool comes up steadily with new solutions to the problem of survival. An ecosystem is a very dynamic system. All of its components, particularly the biological components, are steadily moving and their properties are steadily modified, which is why an ecosystem will never return to the same situation. Furthermore, every point is different from any other point, therefore offering different conditions for the various life forms.

This enormous heterogeneity explains why there are so many species on Earth. There is, so to say, an ecological niche for "everyone," and everyone may be able to find a niche where he or she is best fitted to utilize the resources. *Ecotones,* the transition zones between two ecosystems, offer a particular variability in life conditions, which often results in a particular richness of species diversity. Studies of ecotones have recently drawn much attention from ecologists, because ecotones have pronounced gradients in the external and internal variables, which give a clearer picture of the relation between external and internal variables.

Darwin's theory describes the competition among species and states that species that are best fitted to the prevailing conditions in the ecosystem will survive. Darwin's theory can, in other words, describe the changes in ecological structure and the species composition but cannot directly be applied quantitatively. All species in an ecosystem are confronted with the question: How is it possible to survive or even grow under the prevailing conditions? The prevailing conditions are considered as all factors influencing the species (i.e., all external and internal factors, including those originating from other species). This explains *coevolution,* as any change in the properties of one species will influence the evolution of the other species.

All natural external and internal factors of ecosystems are dynamic—conditions are steadily changing, and there are always many species waiting in the wings, ready to take over, if they are better fitted to the emerging conditions than the species dominating under the present conditions. There is a wide spectrum of species representing different combinations of properties available for the ecosystem. The question is: Which of these species are best able to survive and grow under the present conditions, and which species are best able to survive and grow under the conditions one time step further, two time steps further, and so on? To be able to survive, the species must have genes, or perhaps *phenotypes* (meaning properties), that match these conditions. But the natural external factors and the genetic pool available for the test may change randomly or by chance.

New mutations (misprints produced accidentally) and sexual recombinations (genes are mixed and shuffled) emerge steadily and steadily give new material to be tested toward the question: Which species are best fitted under the conditions prevailing now? These ideas are illustrated in Figure 4.11. External factors are changed steadily and some even relatively fast—partly at random (e.g., the meteorological or climatic factors). The species of the

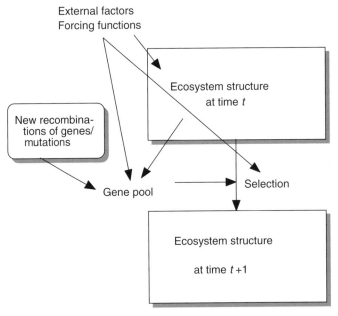

Figure 4.11 Conceptual diagram of how the external factors steadily change the species composition. The possible shifts in species composition are determined by the gene pool, which is changed steadily due to mutations and new sexual recombinations of genes. The development is, however, more complex. The arrow from "external factors" to "selection" accounts for the possibility that species are able to modify their own environment and therefore their own selection pressure. The arrow from "ecosystem structure" to "gene pool" accounts for the possibilities that species can to a certain extent change their own gene pool. (From Jørgensen and Bendoricchio, 2001; copyright 2001; reprinted with permission from Elsevier Science.)

system are selected among the species available and represented by the genetic pool, which again is slowly but surely changed randomly or by chance. The selection in Figure 4.11 includes levels 4 and 5 of hierarchy feedback, illustrated in Table 4.1. It is a selection of the organisms that possess the properties best fitted to the prevailing organisms according to the frequency distribution.

Human-caused changes in external factors (i.e., anthropogenic pollution), have created new problems, because new genes fitted to these changes do not develop overnight, whereas most natural changes have occurred many times previously and the genetic pool is therefore prepared and fitted to meet the natural changes. The spectrum of genes is able to meet most natural changes but not all of the human-made changes, because they are new and untested in the ecosystem.

4.10 SYSTEM THEORIES

Several ecosystem theories were developed over the past few decades that have attempted to describe the essence of the organization and development of systems in general and ecosystems in particular in a thermodynamic sense. Since biological systems have many possibilities to move away from thermodynamic equilibrium, it is therefore crucial in ecology to know which pathways among the possible ones an ecosystem will select for development. That would be the key to describe processes characteristic for ecosystem development. Several thermodynamic theories are summarized by their originators in a collection of papers presented at an international conference, Energy and Ecological Modelling, in 1981 in Louisville, Kentucky (Mitsch et al., 1982). Those and other theories are described in more detail by Jørgensen (2002).

Entropy Minimization and Dissipative Structures

Entropy is disorder. The *second law of thermodynamics* tells us that entropy is increasing in the universe continually and irreversibly. Processes that involve energy transformation will not occur spontaneously unless degradation of energy (disorder) occurs. We know, of course, that life and systems that receive energy in general can reverse this entropy production and make "order out of chaos" (Prigogine, 1982). That entropy production tends to be minimum was proposed by Prigogine (1980), as was his allied theory of *dissipative structures* (Prigogine, 1982). These theories were applied to linear systems at steady nonequilibrium states, not for far-from-equilibrium systems. Prigogine argues that in far-from-equilibrium conditions, new structures called *dissipative structures* can emerge, even in the face of the second law of thermodynamics. As stated by Prigogine (1982): "The equilibrium world is a homeostatic world; fluctuations are damped by the system. However, in far-from-equilibrium situations, fluctuations may grow and invade the system as a whole. Fluctuations may lead to new space–time structures of the system."

Maximum Power Principle

Boltzmann (1905) proposed that "life is a struggle for the ability to perform work," and Lotka (1922) proposed to use maximum power as a goal function to describe ecosystem development. *Power* is defined as energy per unit time and can be expressed as joules per year, or kilowatts. Boltzmann's and Lotka's principles served as the basis for Odum's maximum power principle. The *principle of maximum power* states that systems survive that maximize their energy flow to useful work when in competition with other systems (Odum, 1983). The maximum power concept was first described by Odum and Pinkerton (1955) as "time's speed regulator," analogous to the way that the second law of thermodynamics is referred to as "time's arrow" (Hall, 1995a). If the power is wasted in useless heat or destructive wars, there is often no useful work. But when the system builds structures that feed back and self-design to maximize energy use for useful purposes, that system will survive in competition with other systems. It is also important to note that maximum power does not mean maximum efficiency; it means optimum efficiency that results in highest power flow. A good description of many examples or research that focuses on the maximum power principle is contained in the book *Maximum Power: The Ideas and Applications of H. T. Odum* (Hall, 1995b).

Odum (1996) introduced another goal function to describe energy quality—*emergy*, which is short for "energy memory." Emergy is not the actual content of energy but the energy memory of something or some organism—the energy it took to create that system. It is the energy (originated from the ultimate source of energy—solar energy) that it costs to construct the component considered. If it costs 1000 J of solar energy to create 1 J of electricity, the electricity has a *transformity* of 1000 (Figure 4.12). If it costs 10 units of phytoplankton to construct 1 unit of zooplankton, the energy of zooplankton should be multiplied by 10 to obtain the emergy in zooplankton relative to that of phytoplankton. Jørgensen (1994) showed that there is a good correlation between the exergy index (described below) and emergy when the calculations are based on realistic ecosystem models derived from data and observations.

Exergy Maximization

Exergy is defined as the amount of work (i.e., entropy-free energy) that a system can perform when it is brought into thermodynamic equilibrium with its environment. It is illustrated in Figure 4.13. Exergy is not conserved—this occurs only if entropy-free energy is transferred, which implies that the transfer is reversible. All processes in reality are irreversible, however, which means that exergy is lost (and entropy is produced). Loss of exergy and production of entropy are two different descriptions of the same reality:

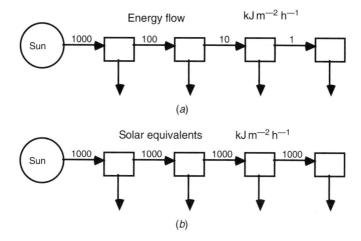

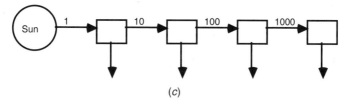

Figure 4.12 Concept of *emergy*: (*a*) possible energy flow through a food chain; (*b*) solar equivalents (emergy) flows illustrated by the same food chain. (*c*) Transformities calculated from diagrams (*a*) and (*b*) can be used to transform energy flow to emergy flow in subsequent studies.

namely, that all processes are irreversible, and we unfortunately always have some loss of energy forms that can do work to energy forms that cannot do work (heat at the temperature of the environment).

Information contains exergy. Boltzmann (1905) showed that the free energy of the information that we actually possess (in contrast to the information we need to describe the system) is $kT \ln I$, where I is the information we have about the state of the system and k is Boltzmann's constant = 1.3803×10^{-23} (J molecules^{-1} deg^{-1}). This implies that 1 bit of information has exergy equal to $kT \ln 2$. Transformation of information from one system to another is often an almost entropy-free energy transfer.

Exergy of the system measures the contrast—it is the difference in free energy if there is no difference in pressure and temperature, as may be assumed for an ecosystem or an environmental system and its environment—against the surrounding environment. If the system is in equilibrium with the surrounding environment, the exergy is of course zero. Since the only way to move systems away from equilibrium is to perform work on them, and since the available work in a system is a measure of the ability, we have to distin-

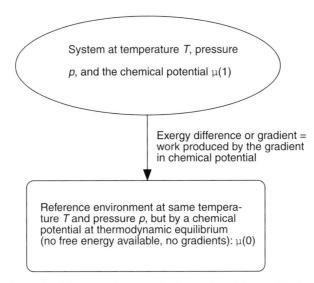

Figure 4.13 *Exergy* is defined as the amount of work that it is possible to extract from a system by bringing it into equilibrium with a reference state (e.g., its environment or as often used in an ecological context). Work capacity (exergy) is found as a quantity (entropy for heat energy, volume for pressure work, and concentrations or number of moles for chemical work) times the difference in the corresponding gradient between the two systems. (From Jørgensen and Bendoricchio, 2001; copyright 2001; reprinted with permission from Elsevier Science.)

guish between the system and its environment or thermodynamic equilibrium alias: for instance, an inorganic soup. Therefore, it is reasonable to use the available work (i.e., the exergy) as a measure of the distance from thermodynamic equilibrium.

Jørgensen and Mejer (1977, 1979), Mejer and Jørgensen (1979), and Jørgensen (1982, 2002) describe the maximization of exergy in the following way. A system that receives a through flow of exergy will have a propensity to move away from thermodynamic equilibrium, and if more combinations of components and processes are offered to utilize the exergy flow, the system has the propensity to select the organization that gives the system as much stored exergy as possible. A flow of energy (exergy) through the system, which means that the system must be open or at least nonisolated, is necessary for its existence. A flow of exergy through the system is also sufficient to form an ordered structure (the dissipative structure of Prigogine).

Le Châtelier's Principle

Another proposed thermodynamic principle may also be considered as an extended version of *LeChâtelier's principle*. Formation of biomass may be described as

energy + nutrients = molecules with more free energy and organization

$$(4.28)$$

If we pump energy into a system, the equilibrium will, according to LeChâtelier's principle, shift toward a utilization of the energy. It means that molecules with more free energy and organization are formed. If more pathways are offered, the pathways that give most relief (i.e., use most energy and thereby form molecules with most embodied free energy) will win according to the proposed tentative law of thermodynamics. The sequence of oxidation of organic matter (see, e.g., Schlesinger, 1997, or Mitsch and Gosselink, 2000) is as follows: by oxygen, by nitrate, by manganese dioxide, by iron(III), by sulfate and by carbon dioxide. It means that oxygen, if present, will always outcompete nitrate, which will outcompete manganese dioxide, and so on.

Ascendency

Ulanowicz (1997) uses *ascendency* to formulate the following hypothesis for development of an ecosystem: In the absence of overwhelming external disturbances, living systems exhibit a natural propensity to increase in ascendency. Ascendency is a measure of the flows in an ecosystem or, rather, our model of the ecosystem. The more through-flow in the ecosystem and the more the network of flow, the higher the ascendency. Jørgensen and Bendoricchio (2001) concluded that ascendancy and exergy indices of models are well correlated. This is not surprising, as the network, which is the basis for the ascendancy calculations, is a result of the ability of the biological components in the network to cooperate and build interrelationships. So, the network, including the transfer of mass or energy from one component to another, reflects indirectly the size of, and information embodied in, the components of the network.

Ecosystem Theories Form a Pattern

The development of ecosystems can be described by three growth forms that are consistent with the ecological succession description by E. P. Odum (1969) as described in Table 4.8:

I. Growth of physical structure (biomass), which is able to capture more of the incoming energy in the form of solar radiation, but also requires more energy for maintenance (respiration and evaporation).

II. Growth of network, which means more cycling of energy and matter.

III. Growth of information (more developed plants and animals with more genes), from *r*-strategists to *K*-strategists, which wastes less energy but also usually carries more information.

The relationships among the above ecosystem theories and their goal functions for the three growth forms have been compared in detail by Jørgensen (2002). It must suffice here to say that while there are differences among the theories, there are also distinct similarities that are worth further investigation.

5

ECOLOGICAL DESIGN
PRINCIPLES

In this chapter we present a number of design principles applicable to ecological engineering that are derived from ecosystem theory discussed in the earlier chapters or provided from the science of ecology. The principles should be applied whenever decisions on the use of ecological engineering or ecosystem restoration have to be taken in environmental management. In an earlier book (Mitsch and Jørgensen, 1989) we suggested 13 ecological concepts to serve as basic design principles of ecotechnology and to provide practical approaches to guiding ecological engineering. From our 1989 list and our observations since then, 19 ecological principles are presented below. We believe that they should be guiding principles for ecosystem restoration as well. All principles were based on system concepts applied to ecology, some of which are described in more detail in Chapter 4. Examples of the application of each of the principles are provided from the entire spectrum of possibilities described in Chapter 3.

Our original list from 1989 was assisted by other publications that have since appeared. One study by Milan Straskraba (1993) presented seven "principles of ecotechnology" which were used to set up 17 rules on how to manage ecosystems. His rules are general and are applicable to environmental management beyond the edges of ecological engineering and ecosystem restoration. Bergen et al. (2001) provided five additional "axioms, heuristics, and suggestions," and Zalewski (2000b) introduces ecohydrology concepts that were also useful. An overall master principle that is fundamental is that ecological engineering and ecosystem restoration need to be based on the science of ecology. While ecological engineering and ecosystem restoration might show that some of our ecological theories are not rigorous enough to survive real-world tests, the principles below are based on ecological concepts that have a strong record of field verification. We must continue to verify and expand these principles; they represent our best attempt from our combined 60 person-years of ecological probing.

Principle 1. Ecosystem structure and functions are determined by the forcing functions of the system.

Ecosystems exchange mass and energy with the environment. There is a close relation between the anthropogenic forcing functions and the state of the agricultural ecosystems, but due to the openness of all ecosystems, the adjacent ecosystems are also affected. Intensive agriculture leads to drainage of surplus nutrients and pesticides to adjacent ecosystems. This is *nonpoint* or *diffuse pollution.* The abatement of this source of pollution requires a wide range of ecotechnological methods.

The use of constructed wetlands and impoundments to reduce the concentrations of nutrients in streams entering a lake ecosystem illustrates application in ecotechnology of this system ecological principle. The forcing functions—nutrient loadings—are reduced and a corresponding reduction of eutrophication should be expected.

We have had much discussion of this principle with our Chinese colleagues, who contend that forcing functions are not as important as the structure of the ecosystem. We agree to the point that components in the ecosystem do have a hierarchy of feedback that alters the function of the ecosystems they are in. But we believe that the overall trajectory of the ecosystem is ultimately decided by forcing functions, particularly if these systems are open to the powers of self-design.

Principle 2. Energy inputs to the ecosystems and available storage of matter are limited.

This principle is based on the conservation of matter and energy. The dominant energy input to ecosystems is solar energy, a very diffuse energy source. We may be able to supplement this energy source with other forms of energy (e.g., fossil fuel), but the only sustainable energy source is solar energy. Use of alternative energy sources (e.g., pumps, plastic liners) may make an ecosystem more predictable, but these subsidies are not sustainable. Similarly, when a wetland is being constructed for nutrient removal, it is often suggested that fertilizer be used to get the plants going because of nutrient limitations in the soil. Consider the paradox here if the wetland is being constructed to remove nutrients.

Principle 3. Ecosystems are open and dissipative systems.

Ecosystems are dependent on a steady input of energy from outside. This principle is an application to ecosystems of the second law of thermodynamics. As ecosystems must obey the conservation principle, of course, they must also obey basic scientific laws, including the basic laws of thermodynamics. The energy input is utilized to cover the energy needs for maintenance, respiration, and evapotranspiration.

Principle 4. Attention to a limited number of factors is most strategic in preventing pollution or restoring ecosystems.

Homeostasis of ecosystems requires accordance between biological function and the chemical composition of forcing functions. The biochemical functions of living organisms define their composition, although these are not to be considered as fixed concentrations but as ranges. Application of the principle implies that the flows of mass through ecological systems should be according to the biochemical stoichiometry. The flow of nutrients through ecosystems is generally in the ratio of $41:7:1$ on average if we accept the Redfield ratio (see Chapter 4). If we expect a wetland plant to take up 10 g of phosphorus, it is probably going to have to fix 410 g of carbon. If this is not the case, the elements in surplus will be exported to adjacent ecosystems and make an impact on the natural balances and processes there. Investigation on a well-managed Danish farm has shown that it is possible to reduce pollution and save resources by a complete material flow analysis of a farm. The results will not only lead to reduction in the emission level of pollutants, but will often also imply cost reductions.

Several restoration projects to reverse eutrophication of lakes and coastal waters have considered this principle. Designing ecological solutions to address the limiting nutrient of the eutrophication of a lake or coastal area is the appropriate approach. Inland lakes are generally phosphorus limited, whereas coastal water eutrophication requires attention primarily to nitrogen. Selecting the appropriate restoration method will further reduce the limiting nutrient and have the optimum results.

Similarly, if an ecosystem is being restored, it is important to focus on the most important resource. If mine land is being restored, the availability of propagules to allow restoration might be the most appropriate limiting factor. If a wetland is being restored, the most appropriate limiting factor might be proper hydrologic conditions.

Principle 5. Ecosystems have some homeostatic capability that results in smoothing out and depressing the effects of strongly variable inputs.

Homeostatic capability is limited. Once it is exceeded, the system breaks down. Several homeostatic mechanisms are known in biology: for instance, the maintenance of pH in our blood and the maintenance of the body temperature of warm-blooded animals. The homeostatic capability may be expressed by means of the concept of ecological buffer capacity (see Chapter 4). As the homeostatic capability is limited, so are the buffer capacities. It is obviously important in environmental management to know and respect the buffer capacities, as otherwise, the ecosystem may change radically and even collapse.

Principle 6. Match recycling pathways to the rates to ecosystems to reduce the effect of pollution.

This principle is related to principle 2. The application of sludge as a soil conditioner illustrates this principle very clearly. The recycling rate of nutrients in agriculture has to be accounted for in any use of sludge. If sludge is applied faster than it can be utilized by the landscape, a significant amount of the nutrients might contaminate the streams, lakes, and/or groundwater adjacent to the agricultural ecosystem. If, however, the influence of the temperature on nitrification and denitrification, the hydraulic conductivity of the soil, the slope of field, and the rate of plant growth are all considered in an application plan for manure, the loss of nutrients to the environment will be maintained at a very low and probably acceptable level. This can be achieved through ecological models to develop a plan for the application of sludge.

Elements are recycled in agrosystems but far less than occurs in nature. For the last couple of decades, animal husbandry has largely become separated from plant production. This has made internal cycling more difficult to achieve, of less economic value, and thus less attractive. We have come to accept losses of the relatively inexpensive, easy-to-apply artificial fertilizers and to compensate for this by increasing their application.

Principle 7. Design for pulsing systems wherever possible.

Very few ecological systems are the same every day. They go through weather seasons, dry and wet seasons and years, floods and droughts, and even seasonal disturbances caused by humans. At least for the pulses brought about by nature, ecological systems have a principle called *pulse stability* or nature's *pulsing paradigm* (H. T. Odum, 1982; W. E. Odum et al., 1995; E. P. Odum, 2000). Coastal marshes have tides twice per day, yet are both productive and stable. Bottomland forests are flooded seasonally, yet are among the most productive forests in any given region. Ecosystems with pulsing patterns often have greater productivity, biological activity, and chemical cycling than do systems with relatively constant patterns.

Ecological engineering of an estuary in Brazil, the Cannaneia, illustrates the recognition of pulsing forces and how it is possible to take advantage of them in ecological engineering. The shores of the islands along the estuary and the coast are very productive mangrove wetlands, and the entire estuary is an important nesting area for fish and shrimp. A channel was built to avoid flooding upstream, where productive agricultural land is situated. The construction of the channel has caused a conflict between farmers, who want the channel open, and fishermen, who want it closed, due to its reduction of the salinity in the estuary (the right salinity is of great importance for mangrove wetlands). The estuary is exposed to tide that is important for maintenance of good water quality with a certain minimum of salinity. The conflict can be solved by use of an ecological engineering approach that takes advantage of the pulsing force (the tide). A sluice in the channel could be constructed to discharge the fresh water when it is most appropriate. The tide would in this case be used to transport the fresh water to the sea as rapidly as possible. The sluice should be closed when the tide is on its way into the estuary. The tidal pulse is frequently selectively filtered to produce an optimal management situation.

Principle 8. Ecosystems are self-designing systems.

The more one works with the self-designing ability of nature, the lower the costs of energy to maintain that system. Much of engineering and landscape management is done to counteract the process of self-design. Traditionally, engineers design structures so that "nothing is left up to nature." For example, the biodiversity of agricultural fields would be significantly higher if pesticides were not used and nature were left to rule on its own.

Self-design implies adaptation at the ecosystem level. The properties of the species in the ecosystem are changed currently in accordance with the prevailing conditions, either for the next generation by dominant heritage of the properties best fitted to the prevailing conditions, or by complete or partial replacement of species with different tolerances. Other species waiting in the wings take over if they have a combination of properties better fitted to the (new) emergent conditions. Self-organization of the ecosystem may be understood as a directional change of the species composition of the ecosystem when its surroundings change, even when that change is a result of human activities.

Whereas self-designing systems are able to implement sophisticated regulations before violent fluctuations or even chaotic events occur, heavily managed systems such as modern agricultures attempts to regulate undesired organisms chemically, using pesticides. This very coarse regulation sometimes causes more harm than anticipated: for instance, when insect predators are affected more than the insects. The conclusion seems clear: Don't eliminate well-working natural regulation mechanisms (i.e., maintain a pattern of nature within ecological systems).

The use of constructed wetlands in lake restoration is an example of the application of this principle of self-design taken from aquatic ecosystems. If we design a wetland to remove the nutrients partially from streams entering the lake, the lake can itself self-design and reduce the level of eutrophication accordingly. The constructed wetland will also continue to self-design. The diversity (complexity) and nutrient removal efficiency will increase gradually, provided that the wetland is not overmanaged and is designed ecologically.

Principle 9. Processes of ecosystems have characteristic time and space scales that should be accounted for in environmental management.

Environmental management should consider the role of a certain spatial pattern for the maintenance of biodiversity. Violation of this principle by drainage of wetlands and deforestation on too large a scale has caused desertification. Wetlands and forests maintain high soil humidity and regulate precipitation. When the vegetation is removed, the soil is exposed to direct solar radiation and dries, causing organic matter to be burned off. Application of excessively large fields prevents wild animals and plants from finding their ecological niches as important components in the pattern of agriculture and more-or-less untouched nature. The solution is to maintain ditches and hedgerows as corridors in the landscape or as ecotones between agricultural and other ecosystems. Fallow fields should be planned as contributors to the pattern of the landscape. The example above on use of the tide to transport fresh water to the sea as rapidly as possible to the sea may also be used to illustrate this principle of using the right time and space scales in ecological engineering.

Principle 10. Biodiversity should be championed to maintain an ecosystem's self-design capacity.

Chemical and biological diversity can contribute to the buffering capacity and self-designing ability of ecosystems. Thereby a wide spectrum of buffer capacities is available to meet the impacts from anthropogenic pollution. Biodiversity plays an important role in buffer capacity and the ability of the system to meet a wide range of possible disturbances by use of the ecosystem's self-designing ability. There are many different buffer capacities, corresponding to any combination of a forcing function and a state variable. Farms that cultivated mixed cultures give a higher yield and are less vulnerable to disturbances: for instance, attacks by herbivorous insects. Mine lands restored with a variety of plant species will be better able to cope with diseases. Wetlands seeded with many species may select one or more of the unanticipated species as its dominant vegetation cover. Restoration of lakes by use of biomanipulation usually increases the biodiversity and some buffer capacities.

Principle 11. Ecotones, transition zones, are as important for ecosystems as membranes are for cells.

Ecological engineering should consider the importance of transition zones. Nature has developed *ecotones* to make soft transitions between two ecosystems. Ecotones may be considered as buffer zones that are able to absorb undesirable changes imposed on an ecosystem from adjacent ecosystems. We must learn from nature and use the same concepts when we design interfaces between human-made ecosystems (agriculture, human settlements) and nature. For example, in wetlands and riparian forests in river lowlands, adsorbing nitrates from agriculture is one of the major issues in environmental management for protecting coastlines (see, e.g., Mitsch et al., 2001). Some countries require buffer zones between human settlements and the coasts of lakes or marine ecosystems (in Denmark it is 50 m or between arable land and streams or lakes). The role of littoral zone protection in lake management is another obvious example. A sound littoral zone with a dense vegetation of macrophytes will be able to absorb contamination before it reaches the lake and will thereby be for a lake as a membrane is for a cell.

Principle 12. Coupling between ecosystems should be utilized wherever possible.

An ecosystem cannot be isolated—it must be an open system, because it needs an input of energy to maintain the system. The coupling of agricultural systems to natural systems leads to transfer of pesticides and nutrients from agriculture to nature, and measures should be taken to (almost) complete utilization of pesticides and nutrients in the agricultural system: for instance, by implementation of a proper fertilization plan accounting for these transfer processes. Ecological management should always consider all ecosystems as interconnected systems, not as isolated subsystems. This means that not only local but also regional and global effects have to be considered.

Principle 13. The components of an ecosystem are interconnected, interrelated, and form a network, implying that direct as well as indirect effects of ecosystem development need to be considered.

An ecosystem is an entity in which everything is linked to everything else. Any effect on any component in an ecosystem is therefore bound to have an effect on all components in the ecosystem either directly or indirectly (i.e., the entire ecosystem will be changed). It can be shown that indirect effects are often more important than direct effects (Patten, 1991). Ecotechnology should attempt to take indirect effects into account; management considering only direct effects often fails. There are numerous examples of the use of pesticides against herbivorous insects that also might have a pronounced effect on carnivorous insects and therefore result in the opposite effect on herbivorous insects than that intended.

Creating a wetland to clean municipal wastewater or agricultural runoff will attract muskrats, geese, and other animals. Building dikes around wetlands invites muskrat damage. Not providing riparian vegetation encourages geese proliferation.

The use of ecotechnology in the abatement of toxic substances in an aquatic ecosystem requires that this principle be considered. The biomagnification of toxic substances through the food chain is a result of the interconnectedness of ecological components. Due to the biomagnification, it is necessary to aim at a far lower concentration of toxic substances in aquatic ecosystems, to avoid undesirably high concentrations of toxic substances in fish used for human consumption.

Principle 14. An ecosystem has a history of development.

The components of ecosystems have been selected to cope with the problems that nature has imposed on ecosystems for millions of years. The high biodiversity of old ecosystems compared with immature ecosystems is another important feature of this principle. The structure of mature ecosystems should therefore be imitated in the application of ecological engineering. An ecosystem with a long history is better able to cope with emissions from its natural environment than is an ecosystem with no history. Conversely, it should be understood that ecological systems do not happen overnight. Attempting to create instant forested wetlands or forested mine land restoration sites will lead to disappointment. Ecosystems take time to develop. The rule of thumb used in the United States—that a wetland creation or restoration should be well on its way to success in five years—is not an ecologically sound practice in many cases (NRC, 2001). Sometimes it takes decades.

Principle 15. Ecosystems and species are most vulnerable at their geographical edges.

When ecological engineering involves ecosystem creation or restoration, the system will have enhanced buffer capacity if the species are in the middle range of their environmental tolerance. Ecosystem manipulation should therefore consider careful selection of the species involved, in accordance with this principle. Ecologically sound planning will use this principle and avoid the use of biological components that are at their geographical edges. This rule is, of course, important for both terrestrial and aquatic ecosystems.

Principle 16. Ecosystems are hierarchical systems and are parts of a larger landscape.

It is important to maintain landscape diversity, such as hedges, wetlands, shorelines, ecotones, and ecological niches. They will all contribute to the health of the entire landscape. Integrated agriculture can more easily follow this principle than can industrialized agriculture, because it has more components to use for construction of a hierarchical structure. In an aquatic landscape, a diversity of habitats, including shallow littoral zones, benthic zones, the epilimnion, and the hypolimnion, have different biogeochemical and food chain roles. Each develops the right conditions with respect to oxygen, pH, nutrients, and temperature to maintain various organisms (the next lower level in the hierarchy) fitted to these zones.

Principle 17. Physical and biological processes are interactive. It is important to know both physical and biological interactions and to interpret them properly.

This principle originated in ecohydrology (Zalewski et al., 1997), in which three hydrological processes are integrated with biota dynamics to achieve new operational strategies. Zalewski and Wagner (2000) give an illustrative example of the utilization of this principle. The regulation of the water level during the fish-spawning period is an effective way to reduce an excessive density of young fish and so to maintain a high concentration of filtering zooplankton. By reducing juvenile fish density at the beginning of the summer to a level less than 5 fish per square meter, the zooplankton concentration stabilizes at 12 to 16 mg L^{-1}. This is sufficient to reduce algae by 80 percent and to avoid toxic blooms.

The example of the proper use of tides to transfer fresh water rapidly to the open sea described above is another illustration of the application of this principle. Regulation of the flow regime to increase biodiversity in wetlands and river floodplains has also been attempted. Flow velocity in a vegetated channel influences macrophyte growth conditions, and submerged plants, in turn, exert an influence on flow characteristics such as velocity, turbulence, and friction. And a natural system does not passively accept changes in hydrology or ecosystem structure. Mitsch and Gosselink (2000) describe the myriad of organisms that regulate wetland systems and their hydrology: alligators, beavers, geese, and muskrats. It is interesting that these "controlling organisms" have been called *ecosystem engineers* by some ecologists (Jones et al., 1994, 1997; Alper, 1998), who were totally unaware that the field of ecological engineering (where humans influence ecosystems) was developing at the same time.

Principle 18. Ecotechnology requires a holistic approach that integrates all interacting parts and processes as far as possible.
This principle is consistent with a holistic view of ecosystems: The system is more than its parts, and ecosystems have emergent properties. Proper use of ecotechnology requires a holistic view of ecosystems. The development of ecosystem models is in most cases compulsory to obtain a sufficient overview of all possible environmental management strategies. This principle is also part of the ecohydrology concept developed by Zalewski (2000b), where the use of holistic and integrated approaches, using all ecosystem properties, is emphasized. Dubnyak and Timchenko (2000) show that oxygen concentration and algal succession are dependent on hydrodynamic processes and the seasonal climatic conditions in reservoirs. It is therefore possible to optimize water quality as a function of time by playing on the interactions between hydrodynamics and the dynamics of ecological processes.

Principle 19. Information in ecosystems is stored in structures.
Structures are a result of the input of energy that is utilized to move away from entropy. Such structures include not only organisms but also the physical structure of the landscape. Size is an important characteristic of structures. Organism size determines many important features of life, such as the rate of development, speed of movement, and the range of areas they inhabit. A certain minimum size of structures surrounding the organisms is necessary to satisfy their needs.

II

APPLICATIONS OF
ECOLOGICAL ENGINEERING

6

LAKE AND RESERVOIR RESTORATION

Most of the restoration methods mentioned in this chapter are applied to reducing eutrophication of lakes and reservoirs. The word *eutrophic* generally means "nutrient-rich." Naumann introduced the concepts of oligotrophy and eutrophy in 1919. He distinguished between oligotrophic lakes, containing little planktonic algae, and eutrophic lakes, containing much phytoplankton. The eutrophication of lakes in Europe and North America has grown rapidly during the last decade, due to increased urbanization and discharge of nutrient per capita. The production of fertilizers has grown exponentially in this century and the concentration of phosphorus in many lakes reflects this growth. The word *eutrophication* is used increasingly in the sense of the artificial addition of nutrients, primarily nitrogen and phosphorus, to water. Eutrophication is generally considered to be undesirable, although it is not always so.

6.1 LAKE AND RESERVOIR TROPHIC STATUS

The green color of eutrophic lakes makes swimming and boating unsafe because of increased turbidity. Furthermore, from an aesthetic point of view, the chlorophyll concentration should not exceed 100 mg m^{-3}. However, the most critical effect from an ecological viewpoint is the reduced oxygen content of the hypolimnion, caused by the decomposition of dead algae. Eutrophic lakes might show high oxygen concentrations at the surface during the summer, but low oxygen concentrations in the hypolimnion, which may cause fish kill (Figure 6.1). On the other hand, increased nutrient concentration may be profitable for shallow ponds used for commercial fishing, as the algae directly or indirectly form food for the fish population.

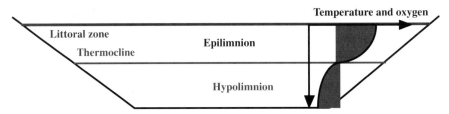

Figure 6.1 Thermal stratification pattern that develops in dimictic lakes in the summer. Epilimnion, hypolimnion, and littoral zone are shown. A plot of temperature and oxygen versus the depth for a typical summer situation for deep temperate lakes is also illustrated.

About 16 to 20 elements are essential for the growth of freshwater plants (Table 6.1). As described in Chapter 4, one of these elements is usually the limiting factor relative to need in ecosystems. In lakes and reservoirs, the present concern about eutrophication usually relates to the rapidly increasing amounts of phosphorus and nitrogen, which are normally present at relatively low concentrations. Of these two elements, phosphorus is considered the major cause of eutrophication of lakes, as it was formerly the growth-limiting factor for algae in the majority of lakes, but its use has increased greatly during the last decades. The importance of phosphorus as the limiting factor is shown in Figure 6.2, where the maximum algal concentration, expressed as μg chlorophyll a per liter, is plotted against the phosphorus concentrations for several lakes. The correlation is obvious.

Nitrogen is a limiting factor in the number of East African lakes as a result of the nitrogen depletion of soils by intensive erosion in the past. Today, nitrogen may become limiting to growth in lakes as a result of the tremendous increase in the phosphorus concentration caused by discharge of wastewater, which contains relatively more phosphorus than nitrogen. While algae uses 4 to 10 times more nitrogen than phosphorus, wastewater generally contains

Table 6.1 Average freshwater plant composition on wet basis

Element	Plant Content (%)	Element	Plant Content (%)
Oxygen	80.5	Chlorine	0.06
Hydrogen	9.7	Sodium	0.04
Carbon	6.5	Iron	0.02
Silicon	1.3	Boron	0.001
Nitrogen	0.7	Manganese	0.0007
Calcium	0.4	Zinc	0.0003
Potassium	0.3	Copper	0.0001
Phosphorus	0.08	Molybdenum	0.00005
Magnesium	0.07	Cobalt	0.000002
Sulfur	0.06		

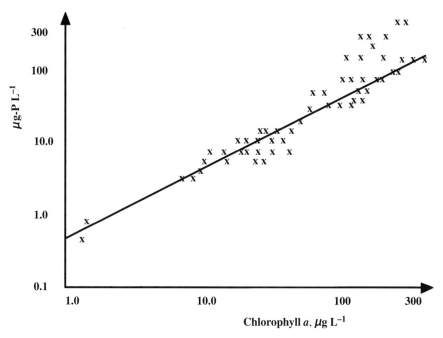

Figure 6.2 Algae biomass (summer maximum) versus orthophosphtate for 56 lakes in England and Denmark.

only three times as much nitrogen as phosphorus. Furthermore, nitrogen accumulates in lakes to a lesser extent than phosphorus, and a considerable amount of nitrogen is lost by denitrification (nitrate to N_2).

The important elements all participate in process cycles. These cycles contain the processes that determine eutrophication. The growth of phytoplankton is the key process in eutrophication, and it is therefore of great importance to understand the interacting processes regulating its growth, which requires that the entire cycle be considered. The cycles of elements in lakes are illustrated for nitrogen and phosphorus in Figure 4.3. The diagrams show, respectively, 22 and 19 processes for the nitrogen and phosphorus cycles in lakes.

Eutrophication Process

The sequence of events leading to eutrophication often occurs as follows. Oligotrophic waters often have an N/P ratio greater than or equal to 10, which means that phosphorus is less abundant than nitrogen relative to the needs of phytoplankton. If sewage is discharged into the lake, the ratio will decrease, since the N : P ratio for municipal wastewater is about 3 : 1, and consequently, nitrogen will be less abundant than phosphorus relative to the needs of phytoplankton. Municipal wastewater typically contains 30 mg-N L^{-1} and 10 mg-P L^{-1}. In this situation, however, the best remedy for excessive algal growth is not necessarily the removal of nitrogen from the sewage, because the mass balance might show that nitrogen-fixing algae would produce an uncontrollable input of nitrogen into the lake.

It is necessary when determining the solution to eutophic lakes and reservoirs first to establish a mass balance for the nutrients. This will often reveal that the input of nitrogen from nitrogen-fixing blue-green algae, precipitation, and tributaries is already contributing too much to the mass balance for any effect to be produced by nitrogen removal from sewage. On the other hand, the mass balance may reveal that most of the phosphorus input (often more than 95 percent) comes from sewage, thus demonstrating that it is better management to remove phosphorus from sewage rather than nitrogen. It is therefore not important which nutrient is limiting but which nutrient can most easily be made to limit algal growth.

Lakes can be classified in accordance with their primary production: the *oligotrophic–mesotrophic–eutrophic series.* Typical oligotrophic lakes are deep, with the hypolimnion larger than the epilimnion. Littoral plants are scarce and the plankton density is low, although the number of species can be large. Due to the low productivity, the hypolimnion does not suffer from oxygen depletion. The nutrient concentration is low and plankton blooms are rare, so the water is highly transparent. An approximate relationship between productivity and transparency is shown in Figure 6.3. Here the transparency (in meters) is plotted versus the maximum productivity (g-C m^{-3} day^{-1}). Although this relationship does not always apply, as the transparency depends not only on the phytoplankton productivity but also on the concentration of inorganic suspended matter (e.g., clay) and the color of the water (very humic rich lakes are brownish), in most lakes the transparency is determined primarily by phytoplankton. Eutrophic lakes are generally shallower and have a higher phytoplankton concentration and thus a generally lower transparency. Littoral vegetation is abundant and summer and spring algal blooms are characteristic.

Another characteristic difference between oligotrophic and eutrophic lakes is the profile of the change in photosynthesis with depth. Most photosynthesis takes place in eutrophic lakes at a depth of 0 to 1 m from the surface (line 3 in Figure 6.4), while oligotrophic lakes have low photosynthesis that is more equally distributed over the depth (line 1 in Figure 6.4). Mesotrophic conditions are midway between these two extremes.

Modeling Eutrophication

These considerations have implied that the eutrophication process can be controlled most easily by a reduction in the nutrient budget. For this purpose a number of eutrophication models have been developed, which take a number of processes into account (see, e.g., Jørgensen, 1976, 1981, 1992; Orlob,

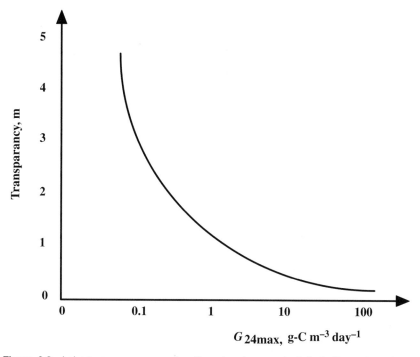

Figure 6.3 Lake transparency as a function of carbon productivity in the water column.

1981; Jørgensen and Bendoricchio, 2001). One of the simplest models to evaluate nutrient limitation in lakes is the *Vollenweider plot* (Vollenweider, 1969), which is much simpler to use than dynamic ecological models. But as it does not consider the dynamics of phytoplankton populations, annual variations, the sediments, and their interaction with the water body, it can give

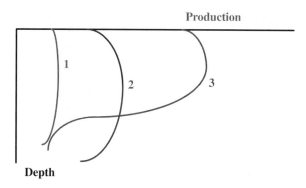

Figure 6.4 Productivity of phytoplankton per unit of volume versus the depth in a series of lakes: (1) oligotrophic, (2) mesotrophic, and (3) eutrophic.

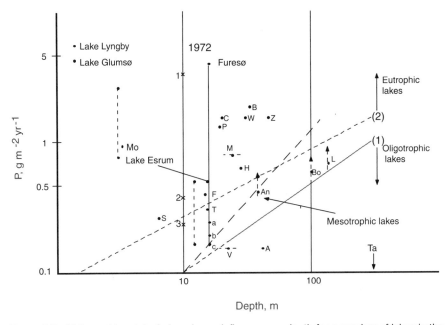

Figure 6.5 Vollenweider plot of phosphorus inflow versus depth for a number of lakes in the world. a, b, and c correspond to removal of 90, 95, and 99 percent of P input, respectively, for Lake Glumsø, 1972 (Jørgensen et al., 1973), Lyngby Lake, 1972, and Esrum Lake, 1972. A, Aegerisee; An, Lac Annecy; B, Baldeggersee; Bo, Bodensee; F, Furesoe (1954); G, Greifensee; H, Halwillersee; L, Lac Leman; M, Lake Mendota; Mo, Lake Moses; P, Pfäffikersee; T, Türlersee; Ta, Lake Tahoe; V, Vänern; W, Lake Washington; Z, Zürichsee.

only a crude picture of the possible control mechanisms in existence. A Vollenweider plot is shown in Figure 6.5, where phosphorus loading in g m^{-2} yr^{-1} is related to the depth of water. The diagram consists of three areas, corresponding to oligotrophic, mesotrophic, and eutrophic lakes.

A similar plot can be constructed for nitrogen, and a comparison of the two diagrams can show approximately whether a possible reduction in the nitrogen loading would be a better management solution than a reduction in the phosphorus loading. Vollenweider (1975) later improved these considerations by taking the input, output, and net loss to the sediment into consideration and by using a correction factor for stratified lakes. In cases where these improvements are required, it may be better to use a dynamic ecological model. Under all circumstances, Vollenweider's plot should be used as a first approximation.

6.2 RESTORATION TECHNIQUES

The most important methods are listed below, and a brief description of their application, advantages, and disadvantages is given for each approach. All applicable methods are listed here; it should be clear that some are more ecological and sustainable than others.

Diversion of Wastewater

Wastewater diversion has been used extensively to rehabilitate lakes, often replacing wastewater treatment. Discharge of effluents into an ecosystem that is less susceptible than the one used at present is, as such, a sound principle, which under all circumstances should be considered, but quantification of all the consequences has often been omitted. Diversion might reduce the number of steps in the treatment but cannot replace wastewater treatment totally, as discharge of effluents, even to the sea, should always require at least mechanical treatment to eliminate suspended matter. Diversion has often been used with a positive effect when eutrophication of a lake has been the dominant problem. Canalization, either to the sea or to the lake outlet, has been used as a solution in many cases of eutrophication. However, effluents must be considered as a freshwater resource. If it is discharged into the sea, effluent cannot be recovered; if it is stored in a lake, after sufficient treatment, of course, it is still a potential water resource. It is far cheaper to purify eutrophic lake water to an acceptable drinking-water standard than to desalinate seawater.

Diversion is often the only possibility when a massive discharge of effluents goes into a susceptible aquatic ecosystem (a lake, river, fjord, or bay). The general trend has been toward the construction of larger and larger wastewater plants, but this is quite often an ecologically unsound solution. Even though the wastewater has received a multistep treatment, it will still have a large amount of pollutants relative to the ecosystem, and the more massive the discharge is at one point, the greater the environmental impact will be. If canalization is a significant part of the overall cost of handling wastewater, it might often turn out to be both a better and cheaper solution to have smaller treatment units with individual discharge points. Although diversion is not considered an ecotechnological method based on sound ecological principles, a number of successful applications of diversion has been reported in the limnological literature. The most frequently cited case of wastewater diversion is probably the restoration of Lake Washington in Seattle, Washington. Wastewater was diverted from the lake to coastal Puget Sound in the 1960s, resulting in immediate improvement in Lake Washington (Figure 6.6). If diversion is accompanied by adding low-nutrient water from other sources, the recovery of the lake will, of course, take place faster.

Removal of Superficial Sediment

Sediment removal can be used to support the recovery process of very eutrophic lakes and of areas contaminated by toxic substances (e.g., harbors). This method can be applied in small ecosystems only with great care. Sedi-

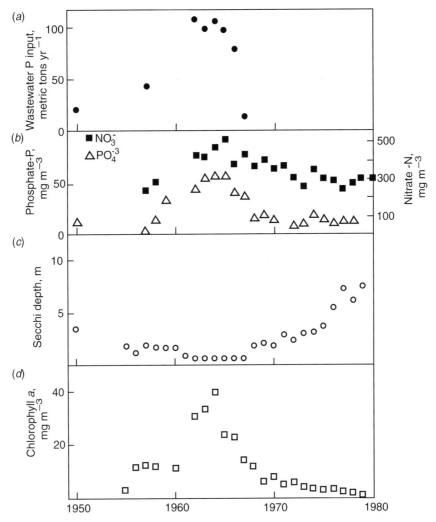

Figure 6.6 Control of eutrophication in Lake Washington, Seattle, Washington for 1950–1980 by diversion of wastewater. Diversion of sewage outflows to nearby Puget Sound occurred over the period 1963–1968. Patterns are shown for (a) wastewater P inflow (b) phosphates and nitrates, (c) Secchi disk reading, and (d) chlorophyll a in top 10 m during July and August. (Data from Edmondson and Lehman, 1981; from Laws, 1993, copyright 1993; reprinted with permission from John Wiley & Sons, Inc.)

ments have a high concentration of nutrients and many toxic substances, including trace metals. If a wastewater treatment scheme is initiated, the storage of nutrients and toxic substances in the sediment might prevent recovery of the ecosystem due to exchange processes between sediment and water. Anaerobic conditions might even accelerate these exchange processes; this is often observed for phosphorus, as iron(III) phosphate reacts with sulfide and forms iron(II) sulfide by release of phosphate. The amount of pollutants stored in the sediment is often significant, as it reflects the discharge of untreated wastewater for the period prior to the introduction of a treatment scheme. Thus, even though the retention time of the water is moderate, it might still take a very long time for the ecosystem to recover.

The removal of sediment can be made mechanically or by use of pneumatic methods. The method is, however, costly to implement and has therefore been limited in use to smaller systems. Perhaps the best known case of removal of superficial sediment occurred in Lake Trummen in Sweden, where 40 cm of the superficial sediment was removed. The transparency of the lake improved considerably, but it decreased again due to the phosphorus in overflows from rainwater basins. Treatment of the overflow after the removal of superficial sediment might give a better result.

Uprooting and Removal of Macrophytes

Uprooting and removal of macrophytes has been widely used in streams and to a certain extent in reservoirs, where macrophytes have caused problems in the turbines. The method can, in principle, be used wherever macrophytes are a significant result of eutrophication. A mass balance should always be set up to evaluate the significance of the method compared with the total nutrient input. Collection of the plant fragments should be considered under all circumstances. Simultaneous removal of nutrients from the effluent should also be considered.

Coverage of Sediment by an Inert Material

Covering sediment with inert material is an alternative to removal of superficial sediment. The idea is to prevent the exchange of nutrients (or perhaps, toxic substances) between sediment and water. Polyethylene, polypropylene, fiberglass screen, or clay is used to cover the sediment surface. The general applicability of the method is limited due to the high cost, even though it might be more moderate in cost than removal of superficial sediment. It has been used in only a few cases, and a more general evaluation of the method is still lacking. Based on our principles of using nature to solve problems, it is marginally ecological to use synthetic materials. Clay is somewhat more natural. Also, this approach treats the symptoms of the problem, not the cause.

Siphoning Hypolimnetic Water from Reservoirs

Figure 6.7 illustrates the use of siphoning and ion exchange of hypolimnetic water. In reservoirs or large ponds, this approach is feasible for reducing the causes of epilimnetic eutrophication and can be used over a longer period and thereby gives a pronounced overall effect. The effect depends on a significant difference between the nutrient concentrations in the epilimnion and

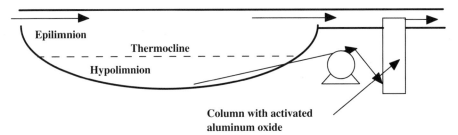

Figure 6.7   Application of siphoning and ion exchange of hypolimnetic water. The dashed line indicates the thermocline. The hypolimnetic water is treated by activated aluminum oxide to remove phosphorus.

the hypolimnion, which is often the case if the lake or reservoir has a pronounced thermocline. This implies, on the other hand, that the method will have an effect only during the period of the year when a thermocline is present (in many temperate lakes from May to October–November), but as the hypolimnetic water might have concentrations five fold or more than the epilimnetic water, it might have a significant influence on the nutrient budget to apply the method anyhow.

As hypolimnetic water is colder and poorer in oxygen, the thermocline will move downward and the possibility of anaerobic zones will be reduced. This might have an indirect effect on the release of nutrients from the sediment. If there are lakes or reservoirs downstream, the method cannot be used, as it removes but does not solve the problem. A possibility in such cases would be to remove phosphorus from the hypolimnetic water before it is discharged downstream. The low concentration of phosphorus in hypolimnetic water (perhaps 0.5 to 1.0 mg L^{-1}) compared with wastewater makes it almost impossible to apply chemical precipitation. However, it will be feasible to use ion exchange, because the capacity of an ion exchanger is more dependent on the total amount of phosphorus removed and the flow than on the total volume of water treated.

Several lakes have been restored by this method, mainly in Austria, Slovenia, and Switzerland, with a significant decrease in the phosphorus concentration as a result. Generally, the decline in total phosphorus concentration in the epilimnion is proportional to the amount of total phosphorus removed by siphoning and to the time the process has been used. The method has relatively low costs and is relatively effective, but the phosphorus must be removed from hypolimnetic water before it is discharged if there are other lakes downstream.

Flocculation of Phosphorus

Either aluminum sulfate or iron(III) chloride can be added to lakes or reservoirs to stimulate the flocculation and subsequent settling of phosphorus from surface waters. Calcium hydroxide cannot be used, even though it is an excellent precipitant for wastewater, as its effect is pH-dependent and a pH of 9.5 or higher is required. The method is not generally recommended, as (1) it is not certain that all flocs will settle and thereby incorporate the phosphorus in the sediment, and (2) the phosphorus might again be released from the sediment at a later stage.

Water Circulation and Aeration

Circulation of water can be used to break down the thermocline. This might prevent the formation of anaerobic zones, and thereby the release of phosphorus from sediment. Aeration of lakes and reservoirs is a more direct way to prevent anaerobic conditions from occurring. Aeration of highly polluted rivers and streams has also been used to avoid anaerobic conditions. In the Danish Lake Hald, pure oxygen has been used instead of air. The water quality of the lake has been improved permanently since the oxygenation started. In most cases, however, the effect was not very great nor as permanent as with other techniques, such as siphoning of hypolimnetic water.

Hydrologic Regulation

Regulation of hydrology has been used extensively to prevent floods. More recently, it has also been considered as a workable method to change the ecology of lakes, reservoirs, and wetlands. If the retention time in a lake or a reservoir is reduced with the same annual input of nutrients, eutrophication will decrease due to decreased nutrient concentrations. The role of the depth, which can be regulated by use of a dam, is more complex. Increased depth has a positive effect on the reduction of eutrophication, but if the retention time is increased simultaneously, the overall effect cannot generally be quantified without the use of a model. The productivity of wetlands is highly dependent on the water level, which makes it highly feasible to control a wetland ecosystem by this method.

Fertilizer Control

Controlling high nutrient water from even getting into a lake or reservoir is, of course, the best action to prevent signs of eutrophication. Fertilizer control can be used in agriculture and forestry to reduce nutrient loss to the environment. Utilization of nutrients by plants depends on a number of factors [temperature, humidity of soil, composition, growth rate of plant (which again depends on a number of factors), chemical speciation of nutrients, etc.]. Models of all these processes are available today on computers, and in the near future, fertilization schemes will be determined by computer on the basis of all the aforementioned information. This will make it feasible to come closer to the optimum fertilization from an economic–ecological point of view.

The occurrence of cyanophyte blooms to a great extent determines the N:P ratio in the lake water. If the ratio is less than 5, at least 50 percent of the bloom is in the form of cyanophytes. By very low ratios (e.g., less than 2), an almost 100 percent cyanophyte bloom may be observed. Adjusting the ratio is possible to a certain extent, as the main source of phosphorus often is wastewater. The phosphorus concentration in treated wastewater can easily be reduced by chemical precipitation to 1 mg L^{-1} and even to 0.1 mg L^{-1}. To avoid cyanophyte blooms, it is important to utilize all the possibilities in hand—fertilizer control, wastewater treatment, and various restoration methods—to obtain the right N:P ratio, which means ≥ 7.

Wetlands or Impoundments as Nutrient Traps

Forming nutrient traps is an appropriate ecological approach to lake restoration, especially wherever nonpoint sources are significant. This method is discussed in detail in Chapter 10.

Calcium Hydroxide Neutralization

Calcium hydroxide is used widely to neutralize low pH values in streams and lakes in areas where acidic rain has a significant impact. Sweden spends about $100 million per year to neutralize acid in streams and lakes.

Algacides

Chemicals such as various copper salts (e.g., copper sulfate) were previously used widely in relatively small lakes, but are now rarely used due to the general toxicity of copper, which accumulates in the sediment and can thereby contaminate a lake for a very long time. The effect of copper on algae varies substantially from species to species. Blue-green algae are generally most sensitive to copper ions. Mitsch and Kaltenborn (1980) performed in situ measurements of metabolism in the euphotic zone of an Illinois lake (Figure 6.8). Although few differences were seen in a treated lake compared to a control lake, during one period about a week after $CuSO_4$ treatments, gross primary productivity appeared to be depressed. By 10 to 14 days after the treatment, the effects of the treatment on metabolism had disappeared.

Shoreline Vegetation

Shading by use of trees at the shoreline is a cost-effective method that can give an acceptable result for small lakes, due to their low area/circumference ratio. It is relatively ineffective in restoring large lakes because of the smaller edge/area ratio.

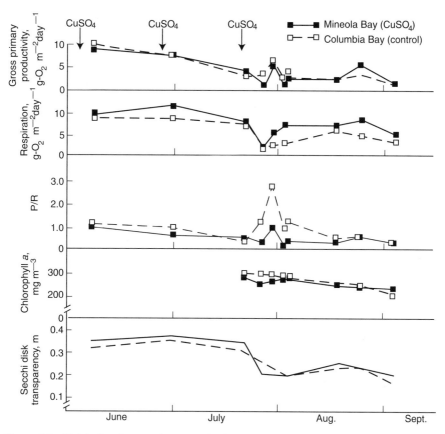

Figure 6.8   Metabolism (gross primary productivity and respiration), P/R rato, chlorophyll, and Secchi disk readings of an Illinois lake epilimnion subjected to three CuSO₄ treatments as indicated. A control bay of similar depth was used as a control area. (From Mitsch and Kaltenborn, 1980.)

Biomanipulation

Biomanipulation can be used as a method of lake restoration if the phosphorus concentration ranges from about 50 to 150 μg L^{-1}, depending on the lake. In this range two ecological structures are possible (Figure 6.9). When the phosphorus concentration initially is low and increases, zooplankton are able to maintain a relatively low phytoplankton concentration by grazing. Carnivorous fish are also able to maintain a low concentration of planktivorous fish, which implies relatively low predation on zooplankton. At a certain phosphorus concentration (about 120 to 150 μg L^{-1}), zooplankton is no longer able to control the phytoplankton concentration by grazing, and as carnivorous fish (e.g., Nile perch or pike) hunt by sight and the turbidity increases, planktivorous fish become more abundant, which involves more pronounced predation on zooplankton. In other words, the structure changed from control by zoo-

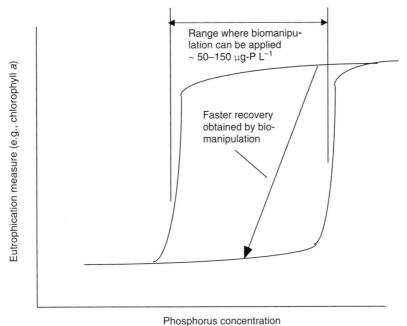

Figure 6.9 Hysteresis relation between nutrient level and eutrophication measured by chlorophyll *a* versus total phosphorus concentration. The possible effect of biomanipulation is shown. An effect of biomanipulation can only be expected in the approximate range 50 to 150 μg-P L^{-1}. Biomanipulation can hardly be applied successfully above 150 μg L^{-1} (see also de Bernardi and Giussani, 1995, and Jørgensen and de Bernardi, 1998).

plankton and carnivorous fish to control by phytoplankton and planktivorous fish. When the phosphorus concentration decreases from a high concentration, phytoplankton and planktivorous fish initially dominate the ecological structure. This structure can, however, be maintained until the phosphorus concentration is reduced to about 50 μg L^{-1}. There are therefore two possible ecological structures in the phosphorus range of approximately 50 to 150 μg L^{-1}. Biomanipulation (de Bernardi and Giussani, 1995) can be used in this range (and only in this range) to make a "shortcut" by removal of planktivorous fish and release of carnivorous fish. If biomanipulation is used above 150 μg-P L^{-1}, some intermediate improvement of the water quality will usually be observed, but the lake will sooner or later get an ecological structure corresponding to the high phosphorus concentration (i.e., a structure controlled by phytoplankton and planktivorous fish). Biomanipulation is a relatively cheap and effective method provided that it is applied in the phosphorus range where two ecological structures are possible. De Bernardi and Giussani (1995) give a comprehensive presentation of various aspects of biomanipulation. Simultaneously, biomanipulation makes it possible to maintain relatively high biodiversity, which does not change the stability of the system, but a higher biodiversity gives the ecosystem a greater ability to meet future, unforeseen changes without changes in the ecosystem function (May, 1977).

There are a number of cases where biomanipulation has been successful, but only if the phosphorus loading was reduced simultaneously and total phosphorus concentrations were below 150 μg L^{-1}. Benndorf (1990) mentions that consistent response to biomanipulation can only be foreseen with a loading of less than about 0.6 to 0.8 g-P m^{-2} yr^{-1}. This statement is in most cases completely in accordance with the results presented in Figure 6.9. The results in this figure have been explained theoretically by catastrophe theory (Bendoricchio, 1988) and by use of exergy as the goal function in a eutrophication model (Jørgensen and de Bernardi, 1998).

CASE STUDY
Bautzen Reservoir, Germany

Experience at the Bautzen Reservoir, Germany, illustrates typical results for biomanipulation used in lakes or reservoirs with too high a phosphorus concentration. The phosphorus loading is calculated to be 7 to 17 g-P m^{-2} yr^{-1}. The reservoir was created in 1973, and the pike population developed rapidly in the reservoir. Sport fishing began in 1976, and the pike population was decimated two years later. Biomanipulation started in 1977 with stocking of pond-raised pike-perch, and the fish catch was restricted simultaneously. Further stocking at rates between 20,000 and 80,000 pike-perch per year occurred in 1980, 1981, 1982, 1984, and 1988. Perch (planktivorous fish) were controlled by the pike-perch but not eliminated. After 1980, no decreases in algal biomass or total phosphorus concentration were observed.

This case study is in contrast to Lake Annone, Italy, and Lake Søbygaard, Denmark, where clear improvements in algal biomass and total phosphorus were observed. The zooplankton concentration increased significantly in both cases and simultaneously with a decrease in the phytoplankton concentration. Both cases were modeled by a structurally dynamic model (Jørgensen and di Bernardi, 1997; Jørgensen, 2002).

Biological Control

Water hyacinths (*Eichhornia crassipes*) and other macrophytes are pests in many tropic lakes and reservoirs. Many methods have been tested to abate this pollution problem. The best method tested up to now seems to be the use of beetles (i.e., a biological control method). The method has had at least partial success in Lake Victoria. Looking at the other view of these macro-phytes, they are themselves sinks for nutrients when the water flows through aquatic plant mats. Floating plants such as water hyacinths can serve as significant nutrient sinks, as shown in Figure 6.10 for a high-nutrient marsh in Florida.

Biological control has also been used as a removal process for heavy metals. Freshwater mussels can be applied for cadmium clearance (Jana and Das, 1997; Das and Jana, 1999). They found that *Lemna trisulca* was able to accumulate as much as 3.8 mg cadmium per gram of dry weight.

Recovery of Submerged Vegetation in Shallow Lakes

A shallow lake can maintain a clear-water state (very small concentrations of phytoplankton) by binding the nutrients to submerged vegetation. This is, however, possible only up to about 250 μg-P L^{-1} (Scheffer et al., 2001). Above this concentration, phytoplankton will take over. When the nutrient concentration afterward is reduced to below 250 μg-P L^{-1}, phytoplankton dominance will be maintained. The submerged vegetation will be reestablished at about 100 μg-P L^{-1}. The two states' submerged vegetation dominance/phytoplankton dominance therefore show hysteresis behavior similar to that of Figure 6.9, but in the phosphorus concentration range 100 to 250 μg-P L^{-1} in this case. This is, furthermore, in accordance with the results of a structurally dynamic model of a shallow lake, Lake Mogan in Turkey (Zhang et al., 2002). This implies that the recovery of a shallow lake that is dominated by phytoplankton by planting a suitable submerged vegetation can take place only in the range 100 to 250 μg-P L^{-1}; above 250 μg-P L^{-1} it is not possible to obtain a long-term effect using this restoration method.

6.3   SELECTION OF A RESTORATION METHOD

It is not possible to give general recommendations as to which restoration method to apply in a specific case. Most restoration problems are associated with eutrophication, and it is necessary in each case to use a eutrophication model to assess the effect of the restoration method and to compare the effects and the costs to decide which method gives most pollution abatement for the money. It is, in other words, necessary to set up a cost–benefit analysis. The following modifications in the eutrophication model must be carried out to account for the effects resulting from application of the restoration method:

1. The forcing functions provide nutrient input, but the hydraulic retention time will be changed.
2. Removal of superficial sediment implies that the sediment contains less phosphorus and nitrogen, which will, of course, change the release rate of these nutrients from the sediment to the water phase.

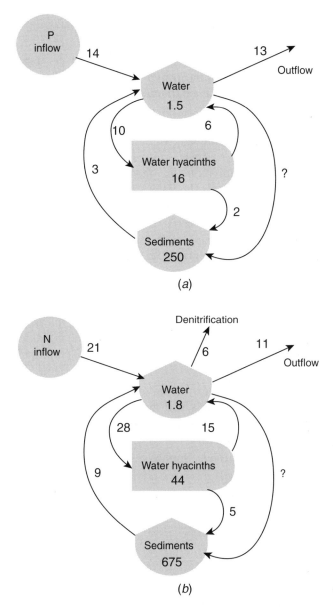

Figure 6.10 Fluxes and storages of (a) phosphorus and (b) nitrogen through a 21-ha water hyacinth (*Eichhornia crassipes*) marsh upstream of a lake in central Florida during summer months. Denitrification was estimated from the nutrient balance. Storage of P and N is in g m^{-2}; flows are in g m^{-2} month^{-1}. (From Mitsch, 1977.)

3. Removal of macrophytes corresponds to the removal of phosphorus and nitrogen in the harvested plants.

4. Coverage of sediment by inert material will have the same effect as modification 2 but will in many cases be more moderate in cost, particularly for deeper lakes.

5. Siphoning of hypolimnetic water corresponds to the removal of more nutrients (the concentration in hypolimnion to replace the concentration in epilimnion) with the outflowing water. It is necessary, of course, to examine what the effect of the higher nutrient concentration will be downstream. If there are other lakes downstream, it is inevitably that the nutrient must be removed, which is possible by a number of methods. For instance, for phosphorus adsorption on activated aluminum oxide and/or precipitation of phosphate with aluminum sulfate, iron(III) chloride, or polyaluminates can be used. This method is obviously applicable only to lakes with a thermocline for at least part of the year.

6. Flocculation of the phosphorus in the water phase implies that the phosphorus is removed once from the water phase to the sediment. Usually, it is necessary to apply this method several times.

7. Circulation and aeration of hypolimnion imply that the release rate of phosphorus and nitrogen is changed from the sediment to the water phase. Aerobic conditions usually imply that the release rate is lower, particularly for phosphorus than under anaerobic conditions.

8. Changes in hydrology mean that the forcing function hydraulic retention time is changed.

9. Fertilizer control and construction of a wetland to cope with nonpoint pollution of nutrients imply that the input of nutrients is reduced. The forcing functions in the model expressing the input of nutrients are changed correspondingly.

10. Use of calcium hydroxide implies a simple change of the lake water pH.

11. Use of algacides will increase the mortality rate of phytoplankton in the eutrophication model. The method is not recommendable as it implies direct discharge of toxic compounds into the lake (e.g., copper sulfate).

12. Shading by the use of trees changes the photosynthetic activity in the lake. The forcing function, solar radiation, in the model is reduced corresponding to the shading effect.

13. Biomanipulation is often a cost-moderate method with a good effect, provided that the phosphorus concentration is in the approximate range 50 to 150 μg-P L^{-1}.

A cost–benefit examination should be carried out prior to the selection of a method. Before a method is considered, it is important to determine whether it can work hand in hand with methods that are simultaneously reducing the loading (compare with the Lake Trummen study presented briefly above, where the superficial sediment was removed). Modeling should be widely used to set up the cost–benefit analysis. Eutrophication models are useful for assessing the benefits of many of the restoration methods presented, focusing on the solution of a eutrophication problem. Below is a case study using a rather complex eutrophication model, but in the practical situation, the model best fitted to the situation and the data available should, of course, be used. The questions to ask of the model are: What is the decrease in primary production? And what is the increase in transparency? For example, biomanipulation often seems to be an attractive and cost-moderate method, but it can be used only in the approximate phosphorus range 50 to 150 μg L^{-1}, whereas removal of sediment often will be very effective in the entire phosphorus range, but it is often prohibitively expensive.

CASE STUDY
Comparison of Lake Restoration Methods for a Danish Lake

Table 6.2 gives the result of a comparison of the effect and cost of five restoration methods for restoration of Lake Glumsø, Denmark, using a model presented in Jørgensen (1976) and Jørgensen and Bendoricchio (2001). The volume of the lake is about 0.5 million cubic meters (depth 2 m and a surface area of about 25 ha). The lake was very eutrophic, but wastewater has been diverted since 1983 to the River Sus downstream of the lake. Before 1983, primary production was about 1000 to 1100 g-C m^{-2} yr^{-1}, while it was reduced to about 500 g-C m^{-2} yr^{-1} during the period 1983–1988. The lake has a re-

Table 6.2 Comparison of restoration approaches for Lake Glumsø, Denmark[a]

Method	Primary Productivity (g-C m^{-2} yr^{-1})	Maximum Chlorophyll a (μg L^{-1})	Investment Cost (millions of dollars)	Annual Operating Cost (thousands of dollars)
Without restoration	500	360	—	—
Coverage of sediment	320	350	1	0
Removal of sediment	320	350	3.5	0
Precipitation of P in lake	460	360	0.6	0
Wetland: removal of nonpoint pollution	210	270	1.0	15
25% reduction of retention time	400	350	0.6	20

[a]Primary productivity and maximum chlorophyll were predicted by a eutrophication model.

tention time of about six months, which implies that the period 1983–1988 corresponds to an exchange of the water 10 times. The transparency in the same period increased from 18 cm at spring and summer blooms to about 60 cm. Maximum chlorophyll *a* concentration was reduced from about 850 μg L^{-1} to about 360 μg L^{-1} as a result of the diversion.

Encouraged by these results, the community considered various restoration methods. The effects were compared by a eutrophication model. The result of this investigation is summarized in Table 6.2, where the simulated effects the third year after restoration are shown. Biomanipulation was not considered because the phosphorus concentration was not sufficiently low.

7

STREAM AND RIVER
RESTORATION

River and stream restoration has become one of the most recent and intriguing aspects of ecological restorations in the landscape. Humans have spent centuries in both the developed and developing worlds trying to tame rivers and bring them under control. This manipulation has been for such reasons as enhancing river transportation, draining landscapes more quickly, providing water for municipal, industrial, and agricultural use, and carrying wastes somewhere downstream. Now the effort to control rivers has almost reversed itself in many developed countries, where dam removals, remeandering, and riparian stabilization are frequently considered.

Except in rare instances, rivers and streams now little resemble their "natural" conditions. Even if river channels have not been altered, flows from cities and control structures such as dams, weirs, and levees have changed the hydrology and hence the ecological conditions of flowing waters in every part of the world. Some classic cases of river changes that have had dramatic and unanticipated effects include the Aswan High Dam on the Nile, the straightening of the Kissimmee River in southern Florida, the concretization of the ephemeral streams of metropolitan Los Angeles, and pollution in industrialized rivers around the world, including dramatic cases in the Ohio River in the eastern United States, and the Rhine, Meuse, Thames, and Ruhr in Europe.

Works specific to the restoration of rivers include the comprehensive U.S. government document *Stream Corridor Restoration* (Federal Interagency Stream Restoration Working Group, 2001), edited books on river restoration (e.g., Boon et al., 2000), the chapter "Rivers and Streams" in the National Research Council report *Restoration of Aquatic EcoSystems* (NRC, 1992), several chapters in *The Rivers Handbook,* Vol. 2 (Calow and Petts, 1994), and

European river restoration paper collections (Hansen, 1996; Hansen and Madsen, 1998; Hansen et al., 1998). A complete description of riparian ecosystems, part of which is included in this chapter, is from Mitsch and Gosselink (2000). Special issues on river restoration have been compiled in *Ecological Engineering* (Nelson et al., 2000; Lefeuvre et al., 2002) and *BioScience* (Hart and Poff, 2002). Classic books on river ecology are by Hynes (1970) and Whitton (1975). The patriarch of river geomorphology is Luna Leopold, and the bible is the classic *Fluvial Processes in Geomorphology* (Leopold et al., 1964). Without knowing Leopold's book, you cannot restore rivers.

7.1 RIVER BASINS

The basic unit of study and restoration of any stream or river starts with its *watershed* (or *catchment* or *drainage basin*). Most larger rivers have three major geomorphic zones: erosion, storage and transport, and sediment deposition (Figure 7.1). The *zone of erosion is* in the headwaters and upper reaches of low-order streams. This zone is at the high altitudes, and if the river basin originates in mountainous areas, the stream course tends to be steep and straight, and the valleys are often V-shaped because they are scoured. The steep banks have narrow riparian zones. Flood frequency and duration vary widely depending on precipitation. If local geology permits, some headwater areas contain fairly extensive flat meadows that may support wetlands at high altitudes. The *zone of storage and transport* occurs below the zone of erosion. These mid-order streams are primarily conduits for sediment, nutrients, and water. They tend to be fairly steep, and their straight V- or U-shaped channels, with some coarse sediment deposition, form a narrow floodplain. Sediments

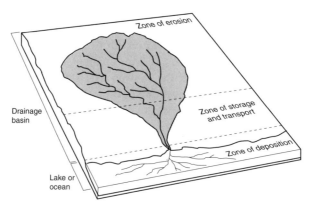

Figure 7.1 Three major geomorphic zones in the fluvial system. (From Faber et al., 1989; redrawn by Mitsch and Gosselink, 2000; copyright 2000; reprinted with permission from John Wiley & Sons, Inc.)

are often scoured during high-energy floods. Flooding is variable and depends on the size of the watershed, the gradient, and local precipitation. The *zone of deposition* is high-order and low-gradient. Sediment deposition is much greater than erosion and transport, and valley slopes are gentle. These two factors lead to the development of broad floodplains and sinuous and meandering stream channels. Sediments grade from coarse at the channel to fine at the periphery of the floodplain. The flooding of the riparian zone tends to be seasonal, characterized by one or a few long-duration spring floods. At their downstream ends, rivers typically debauch into flat, broad valleys, where the channels become braided and the flow is often unconfined. These depositional rivers are characteristic of coastal plains that run into estuaries.

River Channels

Within watersheds, streams and rivers can be classified in terms of stream order and length, while the watersheds can be described in terms of stream density. According to the system designed by Robert E. Horton, *stream order* provides a convenient way of describing the general size of a particular river. It is formally "a measure of the position of a stream in the hierarchy of tributaries" (Leopold et al., 1964). First-order streams have no tributaries, second-order streams have only first-order streams as tributaries, third-order streams have only first- and second-order tributaries, and so on (Figure 7.2).

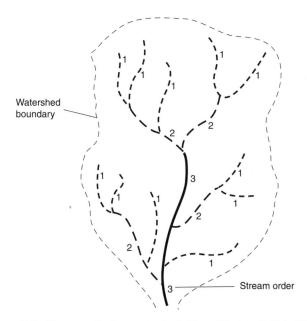

Figure 7.2 Stream order in a watershed. (From Ward and Elliot, 1995.)

River Geomorphology

Leopold (1994) describes rivers as "transporting machines" in which the potential energy at the upstream end is "progressively changed to kinetic form along the channel and the kinetic energy is transformed to heat, doing some work during the process." The work the river does includes both carrying sediments and eroding stream bottoms and banks. The source of energy—the potential energy of elevation—is fed by the solar-based precipitation–evapotranspiration hydrologic cycle.

The importance of the river to the floodplain and the floodplain to the river cannot be overemphasized. If either is altered, the other will change over time because floodplains and their rivers are in continuous dynamic balance between the building and the removal of structure. In the long term, floodplains result from a combination of the deposition of alluvial materials (*aggradation*) and the downcutting of surface geology (*degradation*).

Two major aggradation processes are thought to be responsible for the formation of most floodplains: deposition on the inside curves of rivers (point bars) and deposition from overbank flooding. "As a river moves laterally, sediment is deposited within or below the level of the bankfull stage on the point bar, while at overflow stages the sediment is deposited on both the point bar and over the adjacent flood plain" (Leopold et al., 1964). The resulting floodplain is made up of alluvial sediments (or alluvium) that can range from 10 to 80 m thick. In the lower reaches of the Mississippi River, for example, the alluvium, derived from the river over many thousands of years, generally progresses from gravel or coarse sand at the bottom to fine-grained material at the surface (Bedinger, 1981).

Degradation (downcutting) of floodplains occurs when the supply of sediments is less than the outflow of sediments, a condition that could be caused naturally with a shift in climate or synthetically with the construction of an upstream dam. There are few long-term data to substantiate the first cause, but a considerable number of before-and-after studies have verified stream degradation downstream of dams attributed to the trapping of sediments (Meade and Parker, 1985). In the absence of geologic uplifting, rivers tend to degrade slowly, and "downcutting is slow enough that lateral swinging of the channel can usually make the valley wider than the channel itself" (Leopold et al., 1964). The process is thus difficult to observe over short periods; both aggradation and degradation can only be inferred from the study of floodplain stratigraphy.

The formation of a riparian floodplain and terrace is shown in sequences A to B and C to E in Figure 7.3. When degradation occurs but some of the original floodplain is not downcut, that "abandoned" floodplain is called a *terrace*. Although it may be composed of alluvial fill, it is part of the active floodplain only during peak floods. Aggradation and degradation can alternate over time, as shown in the sequence C to E. A third case, a dynamic steady state, can exist if aggradation resulting from the input of sediments from

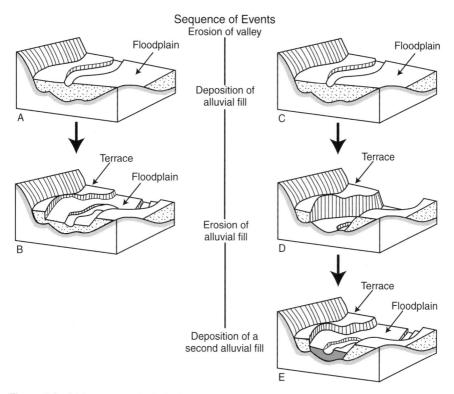

Figure 7.3   Major geomorphologic features of a southeastern U.S. floodplain. (From Leopold et al., 1964, and Brinson et al., 1981.)

upstream is balanced by the degradation or downcutting of the stream. Figure 7.3 demonstrates that the same surface geometry can result from two dissimilar sequences of aggradation and degradation.

7.2   RIPARIAN ECOSYSTEMS

The riparian zone of a river, stream, or other body of water is the land adjacent to that body of water that is, at least periodically, influenced by flooding (Mitsch and Gosselink, 2000). E. P. Odum (1981) described the riparian zone as "an interface between man's most vital resource, namely, water, and his living space, the land." Johnson and McCormick (1979) developed a definition of riparian ecosystems:

Riparian ecosystems are ecosystems with a high water table because of proximity to an aquatic ecosystem or subsurface water. Riparian ecosystems usually occur as an ecotone between aquatic and upland ecosystems but have distinct vegetation and soil characteristics. Aridity, topographic relief, and presence of depositional soils most strongly influence the extent of high water tables and associated riparian ecosystems. These ecosystems are most commonly recognized as bottomland hardwood and floodplain forests in the eastern and central U.S. and as bosque or streambank vegetation in the west. Riparian ecosystems are uniquely characterized by the combination of high species diversity, high species densities and high productivity. Continuous interactions occur between riparian, aquatic, and upland terrestrial ecosystems through exchanges of energy, nutrients, and species.

Riparian ecosystems are found wherever streams or rivers at least occasionally cause flooding beyond their channel confines or where new sites for vegetation establishment and growth are created by channel meandering (e.g., point bars). In arid regions, riparian vegetation may be found along or in ephemeral streams as well as on the floodplains of perennial streams. In most nonarid regions, floodplains and hence riparian zones tend to appear first along a stream "where the flow in the channel changes from ephemeral to perennial—that is, where groundwater enters the channel in sufficient quantity to sustain flow through non-storm periods" (Leopold et al., 1964).

Riparian ecosystems can be broad alluvial valleys several tens of kilometers wide or narrow strips of streambank vegetation in arid regions. Brinson et al. (1981) described the "abundance of water and rich alluvial soils" as factors that make riparian ecosystems different from upland ecosystems. They listed three major features that separate riparian ecosystems from other ecosystem types:

1. Riparian ecosystems generally have a linear form as a consequence of their proximity to rivers and streams.
2. Energy and material from the surrounding landscape converge and pass through riparian ecosystems in much greater amounts than those of any other wetland ecosystem; that is, riparian systems are open systems.
3. Riparian ecosystems are functionally connected to upstream and downstream ecosystems and are laterally connected to upslope (upland) and downslope (aquatic) ecosystems.

7.3 ECOLOGY OF RIVERINE SYSTEMS

Ecologists have reviewed river systems in terms of their ecological function and have developed two different ways of describing flowing water systems. The river continuum concept is related to the general differences in ecology along streams and rivers, going *longitudinally* along the river itself. The concept was developed mostly in low-order streams in the United States. Little attention is paid to lateral connections or to floodplains. The flood pulse concept, on the other hand, has been based on research done in the Amazon River and its tributaries, and it features the importance of seasonal patterns of stream flow and the importance of *lateral* exchange between the river and its floodplain.

River Continuum Concept

The *river continuum concept* (RCC) is a theory developed in the early 1980s to describe the longitudinal patterns of biota found in streams and rivers (Vannote et al., 1980; Minshall et al., 1983, 1985). According to the RCC, most organic matter is introduced to streams from terrestrial sources in head-water areas (Figure 7.4). The production/respiration (P/R) ratio is less than 1 (i.e., the stream is heterotrophic), and invertebrate shredders and collectors dominate the fauna. Biodiversity is limited by low temperatures, low light, and low nutrients. Organic matter is reduced in size as it travels downstream. In river midreaches, more light is available, phytoplankton prospers, and bio-

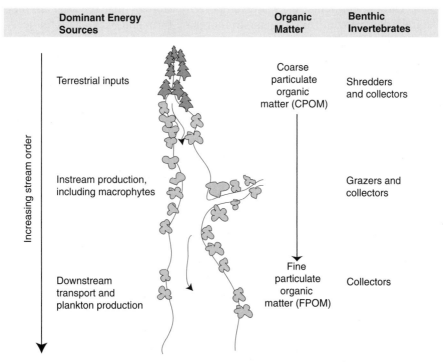

Figure 7.4 River continuum concept (RCC) showing transition from a small first-order stream to a large river. The left column indicates the relative importance of terrestrial, in-stream, or upstream energy sources to the aquatic food chain, while the column on the right indicates the relative importance of different feeding groups of invertebrates. (Redrawn from Johnson et al., 1995.)

diversity is highest. The P/R ratio is greater than 1. Organic matter input from upstream is fine; filter feeders dominate the flora. In braided reaches or where the floodplain is broad, however, the bank habitat is a major source of snags and logs that lead to debris dams that slow water flow and increase stream habitat diversity. The increased input of riparian coarse debris increases food diversity and heterotrophy, and the P/R ratio is less than 1. Terrestrial inputs are probably more nutritious than the fine reworked material from upstream. Even bacteria respond to the concentrations and sources of organic matter. The allelic frequency of metabolic enzymes is correlated with the habitat (McArthur, 1989). This suggests that bacterial flora and genetic selection pressures in a specific habitat (i.e., a floodplain–stream reach) are functions of linkages and interactions between stream and floodplain, including timing, quantity, quality, and source of organic material inputs. Finally, in the highest-order streams, riparian litter inputs are minor, and turbidity reduces primary productivity. Hence the system is heterotrophic again (P/R < 1) and diversity is often low.

There are two major corollaries to the RCC (Johnson et al., 1995). *Nutrient spiraling* or *resource spiraling* refers to the process whereby resources (organic carbon, nutrients, etc.) are temporarily stored, then released as they "spiral" downstream from organic to inorganic form and back again. In head-water streams, nutrient spirals are long. In large high-order rivers, nutrient spirals tend to be tight and short (nutrients are rapidly processed). The importance of backwaters, oxbows, and floodplains to river ecosystem function was virtually ignored in the RCC. The *serial discontinuity concept* developed by Ward and Stanford (1983, 1995) involves the effects that floodplains, dams, and the lateral dimension in general have on the functioning of a river system. Braided streams have the lowest diversity because of shifting sediments typical of these systems, while meandering streams have the highest diversity because of the frequent lateral movement of organisms from bodies of water on the wide floodplain and the spatial heterogeneity of the river–floodplain system.

Flood Pulse Concept

The RCC initially considered the importance of the riparian zone only in an indirect way by noting that small low-order streams are seen to be influenced by shading and abundant contributions of allochthonous organic matter. The importance of backwaters, oxbows, and floodplains to river ecosystem function was virtually ignored. Junk et al. (1989) developed a *flood pulse concept* (FPC) for floodplain–large river systems based on their experience in both temperate and tropical regions of the world (Figure 7.5). They dispute the river continuum concept as a generalizable theory because (1) most of the theory was developed from experience on low-order temperate streams, and (2) the concept is mostly restricted to habitats that are permanent and lotic. In the FPC, the pulsing of the river discharge is the major force controlling biota in river floodplains, and lateral exchange between the floodplain and

Fish Activity		Nutrient Flux	Productivity and Decomposition

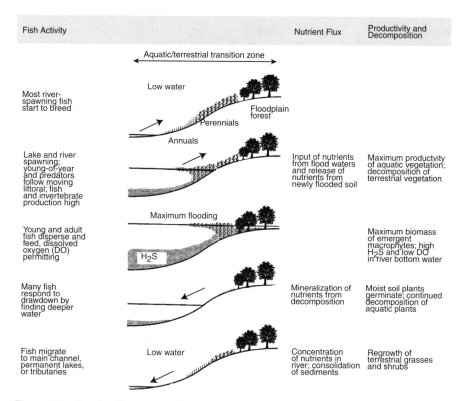

Most river-spawning fish start to breed

Lake and river spawning; young-of-year and predators follow moving littoral; fish and invertebrate production high

Input of nutrients from flood waters and release of nutrients from newly flooded soil

Maximum productvity of aquatic vegetation; decomposition of terrestrial vegetation

Young and adult fish disperse and feed, dissolved oxygen (DO) permitting

Maximum biomass of emergent macrophytes; high H_2S and low DO in river bottom water

Many fish respond to drawdown by finding deeper water

Mineralization of nutrients from decomposition

Moist soil plants germinate; continued decomposition of aquatic plants

Fish migrate to main channel, permanent lakes, or tributaries

Concentration of nutrients in river; consolidation of sediments

Regrowth of terrestrial grasses and shrubs

Figure 7.5 Flood pulse concept (FPC) for a river and its floodplain, illustrating five periods over the wet and dry seasons of a river. (Redrawn from Bayley, 1995.)

river channel and nutrient cycling within the floodplain "have more direct impact on biota than nutrient spiraling discussed in the RCC" (Junk et al., 1989). The FPC thus considers the river–floodplain exchange to be of enormous importance in determining both the productivity of the river and the adjacent riparian zone. Alternating dry and wet cycles optimize the productivity of the littoral zone and the adjacent forest, decomposition of all that is produced, and fish spawning and feeding. Bayley (1995) argues that biological productivity in general, particularly multispecies fish yield, is higher in river–floodplain systems such as that shown in Figure 7.5 over equivalent stable bodies of water.

7.4 RIVER RESTORATION TECHNIQUES

Based primarily on their experience in the rehabilitation of streams and rivers in Denmark, Hansen et al. (1996), describe a classification of river restoration as the following three types (Figure 7.6):

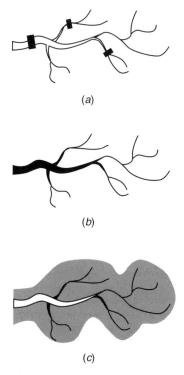

Figure 7.6   Three types of river restoration in terms of overall objectives of the project: (a) type 1, watercourse reaches; (b) type 2, continuity between reaches; (c) type 3, rehabilitation of river valleys. (From Hansen et al., 1996.)

1. *Rehabilitation of watercourse reaches.* The project is designed to improve conditions on short reaches of streams and rivers.
2. *Restoration of continuity between watercourse reaches.* The project is involved in restoring the passage of the stream along the river or streams course. The key is restoring free passage of the waterways.
3. *Rehabilitation of river valleys.* Projects include restoring long reaches of streams and the river valley as one hydrologic entity.

Specific types of projects under this classification are listed in Table 7.1.

The actual practice of river restoration has been defined best in projects that attempt to correct ecological problems that have been caused by previous human activity. We describe here several of the more prevalent restoration activities.

Removing Dams and Stream Impediments

One of the most obvious approaches to restoring rivers is to restore its free-flowing condition by the removal of dams and other stream impediments

Table 7.1 Examples of river restoration

Type 1: Rehabilitation of watercourse reaches
    Reach meandered
    Culverted reach opened to create better habitats
    Two-step cross-sectional profile created
    Lake established/reestablished in connection with the watercourse
    Ochre sedimentation basin established in connection with the watercourse
    Stones or gravel laid out
    Artificial fish hiding places established
    Other solid objects laid out
    Current concentrators established
    Sand traps constructed
    Trees and bushes planted
    Trees and bushes removed
    Artificial bed and/or bank established
    Artificial bed and/or bank removed
    Fences, watering places, etc.
Type 2: Restoration of continuity between watercourse reaches
    Obstruction replaced by riffle
    Obstruction replaced by meanders
    Bypass riffle established at preserved obstruction
    Riffle established at preserved obstruction
    Culverted reach opened to create free passage
    Culvert falls evened out (drop manhole removed, etc.)
    Greater water depth and/or current breakers in underpass culverts
    Falls evened out at culvert outlet/bridge
    Fish ladder/fish sluice established/removed
    Stream formerly dried up periodically restored completely
    Stream formerly dried up periodically restored partly
    Water pumped into stream to maintain flow in reach dried up periodically
    Otter pass established
Type 3: Rehabilitation of river valleys
    Water table and flooding frequency increased by:
    • Remeandering the watercourse
    • Raising the bed
    • Terminating drains into meadows or wetlands
    • Establishing a dam
    • Meadow trickling
    • Narrowing the watercourse
    Lakes/ponds/wetlands established/reestablished in the river valley
    Vegetation management in the river valley

Source: After Hansen et al. (1996).

constructed by humans. Although impediments on rivers were often for good reasons—transportation, flood control, mills, cooling ponds, recreation, and electrical energy production—these uses have long-since been rendered unnecessary. In the meantime, the ecological effects of dams—changing the energy patterns of the river downstream, disrupting fish migration patterns, inundating productive upland, altering sediment dynamics, and even disrupting aesthetics—have led many groups around the world to call for their removal. Dams can adversely affect streams by changing channel morphology, flow regimes, and sediment transport; modifying water chemistry; modifying algal and macroinvertebrate communities; and disrupting resident and migratory fish communities. Rivers are longitudinally linked systems with processes occurring in the upper reaches affecting downstream reaches. In addition, processes occurring in the downstream reaches can affect upstream reaches as denoted by biophysical legacies (e.g., reduced gene flow, changes in community structure, and alteration of nutrient cycling) (Pringle, 1997).

It is a natural instinct of humans to build dams to store water. It is estimated that there are 45,000 "large" dams worldwide (WCED, 2000; a large dam is more than 15 m high or an impoundment of more than 3 million cubic meters). By a different definition of "large" (>1.8 m high or an impoundment >18,500 m³), the U.S. Army Corps of Engineers estimates that there are 76,500 regulated dam structures in the United States. And this number does not count an estimated 2 million small dams in the United States that are too small to be on the federal list (Poff and Hart, 2002).

Dam removal in the developed world has increased exponentially in the past decade, with an estimated 180 dams removed in the 1990s in the United States and another 30 removed in 2001 alone (Poff and Hart, 2002). Most of these dam removals are done with little scientific study on the effects. When scientific studies are done, they are usually piecemeal in scope, with a "university department" rather than a holistic approach. One scientist will study fish, another one will study invertebrates, and a third might take water samples (biology, entomology, and geology departments respectively). Hart et al. (2002) point out that past studies rely on qualitative observations rather than quantitative measurements, there is inadequate replication (must be pseudo-replication at one site unless several dams are removed at the same time!), and cause and effects are often wrongly hypothesized, due to lack of study of abiotic factors such as sediment transport and water temperature.

Restoration of rivers by dam removal instantly changes hydrologic features, with changes of vegetation, fish, and invertebrate communities occurring over a much longer period. Effects will occur at the dam site, downstream of the dam, and upstream of the original pool. Some of these changes are summarized in Figure 7.7.

Very few studies have been completed on enough studies to degree categorically that dam removal techniques are always beneficial. In fact, some ecologists and resource managers who favor lentic environments have argued

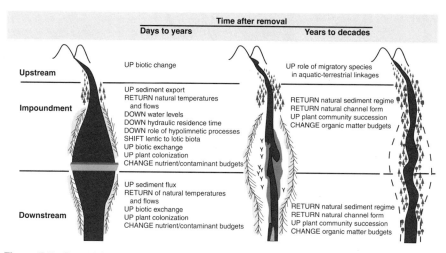

Figure 7.7   Potential upstream, impoundment, and downstream effects of dam removal. Effects are also shown for the short term (days to years) and the long term (years to decades). (Redrawn from Hart et al., 2002.)

against dam removal. The process has to be undertaken with complete studies of all stream and river impacts that would be affected: upstream, downstream, and in the social setting in which the dam and impoundment are found. It is also important to note that a dam removal is often limited in its spatial impact and needs to be done in the context of what benefits it provides to the watershed as a whole. Dam removal can make urban rivers even flashier than they normally are, often causing increased flooding downstream. With these precautions noted, the removal of a dam from a river can have important benefits.

CASE STUDY
Dam Removal, Manatawny Creek, Pennsylvania

The Academy of Sciences in Philadelphia is conducting a study of the ecological impacts of dam removal with a previously funded dam removal and riparian restoration project on the Manatawny Creek near Pottstown in southeastern Pennsylvania (Figure 7.8). Images showing the dam removal and immediate aftermath in the vicinity of the dam are shown in Figure 7.9. The project has four components:

1. Removing the dam
2. Restoring the newly established riparian corridor within the former impoundment area

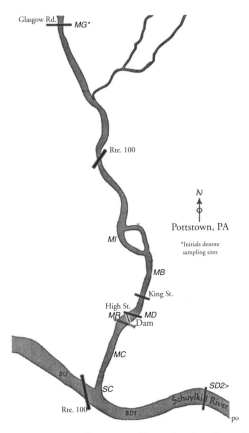

Figure 7.8 Manatawny Creek near Pottstown, Pennsylvania, where studies are being undertaken on the effects of dam removal on a small stream. Figure 7.9 illustrates visual changes before and after dam removal. (Map reproduced courtesy of the Patrick Center for Environmental Research, Academy of Natural Sciences, Philadelphia, PA.)

3. Conducting a comprehensive, interdisciplinary study of the physical, chemical, and biological changes in the Manatawny Creek watershed to measure and evaluate project success

4. Educating the public and providing technology transfer of dam removal as a watershed restoration method

Manatawny Creek drains a 238-km^2 watershed and is a tributary to the Schuylkill River and then the Delaware River. Pottstown, with an estimated population of 22,000, is the major urban center in the watershed. The major land uses in the watershed are forest and open (56 percent), agriculture (41 percent), and urban development (3 percent). The low-head dam on the Manatawny was constructed around 1850, about 500 m upstream of the confluence with the Schuylkill

Figure 7.9 Visual effects of the removal of the Manatawny Creek dam in Pennsylvania: (a) dam prior to removal; (b) dam site soon after its removal (November 2000). View from King Street bridge (see the location in Figure 7.8) (c) before dam removal; (d) immediately after dam removal; (e) 21 months after dam removal (April 2002). (Courtesy of Karen L. Bushaw-Newton, American University; reprinted with permission.)

River (Figure 7.8). The stone and concrete dam was approximately 2.5 m in height and 30 m in length (Figure 7.9a). Water was impounded about 800 m upstream of the dam, and no fish ladders or other structures were ever constructed to allow fish migration. In 1999, the Pennsylvania Department of Environmental Protection designated the dam an "orphan" and it was removed in August to November 2000. The Patrick Center of the Academy of Natural Sciences in Philadelphia is conducting experiments and analyzing several components of the creek ecosystem by comparing before, during, and after dam removal at various sites upstream and downstream of the dam. The fields that are being investigated include geomorphology, streamwater and sediment chemistry, food webs, dissolved organic matter, algae, freshwater mussels, macroinvertebrates, fish, and riparian vegetation.

Channel Restoration

For many reasons, particularly for rapid drainage of a landscape, streams and rivers have been changed from meandering patterns to straight channels. We now realize that this kind of change has caused extensive pollution problems by itself but has also created systems that are difficult to maintain. In several locations in the world, straight rivers are being restored in some manner back to their original shapes or to shapes that are more natural in energy dissipation.

Rivers are rarely straight and are often characterized as being meandering. Meandering streams have sinusoidal patterns of wavelengths and radii of curvature (Figure 7.10a). Even straight channels have sine patterns to the

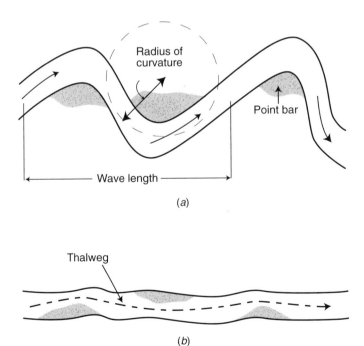

Figure 7.10 Channels with various degrees of sinuosity: (a) meandering stream; (b) straight stream with slight meanders of the thalweg. (From Leopold, 1994.)

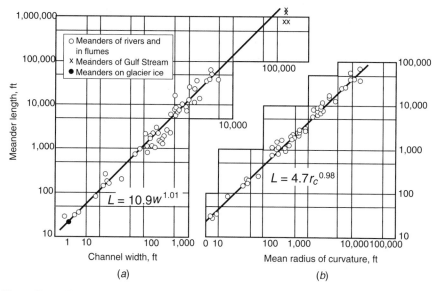

Figure 7.11 Relationships between meander length and (a) channel width and (b) mean radius of curvature. (From Leopold, 1994.)

thalweg, the path of highest velocity and deepest water pathway of streams (Figure 7.10b). Point bars tend to develop on the concave parts of the meanders, and greatest bank erosion tends to occur in areas where the thalweg comes in closest contact with the riverbank. The wavelength of river meanders is described as a remarkably similar linear pattern from rivers throughout the world (Figure 7.11):

$$L = 11w \tag{7.1}$$

and the radius of curvature is approximated by the equation

$$L = 5r_c \tag{7.2}$$

where L is the wavelength of the meander, w the stream width, and r_c the radius of curvature of the stream.

CASE STUDY
Restoring the Kissimmee River, Florida

The Kissimmee River basin is the northern portion of the Kissimmee River–Lake Okeechobee–Everglades (KOE) ecological system in

Florida. The river flows south from Lake Kissimmee into Lake Okee-chobee. The Kissimmee River originally meandered 166 km to Lake Okeechobee, with 94 percent of the floodplain inundated over 50 percent of the time. From 1961 to 1971, the 166-km-long river was channelized 9 m deep into a basically straight 90-km-long 100-m-wide canal (Figure 7.12; see also Chapter 1). Named the C38 canal,

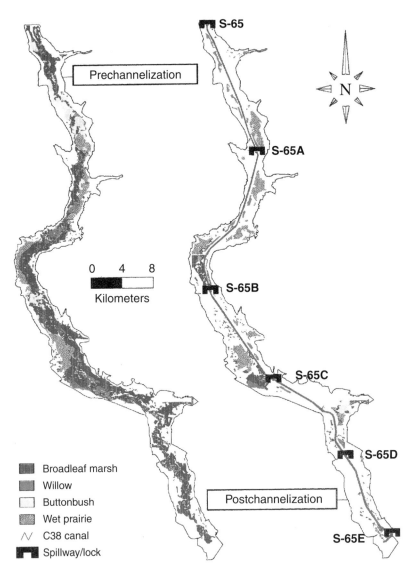

Figure 7.12 Channelization of Kissimmee River and subsequent change in riparian ecosystems. (From Toth et al., 1995; copyright 1995; reprinted with permission from Blackwell Publishing Ltd.)

it controlled the passage of floodwater using six flow-control structures. Transformation of the river floodplain ecosystem into a series of deep impounded reservoirs drained approximately 12,000 to 14,000 ha of floodplain wetlands, degrading fish and wildlife values within the river. The physical effects of channelization, including alteration of the system's hydrologic characteristics, largely eliminated river and floodplain wetlands and degraded fish and wildlife values of the Kissimmee River ecosystem (Table 7.2). The floodplain at the lower end of each pool remained inundated, but prechannelization water-level fluctuations were eliminated. Low- and no-flow regimes in remnant river channels resulted in encroachment of vegetation, especially floating exotics such as *Pistia stratiotes* (water lettuce) and *Eichhornia crassipes* (water hyacinth) to the river channel. Thick accumulations (up to 1 m) of organic matter increased the biological oxygen demand of the system. Low- and no-flow regimes in the canal and remnant river channels resulted in chronically low dissolved oxygen levels and sport fish species such as largemouth bass were largely replaced by species tolerant of low-dissolved-oxygen regimes, such as Florida gar and bowfin. Invertebrate taxa typical of many large river systems (e.g., certain caddis flies and mayflies) were replaced by species common to lentic systems (Toth, 1995).

By the early 1970s, floodplain utilization by wintering waterfowl declined by over 90 percent. Wading bird populations, a highly visible component of the historic system, declined and were largely replaced by *Bubulcus ibis* (cattle egret). Stabilized water levels and reduced flow also eliminated river–floodplain exchanges. Influx of organic matter, invertebrates, and forage fishes to the river from the floodplain during periods of water recession was eliminated. Stabilized water levels also largely eliminated adult spawning foraging

Table 7.2 Wetland changes due to channelization of the Kissimmee River in southern Florida[a]

Wetland Type	Prechannelization (ha)	Postchannelization (ha)	Percent Change
Marsh	8,892	1,238	−86
Wet prairie	4,126	2,128	−48
Shrub–scrub wetland	2,068	1,003	−51
Forested wetland	150	243	+62
Other	533	919	+72
	15,769	5,531	−65

Source: After Toth et al. (1995).

[a]Channelization took place between 1962 and 1971 and transformed a 166-km meandering river into a 90-km-long 10-m-deep 100-m-wide canal.

habitat, as well as larval and juvenile refuge sites for fish on the and floodplain. In 1976, the Florida legislature passed the Kissimmee River Restoration Act, which sparked several major restoration and planning projects. Plans to restore the Kissimmee River will cost eight times that of initial straightening and will take over 15 years to complete. In current plans for restoration, the upper basin project would employ flow-control structures to raise water levels, and the lower basin project would fill in channels connecting Lake Kissimmee to Lake Okeechobee, forcing water to flow through its original path. Among the reasons for the restoration were the loss of fluctuating water levels in the river and hence flooding of wetland areas and the deterioration of water quality in Lake Okeechobee as a result of the straightening of the river.

An initial demonstration project developed in the late 1980s for trial restoration of the Kissimmee River had four major components:

1. Development of a pool stage fluctuation schedule to reestablish seasonal water fluctuations in a wetland area of 1100 ha.
2. The building of three notched weirs across C38 to simulate the effects of backfilling by diverting flow back into the original winding river and floodplain.
3. The creation of a flow-through marsh system.
4. Hydrologic and hydraulic modeling studies to evaluate the engineering feasibility of the backfill, flood control potentials, and sedimentation issues.

As expected, the demonstration project established prechannelization flooding patterns over a limited portion of the floodplain. Stage fluctuations and backwater effects resulted in a flooding frequency approximately 25 percent of the historic frequency and increased frequency for the floodplains overall. Dead and decaying organic plant matter was washed into C38, restoring the historic shifting sand substrate in the remnant channels. Plant communities showed that they can reestablish themselves under appropriate hydrological conditions in the floodplain. Invertebrates, fishes, water birds, and waterfowl all showed positive responses to the demonstration project. In all cases, the number of species and quantities of each species increased. The results of the demonstration project confirmed the feasibility of restoring the structure and function of the Kissimmee River ecosystem.

Based on these findings, environmental restoration goals and objectives for the Kissimmee River continue to be formulated. Despite the interest in complete restoration, restoration of the entire river to prechannelization conditions is almost impossible given the land development that has occurred in the basin since channelization. The project's immense cost is caused not only by the labor involved but also by the cost for land acquisition. Owners of land within the Kissimmee watershed, where restoration projects will occur, are being "compensated handsomely" for their land.

CASE STUDY
Skjern River Restoration Project, Denmark

River restoration has flourished in Denmark since the early 1990s. The Skjern River in west central Jutland drains more than 11 percent of Jutland, and the flow of water in the Skjern is the largest in Denmark. In Denmark's largest drainage project ever, 4000 ha of wet meadow was converted into arable land, and the lower Skjern was straightened to a fraction of its former meandering self. By the late 1980s, the river was essentially a straight line to the Ringkobing Fjord on the North Sea, eliminating thousands of hectares of marshland, meadows, and river habitat (Figure 7.13). The channelized river was diked, canals were built, and pumps were installed to hasten the downstream movement of water from the land. This public works project cost 30 million Danish krone (DKr) (about $3.6 million) and was considered a success by the agricultural community at first, as grains could now be grown in the formerly wet region. But the environment was paying a heavy price with this artificial river. The self-cleansing ability of the river was lessened, the downstream fjord was becoming polluted with nutrients and sediments, and the land that was draining began to subside due to peat oxidation and loss of water—up to 1 m or more in some locations. The human interference on this river has been described as "some of the most severe in northern Europe" [Danish Ministry of Environment and Energy (DMEE), 1999].

A few short years after the drainage, it appeared that another drainage project might be necessary (DMEE, 1999). Meanwhile, the Danish parliament passed a public works act in 1998 with a huge majority that called for restoration of the lower Skjern River and earmarked about DKr 254 million for this project. The project is being implemented in three phases for three reaches of the river. When the restoration is complete, 2200 ha of meadows and wetlands—half of the area drained in the 1960s—will be restored. The river restoration in this case called for the following:

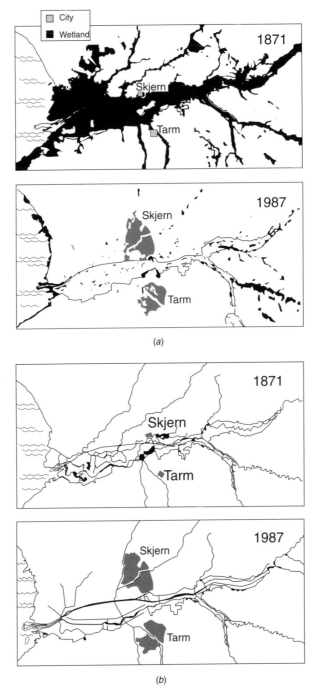

Figure 7.13 Conversion of (a) meadows and marshlands and (b) watercourses along the Skjern River in western Denmark between 1871 and 1987. (From the Danish Ministry of Environment and Energy, 1999.)

1. Restore the meanders of the river wherever possible (Figure 7.14).
2. Remove dikes along the river to allow adjacent meadows to be flooded once again.
3. Define the project area by dikes far away from the river to prevent flooding of farmland outside the project area.

Restoring and Creating Floodplain Ecosystems

The floodplain of a river must be considered as important a part of a river system as the stream channel itself. In the eastern United States, rivers tend to flood their floodplains with a frequency of about two out of three years (Leopold et al., 1964). A typical broad floodplain in climates with adequate moisture (Figure 7.15) contains several major features:

1. The *river channel* meanders through the area, transporting, eroding, and depositing alluvial sediments (see Section 7.1).
2. *Natural levees* adjacent to the channel are composed of coarse materials that are deposited when floods flow over the channel banks. Natural levees, sloping sharply toward the river and more gently away from the floodplain, are often the highest elevation on the floodplain.
3. *Point bars* are areas of sedimentation on the convex sides of river curves. As sediments are deposited on the point bar, the meander curve

(a)   (b)

Figure 7.14 Skjern River restoration in Denmark in July 2001 showing (a) a straightened river channel that is to be replaced and (b) a recently constructed meander. (Photos by W. J. Mitsch.)

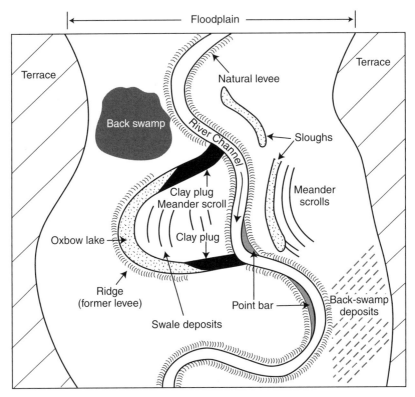

Figure 7.15 River floodplain features. (After Leopold et al., 1964; redrawn by Mitsch and Gosselink, 2000; copyright 2000; reprinted by permission of John Wiley & Sons, Inc.)

of the river tends to increase in radius and migrate downstream. Eventually, the point bar begins to support vegetation that stabilizes it as part of the floodplain.

4. *Meander scrolls* are depressions and ridges on the convex side of bends in the river. They are formed from point bars as the stream migrates laterally across the floodplain. This type of terrain is often referred to as *ridge and swale topography*.

5. *Oxbows* and *oxbow lakes* (*billabongs* in Australia) are bodies of permanently standing water that result from the cutoff of meanders. Deepwater swamps or freshwater marshes often develop in oxbows in these cutoffs, with substantial connections to the river during flood periods and isolation from the river during low-flow periods.

6. *Sloughs* are areas of dead water that form in meander scrolls and along valley walls. Deepwater swamps can also form in the permanently flooded sloughs.

7. *Back swamps* are deposits of fine sediments that occur between the natural levee and the valley wall or terrace.

8. *Terraces* are abandoned floodplains that may have been formed by the river's alluvial deposits but are not connected hydrologically to the present river.

Ecological restoration can include goals of restoring any number of ecosystem functions, as illustrated in Figure 7.16. Restoration on the floodplain can take place as a way to (1) enhance the ecological character of the floodplain itself, (2) provide a place for floodwaters to go to desynchronize flooding downstream, (3) improve the quality of both water entering the floodplain from adjacent uplands (transversely) and the flooding river water itself (longitudinally), (4) provide organic carbon as a base for the river food chain itself and macrodetritus as habitat structure, and (5) subsidize stream flow during periods of low flow. Three case studies of riparian ecosystem creation and restoration are described here: two at one site in central Ohio and one in central France.

Restoring the floodplain dynamics of a riverine system is just as important as restoring the channel dimensions of a river and sometimes more important. Two projects at the Olentangy River Wetland Research Park in Columbus, Ohio, illustrate ecosystem creation and restoration projects that illustrate ecotechnology approaches that rely on self-design: (1) creation of riparian oxbow wetland and (2) restoration of bottomland hardwood forest.

CASE STUDY
Oxbow Creation, Midwestern United States

A 3-ha riparian wetland (locally called a *billabong*) was constructed in the summer of 1996 as a freshwater oxbow wetland that receives water periodically from Olentangy River flooding in central Ohio (Figure 7.17). When the river level is higher than the wetland water level, water flows into the wetland through a check valve that prevents backflow when the river stage drops. After the water flows through the wetland, it eventually returns to the Olentangy River through a control weir at the outlflow.

The following equation was used to determine hydrologic budgets for this wetland:

$$\frac{\Delta V}{\Delta t} = S_{\text{in}} + S_{\text{out}} + G_{\text{in/out}} + P - \text{ET} \qquad (7.3)$$

where $\Delta V/\Delta t$ is the change of water volume in wetland over time, S_{in} the river inflow, S_{out} the surface outflow, $G_{\text{in/out}}$ the groundwater exchange, P the precipitation, and ET the evapotranspiration.

Most parameters (water volume, precipitation, evapotranspiration, and groundwater) in equation (7.3) can be estimated from simple field

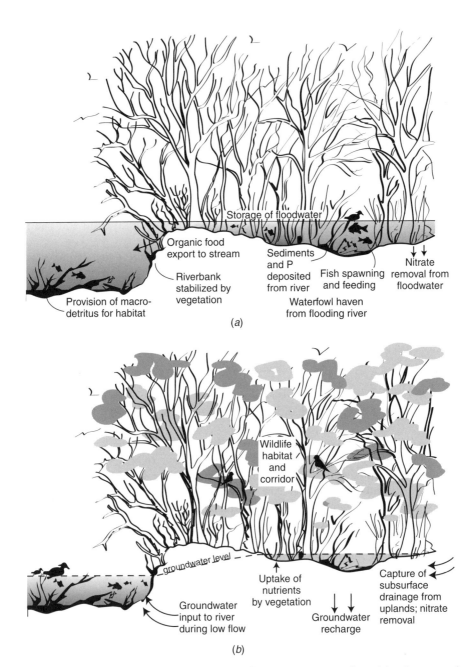

Figure 7.16 Functions and values of riparian forest ecosystems adjacent to streams and rivers during (a) the nongrowing season and (b) the growing season. (From Mitsch and Gosselink, 2000; copyright 2000; reprinted with permission from John Wiley & Sons, Inc.)

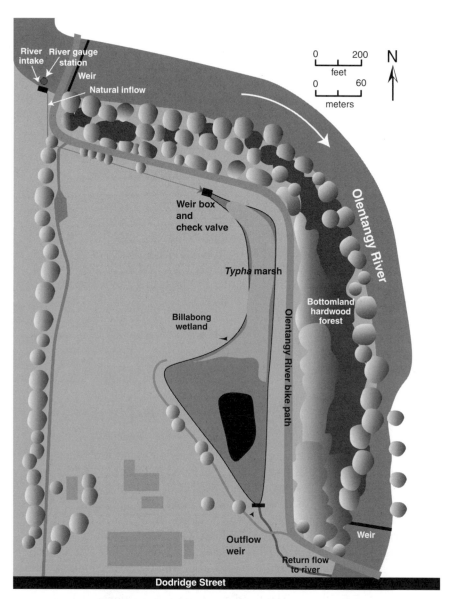

Figure 7.17   A created 3-ha riparian oxbow wetland (referred to as a billabong wetland) and a 5-ha bottomland hardwood forest under restoration at the Olentangy River Wetland Research Park on the Ohio State University campus, Columbus, Ohio.

instruments and subsequent calculations. Surface inflow and surface outflow can be estimated by the equations presented here (Wang et al., 1997). Surface inflow is a function of the river stage (L_r) relative to the water level of the wetland basin (L_m) and is estimated by

$$S_{in} = 1.8447ah^{1/2} \qquad (7.4)$$

Here

$$a = 0.7372D + 1.387D^2 - 0.39993D^3$$

$$= \text{cross-sectional area of the inflow pipe (ft}^2)$$

$$= \max(0, \min(L_r - L_m, L_r - 0.6D - 723.5))$$

$$h = L_r - L_m$$

where D is the water depth in the pipe at the weir box (ft) [$= \min(d, \max(0, L_r - 723.5))$]; L_r, the river water level (ft above MSL), $= 723 + L_s$ [L_s is the staff gauge reading (ft)]; and L_m is the wetland water level (ft above MSL); h is the head causing flow. The surface outflow is defined as

$$S_{out} = 10.16h_0^{1.436} \qquad (7.5)$$

where

$$h_0 = \begin{cases} L_m - 724.2 & L_m > 724.2 \\ 0 & L_m \leq 724.2 \end{cases}$$

A pattern over one year of the river elevation and the water elevation in this floodplain wetland is illustrated in Figure 7.18. The wet periods (Table 7.3) are significantly longer in most of the basin than the critical number of days that are necessary for an area to be formally a jurisdictional wetland, which ranges from 7 to 21 days in the growing season (NRC, 1995). General water-level trends of the billabong follow a pattern of high water levels in the late winter and early growing season (Figure 7.19a), followed by drier conditions in late summer and fall (Figure 7.19b), a pattern that is typical of midwestern U.S. wetlands. The flood events provide nutrients, sediments, and introduction of seeds and small organisms to the wetland ecosystem. Thus this wetland basin should provide adequate habitat for wetland organisms as well as suitably saturated soil for wetland biogeochemical processes to take place.

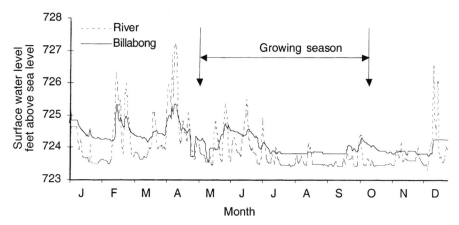

Figure 7.18 Patterns of a water stage in riparian wetland (solid line) and the adjacent Olen-tangy River stage (dashed line) in central Ohio for 2001. Note the water-level pulses in wetland from February to June but little river flooding from July to early December. There is more than sufficient flooding during the growing season for this created wetland to be a jurisdictional wetland in the United States.

Water quality also appears to be considerably enhanced, particu-larly during the flood pulse events measured in 2001 (Table 7.4). Soluble reactive phosphorus decreased by an average of 74 percent during flood events and nitrate-nitrogen decreased by 77 percent. The development of vegetation communities in the billabong reflects this nutrient gradient from inflow to outflow. Near the inflow, very large stands of the high-nutrient-loving *Typha* are found (see Figure 7.16). The water then passes through a *Typha* stand that diminishes in in-tensity by the middle of the wetlands. A nice variety of sedges and more desirable plants are found along the edges of the wetland in its outflow half, partially the result of intensive planting in 1997.

Table 7.3 Summary of wet days and wet area for two different definitions of growing seasons in 2000 for the oxbow wetland at the Olentangy River Wetland Research Park in Columbus, Ohio

	Number of Wet Days per Wet Area (%) of:		
Growing Season	68	72	82
April 20–October 19 2000	140	91	0
May 1–October 31 2000	140	75	0

Source: After Zhang and Mitsch (2001).

(a)

(b)

Figure 7.19 Water levels in Olentangy River riparian wetland during (a) the wet season and (b) the dry season. (Photos by W. J. Mitsch.)

Table 7.4 Water quality measurements in the oxbow wetland created at Olentangy River Wetland Research Park, 1999–2001, when flooded river water is flowing

Parameter	Year	Inflow[a]	Outflow[a]	Removal (%)
Soluble	1999[b]	36 ± 3 (19)	24 ± 2 (89)	33
reactive P	2000[b]	62 ± 6 (42)	26 ± 3 (60)	56
(μg-P L^{-1})	2001[c]	46 ± 5 (39)	11 ± 2 (19)	74
NO$_3$ + NO$_2$	1999[b]	5.15 ± 0.27 (19)	3.43 ± 0.16 (102)	33
(mg-N L^{-1})	2000[b]	4.60 ± 0.30 (43)	4.05 ± 0.69 (69)	12
	2001[c]	4.07 ± 0.28 (40)	0.92 ± 0.19 (19)	77
Turbidity	1999[b]	94 ± 10 (16)	42 ± 3 (137)	55
NTU	2000[b]	146 ± 14 (27)	74 ± 9 (50)	49

Source: After Mitsch and Zhang (2002).

[a]Average ± standard error (number of samples).
[b]Manual sampling during flood events.
[c]Autosampler sampling during spring flood events.

CASE STUDY
Bottomland Hardwood Forest Restoration,
Midwestern United States

In attempts to prevent flooding, artificial levees were frequently constructed along rivers to keep the river water purposefully in the stream channel and not on the floodplain. This practice, although keeping water off the floodplain, simply shifts the flooding to a downstream location. Partial restoration of a 5-ha bottomland hardwood forest by levee removal has been carried out on the Olentangy River in Ohio (Figures 7.17 and 7.20). The bottomland forest restoration, a cooperative project with the Ohio Department of Transportation (ODOT), began as a project to mitigate the loss of 2 ha of forested wetland loss due to a highway project in Columbus. The restoration project began with an evaluation of river stage information that had been collected for several years near the forest site. Figure 7.21a illustrates that putting notches to floodplain level in the artificial levee would have allowed approximately 11 overbank flooding events for a 31-month period during 1994–1996. With the artificial levee, there was only one flood during that period that equaled the top of the levee and that flood was estimated to have a recurrence interval of 100 years. It was estimated that there would be three to six floods per years into the bottomland forest if the levee were removed.

Removing the entire levee was not possible because of cost and environmental considerations. The actual restoration began with ODOT personnel cutting four notches in the bottomland artificial

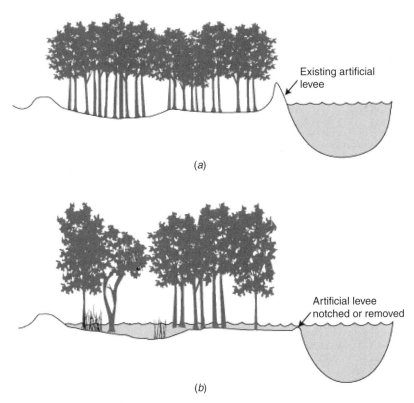

Existing artificial
levee

(a)

Artificial levee
notched or removed

(b)

Figure 7.20 Concept of bottomland forest restoration carried out in 2000 at the Olentangy River Wetland Research Park in central Ohio: (a) situation where a forest is isolated by an artificial levee that prevents full flooding prior to restoration; (b) situation after restoration, where a dike is removed or breeched, allowing bottomland flooding.

levee in June 2000. The restoration openings, created by small earth-moving equipment, were approximately 6 m wide and 2 m high. The notches were cut to the base of the floodplain natural levee to ap-proximate the bottomland forest's pre-levee hydrology. Seven inde-pendent flood events occurred over a 25-month period from June 2000 through June 2002; that is, there was significantly more flooding after the notches were cut (Figure 7.21b). Floods resulted in overbank flooding through the notches with major exchanges of water, sedi-ments, and presumably plant propagules from the river to the bottom-land during the rising-stage period and from the bottomland to the river during falling river stages (Figure 7.22). This project is still in the early stages of development, and this "quadruple bypass surgery" will take several decades to change the character of the vegetation and hence the ecology of the forest. The river is also benefitting because of brief but important exchanges of material with the flood-

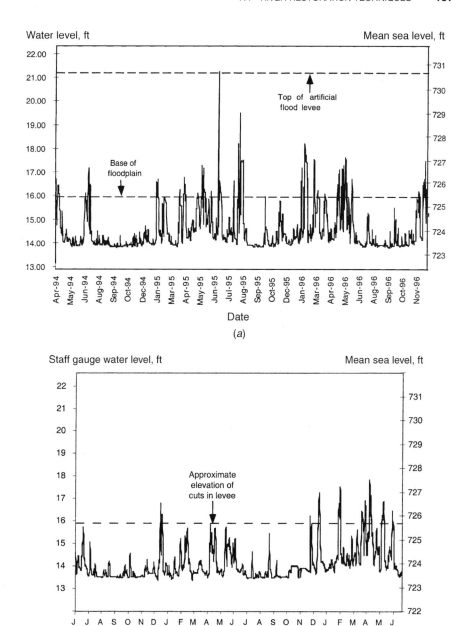

Figure 7.21 River stage of the Olentangy River adjacent to the Olentangy River Wetland Research Park and adjacent to restored bottomland. (a) Planning data prior to restoration. Only one flood in this 31-month period (estimated to be the 100-year flow) equaled the elevation of the artificial levee. Also shown is the base level of the floodplain (1.7 m below the levee height). If the levee had been notched then, 11 separate bottomland overbank flooding events would have occurred in the floodplain. (b) Data after restoration. Levee was breached in June 2000 and seven independent bottomland flood events occurred in the 25-month period after the levee breeching.

(a)

(b)

Figure 7.22 (a) Flood condition in bottomland hardwood forest after dike breeching; (b) dike breech showing the flow of water from river to bottomland. (Photos by W. J. Mitsch.)

plain. Humans benefit because of flood relief downstream, ecological enhancement of the forest and river, and possibly improved water quality and food web support in the river. Trees growing on the levee were destabilized by the notching, and reefs of large woody debris have developed since the cutting in the river at each of the cut locations, due to fallen trees. These provide additional fish habitat in the river.

CASE STUDY
Restoring Flood Pulses to a Former River Channel in France

A long-term study on the restoration of an abandoned river channel adjacent to the Rhône River in the Brégnier–Cordon plain of central France (Figure 7.23) has been conducted since 1993 (Henry and Amoros, 1995; Henry et al., 2002). The river channel had been subjected to drainage and rapid terrestrialization since a dam was built upstream in the early 1980s. The channel was first dredged to remove fine organic sediments to expose a gravel bottom and increase groundwater connectivity with the adjacent river. Flood pulses were also reintroduced from the main channel (Figure 7.24). Results of five-years of study using principal-components analysis for aquatic vegetation in the restored basins compared to nearby reference channels have shown a marked difference between the two (Figure 7.25). Eutrophic species that dominated the disturbed site since construction of the dam (e.g., *Ceratophyllum demersum, Lemna* spp.) have been partially replaced by mesotrophic species such as *Berula erecta, Callitriche platycarpa,* and *Groenlanda densa.* There is clearly a reversal of eutrophication in these back channels with the increase in groundwater supply and organic sediment removal. A question remains as to this reversal in eutrophication in the long term, as organic sediments will redevelop and high-nutrient inorganic sediments will continue to flood from the adjacent river (Henry et al., 2002).

7.5   MEASURING RIVER RESTORATION SUCCESS

There are, of course, a number of ways to measure the success of restored riverine systems; all have shortcomings and advantages. The first and easiest method involves a *before–after study,* where conditions prior to and after the restoration are compared. The problem with this approach is that it cannot be proved that the changes are due to the restoration; they might have happened

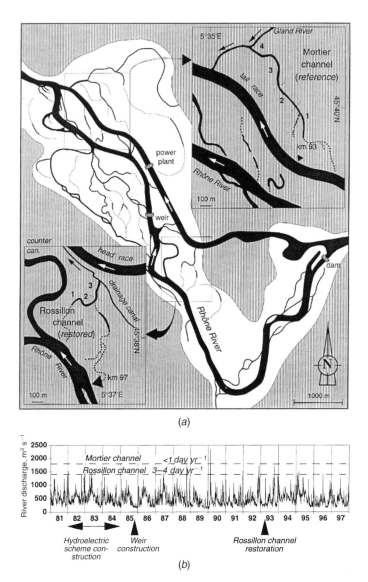

(a)

(b)

Figure 7.23 Brégnier–Cordon section of Rhône River in central France: (*a*) location of restored Rossillon channel and reference Mortier channel, which has characteristics of the Rossillon channel prior to restoration; (*b*) river discharge at which the two sites are flooded by the Rhône River. (From Henry et al., 2002; copyright 2002; reprinted with permission from Elsevier Science.)

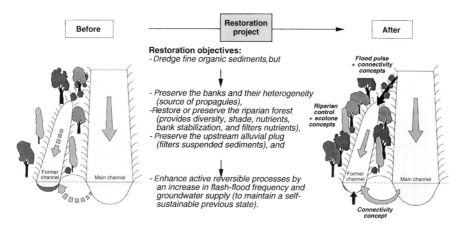

Figure 7.24 Conceptual diagram and objectives of the Rhône River channel restoration. (From Henry et al., 2002; copyright 2002; reprinted with permission from Elsevier Science.)

without the restoration. Kolka et al. (2000) argue that without a reference system, this type of before–after comparison simply assesses change in the state of the ecosystem, not success or failure, and that it may even be biased toward predicting success. A second method involves comparison to an *unrestored reference*—comparing a "restored" stream or river reach with a stream or river reach that is initially in the same deteriorated condition as is the target stream or river. If the restoration is successful, a comparison of the two systems will show an eventual divergence, with the restored system showing better biological, chemical, and physical indicators than the unrestored control. This is similar to comparing a diseased person who is being treated for that disease with a person or population that continues to have that disease. This is the approach taken by Henry et al. (2002) in their river restoration project on the Rhône River in France.

A third approach is to measure *physical, chemical,* or *biological indices* in the restored stream and compare these with known "standards" of the same indices (Stein and Ambrose, 1998). This is the essence of using water quality standards or biological indices as metrics of stream and river health. This is similar to taking blood pressure or body temperature measurements of a patient to determine the person's relative health. In everyday activity, this is the most common approach for determining human health and probably the most common way for estimating ecological health.

In what is becoming the most accepted method for evaluating restored ecosystems, the restored system is compared with an unaffected *reference system* that reflects the best that nature can do in that region; in this case the control is a system that is desired, and various ecological indices are used for the comparison. This comparison can be made both before and after the restoration. This last approach is similar to the approach most often used for

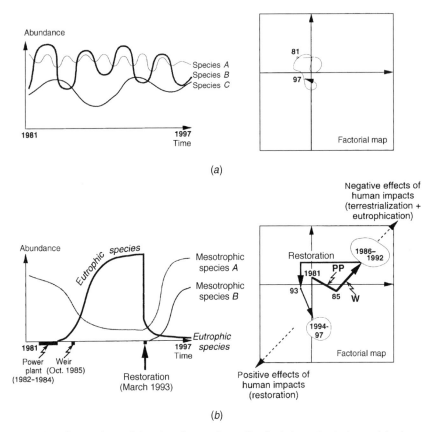

Figure 7.25 Comparison of the abundance of aquatic plant dynamics between (a) reference (not restored) and (b) restored back channels of the Rhône River. PP, hydroelectric power plant installation; W, weir construction. Restoration by removing organic sediments and increasing groundwater inflow occurred in a restored channel in 1993. (From Henry et al., 2002; copyright 2002; reprinted with permission from Elsevier Sceince.)

wetland mitigation evaluation (see Wilson and Mitsch, 1996; Rheinhardt et al., 1997; Stein and Ambrose 1998; NRC, 2001). Because river reaches are longitudinally connected systems that are not independent, river restoration projects then need to be compared in this method with similar high-quality streams and rivers elsewhere. It is often quite difficult to find unaffected streams and rivers, particularly systems with similar stream order and geomorphology. A great deal of professional judgment is needed in the end to evaluate the effectiveness of restoration projects properly. Indices and references only assist that judgment.

8

WETLAND CREATION AND RESTORATION

Loss rates of wetlands around the world and subsequent recognition of wetland values have stimulated restoration and creation of these systems around the world. The creation of wetlands in previously dry and/or nonvegetated areas and the restoration of wetlands where they once were found are exciting opportunities for reversing the trend of decreasing wetland resources and for providing aesthetic and functional units to the landscape. Wetland creation and restoration can range from the relatively simple building of farmland freshwater marshes by plugging existing drainage systems to the restoration of more extensive wetlands such as those found in the Florida Everglades or Danube Delta in Europe. Knowledge of the principles and practices of wetland creation and restoration is one of the qualifications required of those who would like to pursue this profession. Among the necessary knowledge required is to know how natural wetlands function, a topic much too elaborate for us to cover in one chapter. If you are interested in this field, our advice is first to become an expert in wetland ecology.

The material in this chapter is partially from the book *Wetlands,* 3rd edition (Mitsch and Gosselink, 2000). That book should be consulted for more details on specific wetland types and wetland function. Some early publications on the subject in the early 1990s remain popular (Kusler and Kentula, 1990; Kentula et al., 1992; NRC, 1992). There are now several reviews on creation and restoration of specific types of freshwater wetlands, including marshes (Galatowitsch and van der Valk, 1994; Hammer, 1997; Mitsch and Bouchard, 1998), forested wetlands (Clewell, 1999), and peatlands (Price et al., 1998). Middleton (1999) describes wetland restoration in the context of the importance of flood pulses and disturbance ecology, and Streever (1999) provides regional overviews and case studies from around the world. A critique of the policies and techniques of wetland creation and restoration in the United States was published as NRC (2001). Wetland restoration associated with coastal regions is covered in Chapter 9; wetlands created primarily for water pollution control are described in Chapter 10.

8.1   GLOBAL WETLAND EXTENT AND LOSSES

Wetlands are known by terms such as *bogs, fens, marshes, swamps,* and *mires.* They are found in every climate and continent and are particularly dominant in both the tropic and boreal zones of the world (Figure 8.1). There are an estimated 7 to 9 million square kilometers of wetlands in the world (5 to 6 percent of the land surface; Mitsch and Gosselink, 2000), with well over 90 percent of those wetlands as inland freshwater systems. Freshwater wetlands such as marshes, swamps, and peatlands are the types of wetlands emphasized in this chapter.

The rate at which wetlands are being lost on a global scale is unknown. There are too many vast areas of wetlands where accurate records were not kept, and many wetlands in the world have been drained over the history of humankind. It is probably safe to assume that (1) we are still losing wetlands at a fairly rapid rate globally, and (2) we have lost as much as 50 percent of

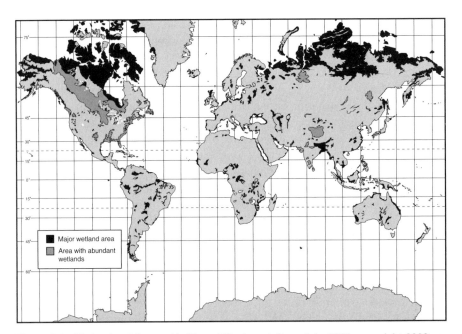

Figure 8.1   Wetlands of the world. (From Mitsch and Gosselink, 2000; copyright 2002; reprinted by permission of John Wiley & Sons, Inc.)

the original wetlands on the face of the Earth (Dugan, 1993). There are a number of areas where the loss rate has been documented (Table 8.1). The estimate of about 50 percent loss of wetlands since European settlement in the lower 48 states of the United States is fairly accurate. Similarly, the 90 percent loss of wetlands in New Zealand is well documented. The loss rate of 60 percent from China in Table 8.1 is based on the present estimate of 250,000 km² of natural wetlands in the country out of a total of 620,000 km² of total plus artificial wetlands (rice paddies, etc.) (Lu, 1995).

8.2 DEFINITIONS

Several terms are frequently used in connection with the creation and restoration of wetlands. Precise definitions are important, and confusion about the exact meaning of wetland creation, restoration, and related terms is common (Lewis, 1990a). Bradshaw (1996) concurs that "we must be clear in what is being discussed." *Wetland restoration* refers to the return of a wetland from a disturbed or altered condition caused by human activity to a previously existing condition. The wetland may have been degraded or altered hydrologically, and restoration then may involve reestablishing hydrologic conditions to reestablish previous vegetation communities. *Wetland creation* refers to the conversion of a persistent upland or shallow-water area into a wetland

Table 8.1 Percent loss of wetlands in various geographic locations in the world

Location	Percent Loss	Reference
North America		
United States	53	Dahl, 1990
Canada		National Wetland
Atlantic tidal and salt marshes	65	Working
Lower Great Lakes–St. Lawrence River	71	Group, 1988
Prairie potholes and sloughs	71	
Pacific coastal estuarine wetlands	80	
Australasia		
Australia	>50	Australian Nature
Swan coastal plain	75	Conservation
Coastal New South Wales	75	Agency, 1996
Victoria	33	
River Murray basin	35	
New Zealand	>90	Dugan, 1993
Philippine mangrove swamps	67	Dugan, 1993
China	60	Lu, 1995
Europe	>90	Estimate

Source: Data from Mitsch and Gossellink (2000).

by human activity. *Wetland enhancement* refers to a human activity that in-creases one or more functions of an existing wetland. One type of created wetland, a *constructed wetland,* refers to a wetland that has been developed on a former upland environment to create poorly drained soils and wetland flora and fauna for the primary purpose of contaminant or pollution removal from wastewater or runoff (Hammer, 1997). This kind of wetland is also referred to as a *treatment wetland.* Wetlands created specifically for waste-water treatment are discussed in Chapter 10.

8.3 REASONS FOR CREATING AND RESTORING WETLANDS

Significant efforts now focus on the voluntary restoration and creation of wetlands. Part of the interest in wetland creation and restoration stems from the fact that we are losing or have lost so much of this valuable habitat. Often, interest is less voluntary and more in response to government policies such as "no net loss" in the United States that require the replacement of wetlands for those unavoidably lost. New Zealand, which has lost 90 percent of its wetlands, has major efforts to restore marshes and other wetlands in the Wai-kato River basin on North Island and in the vicinity of Christchurch on South Island. In southeastern Australia, restoration of the Murray–Darling water-sheds, particularly the riverine billabongs, has become a major undertaking, while coastal plain wetland restoration and creation are occurring in south-western Australia. Now wetland restoration and creation are being proposed or implemented on very large scales to prevent more deterioration of existing wetlands (Everglades in Florida), to mitigate the loss of fisheries (Delaware Bay in the eastern United States), to reduce land loss (Mississippi Delta in Louisiana), and to solve serious cases of overenrichment of coastal waters (Baltic Sea in Scandinavia; Gulf of Mexico fed by the Mississippi River basin in the United States.

Replacing Wetland Habitat

Wetland protection regulations in the United States and now elsewhere have led to the practice of requiring that wetlands be created, restored, or enhanced to replace wetlands lost in developments such as highway construction, coastal drainage and filling, or commercial development. This is referred to as the process of mitigating the original loss, and these "new" wetlands are often called *mitigation wetlands.* Perhaps a more appropriate term should be *replacement wetland.* One such riverine replacement wetland was described in Chapter 7. Replacement wetlands are designed to be at least the same size as the lost wetlands, and a *mitigation ratio* is often applied so that more wetlands are created and/or restored than are lost. For example, a mitigation ratio of 2:1 means that 2 ha of wetlands will be restored or created for every hectare of wetland lost to development. Considerable controversy exists, for example, in the United States, on the question as to whether wetland loss can be mitigated successfully or if it is essentially impossible (NRC, 2001).

On paper, the U.S. Army Corps of Engineer's implementation of "no net loss" appears to be working in the United States (Figure 8.2). An estimated 8100 ha yr^{-1} of wetlands and associated uplands in the United States from 1993 through 2002 were gained due to enforcement of the Clean Water Act through wetland mitigation. This number is the result of the issuing of permits for the destruction of 9700 ha yr^{-1} of wetlands and the creation, restoration, enhancement or preservation of 17,800 ha yr^{-1} of wetlands and associated uplands. There are two reasons why one should not be so euphoric about this net gain of wetlands. First, it is impossible to tell from these general numbers just how successful this wetland trading has been, as few statistics exist on what functions were lost versus what functions were gained (NRC, 2001). Second, the estimated gain of 81,000 ha over ten years does not make much of an impact on the loss of 47 million hectares of wetlands that occurred from presettlement time to the 1980s in the United States.

Conservation programs are now in place to encourage individual farmers in the United States to restore wetlands on their land. Both the Conservation Reserve Program (CRP) and the Wetlands Reserve Program (WRP) under the U.S. Department of Agriculture have led to significant areas of wetlands being restored or protected. CRP guidelines, announced in 1997, give increased emphasis to the enrollment and restoration of *cropped wetlands,* that is, wetlands that produce crops but serve wetland functions when crops are not being grown. The CRP also encourages wetland restoration, particularly through hydrologic restoration. In the CRP, participants voluntarily enter into contracts with the U.S. Department of Agriculture (USDA) to enroll erodible and other environmentally sensitive land in long-term contracts for 10 to 15 years. In

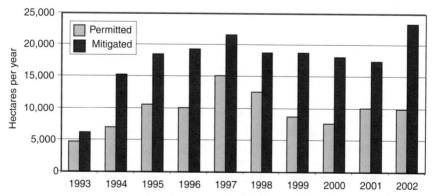

Figure 8.2 Approximate area of wetlands lost permitted and gained (mitigated) through U.S. Army Corps of Engineers' Section 404 dredge and fill permit program, 1993–2002, in the United States. "Permitted" refers to the approximate area of wetlands lost through the permit program. "Mitigated" refers to the area of restored, created, enhanced, and preserved wetlands and associated uplands required for the permit. (Data from U.S. Army Corps of Engineers.)

exchange, participants receive annual rental payments and a payment of up to 50 percent of the cost of establishing conservation practices. The Wetland Reserve Program (WRP) is a newer voluntary program offering landowners the opportunity to protect, restore, and enhance wetlands on their property. The USDA Natural Resources Conservation Service (NRCS) provides technical and financial support to help landowners. The WRP options to protect, restore, and enhance wetlands and associated uplands include permanent easements, 30-year easements, or 10-year restoration cost-share agreements. As of November 2002, approximately 517,000 ha of wetlands and adjacent uplands have been enrolled in the WRP in the United States, with the most intense activity in the lower Mississippi River basin, California, and Florida.

There is less experience at forested wetland restoration and creation than in herbaceous marshes, despite that fact that these wetlands have been lost at alarming rates, particularly in the southeastern United States. Forested wetland creation and restoration are different from marsh creation and restoration because forest regeneration takes decades rather than years to complete and there is more uncertainty about the results.

CASE STUDY
Mitigation Wetland in Central Ohio

An example of a mitigation or replacement wetland that was followed with extensive study for five years in central Ohio is described here. This case study is typical of wetland mitigation in that five years' worth of data were collected after wetland construction; this is the standard period in the United States that is used to determine wetland "success." It is different in that models were used to estimate its water quality function and its forest cover after 50 years. The aerial photograph in Figure 8.3 shows the wetland mitigation site four years after construction; one of the goals of that project was replacement of a forested wetland, yet five years of monitoring is hardly adequate to determine whether that effort would be successful.

The watershed of the wetland, estimated at 260 ha, contains an industrial park, agricultural land, and private homes, but otherwise is within the beltway of Columbus, Ohio. The outflow from the wetland flows into a channelized stream. The hydrology of the area has been greatly altered by highway construction, stream channelization, tiling of farmland, and urban development. The wetlands that were lost were approximately 3 ha of degraded wetlands located in corn–soybean fields and along a channelized stream immediately north of the study site. Little was documented on the function of the lost wetlands. The wetland created was excavated in the fall of 1991 and planted in the spring of 1992. The wetland, built in adjacent corn-

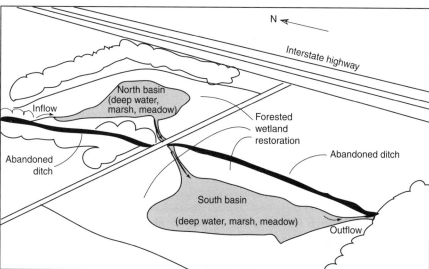

Figure 8.3 On-site mitigation for small wetlands in urban or suburban areas is common in the United States. This 6.1-ha wetland was constructed in Columbus, Ohio, in 1992 to mitigate the loss of about 3 ha of degraded wetland nearby. The photo is from 1996, four years after wetland restoration. The long-term goal called for the development of forested wetlands surrounding the marshes; small planted samplings can be seen as small dots in the photo around both basins. (From Mitsch and Gosselink, 2000; copyright 2000; reprinted with permission from John Wiley & Sons, Inc.)

fields, consists of two basins (north and south; left and right in Figure 8.3), for a total size of 6.1 ha. While the marshes shown in Figure 8.3 are well developed after five years, the area that is to be forested shows only tiny dots of small seedlings barely discernible in the photograph but hardly comparable to adjacent forest canopies in the picture.

The most important predictor of future wetland success is the hydroperiod. Water-level records for four years (Figure 8.4) show a wetland with a pulsing hydrograph typical of both low-order streams and streams found in urban areas. The frequency of stream pulses appears to have increased in 1996 compared to data from previous years, but magnitudes do not appear to be greater; changes in hydrology possibly indicate increased urban development in the area, a major concern when mitigation wetlands are placed in urban areas. After five years, the wetland appeared to be productive and continued to develop and respond as a natural wetland should. Plant cover and diversity remained healthy, and some of the vegetation planted, along with several colonizing species, respond to hydrologic variation and animal herbivory by musrats. Changes in plant richness over the five years of monitoring (Figure 8.5) are dramatic in this wetland. The survey found 45 taxa in the first year, 10 of which were introduced woody plants (trees or shrubs). Surveys in 1993 and 1994 showed a doubling in plant richness, with 81 and 78 taxa, respectively, seen in those years. Both years were characterized by generally low water levels that would have allowed a dramatic increase in the number of species in the wetland. In 1995 the number of taxa dropped to 66, possibly due to a relative lack of attention to botanical detail but also due possibly to the much higher water levels that year. Then with lower water levels in the growing season, coupled with a very wet spring, 101 species were observed in 1996. This five-year vegetation pattern is coupled to observed hydrologic conditions and suggests a healthy, yet dynamic system. Increased stream flow and exceptionally flashy stream conditions increased in latter years, due possibly to increased surface runoff from development. Changes in the plant community over that time reflect changing hydrologic conditions due to increased stream flow into the wetlands and possible plant harvesting/herbivory by muskrats. This resulted in the elimination of some cattails (*Typha* spp.) and bulrushes in the basins, but the vegetation was replaced rapidly and diversity remained high. Vegetation grades well from upland to wetland across the transects. Wildlife use has also been good for a wetland located in an urban area with a consistent increase in richness of bird species over the five-year monitoring period (Table 8.2).

Most of the hundreds of trees introduced that survived the initial two years of flooding still appeared to be alive after five-years, but

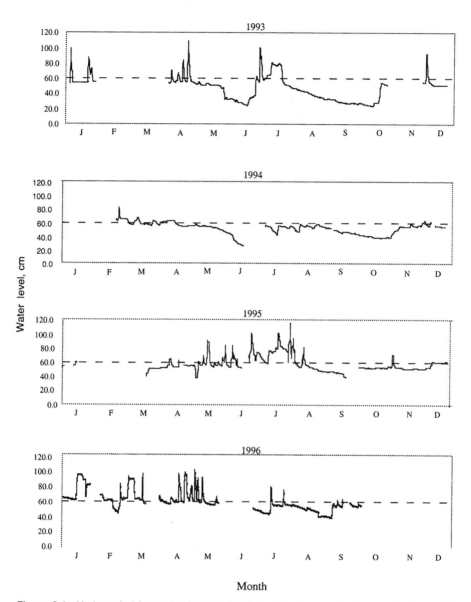

Figure 8.4 Hydroperiod (water level versus time) for mitigation wetland shown in Figure 8.3 for four years.

few new tree propagules were observed in that short a time. It is very unlikely that the sedge meadow/forested wetland portion of these basins will have a closed canopy for decades. Computer simulations suggested that the basal area of this created forest will approximate that of natural wet forests only after 50 years (Niswander and Mitsch,

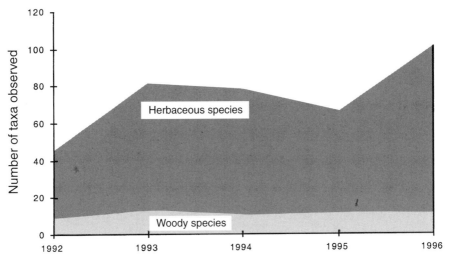

Figure 8.5 Plant species richness for mitigation wetland shown in Figure 8.3 for five-year monitoring study. Tree species are shown separately.

1995). Most forested wetland restoration is carried out on sites where hydrology and soils are largely intact and involves primarily planting appropriate vegetation.

CASE STUDY
Florida Forested Wetland Restoration

Forested wetland creation that involves the engineering of an entire wetland setting has been attempted in the phosphate mining region of central Florida (M. T. Brown et al., 1992; Clewell, 1999). Details of the mine restoration are described in Chapter 12. One of the largest forested wetland creation projects of this type consisted of building 61 ha of marsh and forested wetlands at a phosphate mine reclamation site. Fifty-five thousand trees, representing 12 wetland species, were planted (Erwin et al., 1984). The survival rate of the seedlings for the first year was 77 percent. In one of the longest documented histories of a forested wetland restoration, Clewell (1999) describes a 1.5-ha forested wetland planted in Florida in 1985 after extensive hydrologic and soil restoration of the phosphate mine site. The state of Florida specified that this wetland have a minimum density of 988 trees per hectare; 11 years after planting, there were 697 trees per hectare over 10 cm in diameter, with domination by planted cypress

Table 8.2 Bird species observed at central Ohio mitigation wetlands over the five-year required monitoring period, 1992–1996

Scientific Name	Common Name	1992	1993	1994	1995	1996
Agelaius phoeniceus	Red-winged blackbirds	×	×	×	×	
Aix sponsa	Wood duck					×
Anas discors	Blue-winged teal		×	×		
A. platyrhynchos	Mallard	×	×	×	×	×
Ardea herodias	Great blue heron		×	×	×	×
Bombycilla cedorum	Cedar waxwing		×			
Branata canadensis	Canada goose	×	×	×	×	×
Buteo jamaicensis	Red-tailed hawk		×	×		×
Butorides virescens	Green heron			×	×	×
Cardinalis cardinalis	Northern cardinal		×	×	×	×
Cardoparus mexicanus	House finch					×
Casmerodius albus	Great egret		×			
Cathartes aura	Turkey vulture		×	×		
Chaetura plagica	Chimney swift					×
Charadrius vociferus	Killdeer		×			×
Circus cyaneus	Marsh hawk					×
Colaptes auratus	Common flicker					×
Corvus brachyrnchos	Crow		×	×		×
C. corax	Northern raven			×		
Cyanocitta cristata	Blue jay		×	×		×
Epidonax sp.	Epidonax flycatcher					×
Falco sparverius	American kestrel		×			
Fulica americana	American coots				×	×
Hirundo rustica	Barn swallow		×			×
Hirundo/Iridoprocne	Swallows				×	
Iridoprocne bicolor	Tree swallow		×			×
Junco hyemalis	Northern junco					×
Lanius sp.	Shrike		×			
Larus delawarensis	Ring-billed gull		×			
Megaceryle alcyon	Belted kingfisher	×				×
Melospiza georgiana	Swamp sparrow	×	×	×		
M. melodia	Song sparrow			×	×	×
Parus bicolor	Tufted titmouse					×
Passerina cyanea	Indigo bunting					×
Philohela minor	Woodcock		×			
Quiscalus quiscalus	Common grackle	×				
Rallus limicola	Virginia rail		×			
Spinus tristis	American goldfinch			×	×	×
Sturnus vulgaris	European starling					×
Turdis migratorius	Robin		×		×	×
Zenaida macroura	Mourning doves		×	×	×	×
Total taxa observed[a]		6	23	16	12	27

[a]Different individuals did the survey each year, so identification skills varied from year to year.

Taxodium spp. and volunteer willows *Salix caroliniana*. By comparison, 10 reference forested wetlands in the region had a density of 469 trees per hectare, considerably fewer than the stated goal and less than the density of the restored wetland. The basal area of the restored wetland after 11 years was 8.3 m^2 ha^{-1}, about 38 percent of that of the reference forests.

CASE STUDY
Restoring the Florida Everglades

The restoration of the Florida Everglades, the largest wetland area in the United States, is actually several separate initiatives being carried out in the 4.6-million-hectare Kissimmee–Okeechobee–Everglades (KOE) region in the southern third of Florida (see Figure 1.1). Overall, the Everglades restoration, as now planned by the U.S. Army Corps of Engineers, will cost almost $8 billion and will be carried out over 20 years or more. Problems in the Everglades have developed because of (1) excessive nutrient loading to Lake Okeechobee and to the Everglades itself, primarily from agricultural runoff; (2) loss and fragmentation of habitat caused by urban and agricultural development; (3) spread of *Typha* and other invasives and exotics to the Everglades, replacing native vegetation; and (4) hydrologic alteration due to an extensive canal and straightened river system built by the U.S. Army Corps of Engineers and others for flood protection, and maintained by several water management districts.

Everglades restoration also involves halting the spread of high-nutrient cattail (*Typha domingensis*) through the low-nutrient sawgrass (*Cladium jamaicense*) communities that presently dominate the Everglades. Since the main cause of the spread of cattails is nutrients, especially phosphorus emanating from agricultural areas in the basin, 16,000 ha of created wetlands, called stormwater treatment areas (STAs), are planned for phosphorus control from the agricultural area. A prototype of the STAs, a 1500-ha site called the Everglades Nutrient Removal Project, has operated since mid-1994. Early results from this testing of the concept of creating and restoring wetlands to protect downstream wetlands from nutrient enrichment are encouraging. Experimental wetland basins are consistently decreasing phosphorus to levels below 10 μg-P L^{-1}.

Watershed Restoration with Wetlands

Lines often blur between wetlands created and restored for habitat restoration and those restored for water quality improvement or flood control. In fact, most wetlands are restored or created for two or more of these values. For example, a large-scale wetland and riparian forest restoration and creation, on the order of millions of hectares, is being proposed to help solve a major coastal pollution problem in the Gulf of Mexico (Mitsch et al., 2001; see the following Case Study). Restoration done for water quality improvement would also have the major advantages of habitat restoration and flood mitigation in addition to water quality improvement. Similar wetland/watershed restorations are being discussed to alleviate pollution in Chesapeake Bay in the eastern United States and the Baltic Sea in Scandinavia.

CASE STUDY
Solving the Gulf of Mexico Hypoxia

A hypoxic zone has developed off the shore of Louisiana in the Gulf of Mexico where hypolimnetic waters with dissolved oxygen below 2 mg-O_2 L^{-1} now extend over an area of 1.6 to 2.0 million hectares (Rabalais et al., 1996, 1998, pers. commun.; Figure 8.6). Nitrogen, particularly nitrate-nitrogen, is the most probable cause; 80 percent of the nitrogen input is from the 3-million square-kilometer Mississippi River basin (41 percent of the lower 48 states of the United States). The control of this hypoxia is important because the continental shelf fishery in the Gulf is approximately 25 percent of the U.S. total.

Figure 8.6 Mississippi River Basin in the central United States and general location of Gulf of Mexico hypoxia. (From Mitsch et al., 2001.)

A number of approaches are being considered for controlling nitrogen flow into the gulf; many of them involve large-scale modifications of land-use practices in the midwestern United States. Among the options are modifying agricultural practices (e.g., reduced fertilizer use or alternative cropping techniques); tertiary treatment (biological, chemical, physical) of point sources; landscape restoration (e.g., riparian buffers and wetland creation; see Figure 1.3) to control nonpoint-source pollution from farmland; stream and delta restoration; and atmospheric controls of NO_x. The approach that appears to have the highest probability of success with a minimum impact on farming in the midwestern United States is landscape restoration. It has been suggested 2 million hectares of restored and created wetlands and about 7.7 million hectares of restored riparian buffers would be necessary to provide enough denitrification to substantially reduce the nitrogen entering the Gulf of Mexico (Mitsch et al., 1999, 2001). Restoring the Gulf of Mexico requires restoring 3 percent of the Mississippi River basin. Interestingly, Hey and Phillipi (1995) found that a similar wetland restoration effort would be required in the upper Mississippi River basin to mitigate the effects of very large and costly floods such as the one that occurred in the summer of 1993.

Peatland Restoration

Peatland restoration has been a relatively new type of wetland restoration. Early attempts with peatlands occurred in Europe, specifically in Finland, Germany, the United Kingdom, and the Netherlands. Increased peat mining in Canada and elsewhere has led to increased interest in understanding if and how mined peatlands can be restored. When peat surface mines are abandoned without restoration, the area rarely returns through secondary succession to the original moss-dominated system (Quinty and Rochefort, 1997). There is promise that restoration can be successful (Lavoie and Rochefort, 1996; Wind-Mulder et al., 1996; Quinty and Rochefort, 1997; Price et al., 1998) but because (1) surface mining causes major changes in local hydrology, and (2) peat accumulates at an exceedingly slow rate, restoration progress will be measured in decades rather than years. In the 1960s and 1970s, block harvesting of peat was replaced by vacuum harvesting in southern Quebec and in New Brunswick, necessitating the development of different restoration techniques. Whereas traditional block cutting of peat left a variable landscape of high ground and trenches, vacuum harvesting leaves relatively flat surfaces bordered by drainage ditches. Abandoned block-cut sites appear to revegetate with peatland species more easily than do vacuum-harvested sites, and the latter can remain bare for a decade or more after mining (Rochefort and Campeau, 1997).

8.4 WETLAND CREATION AND RESTORATION TECHNIQUES

General Principles

Two basic principles should be the starting point for anyone involved in wetland restoration and creation:

1. Understanding wetland ecology and its principles (e.g., hydrology, biogeochemistry, adaptations, succession) is essential to create and restore wetlands successfully as part of a natural landscape.
2. Resisting the ever-present temptation to overengineer wetlands by attempting either to channel natural energies that cannot be channeled or to introduce species that the landscape and climate do not support.

In all situations of wetland creation and restoration, human contributions to the design of wetlands should be kept simple and should strive to stay within the bounds established by the natural landscape, without reliance on dams, dikes, weirs, pumps, and other technological approaches that invite failure. As stated by Boule (1988): "Simple systems tend to be self-regulating and self-maintaining." Some general principles of ecotechnology that apply to the creation and restoration of wetlands are:

1. Design the system for minimum maintenance and a general reliance on self-design.
2. Design a system that utilizes natural energies, such as the potential energy of streams as natural subsidies to the system.
3. Design the system with the hydrologic and ecological landscape and climate.
4. Design the system to fulfill multiple goals, but identify at least one major objective and several secondary objectives.
5. Give the system time.
6. Design the system for function, not form.
7. Do not overengineer wetland design with rectangular basins, rigid structures and channels, and regular morphology.

Defining Goals

The design of an appropriate wetland or series of wetlands, whether for the control of nonpoint source pollution, for a wildlife habitat, or for wastewater treatment, should start with the formulation of the overall objectives of the wetland. One view is that wetlands should be designed to maximize ecosystem longevity and efficiency and minimize cost. The goal, or a series of goals, should be determined before a specific site is chosen or a wetland is designed. If several goals are identified, one must be chosen as primary.

Placing Wetlands in the Landscape

In some cases, particularly when sites are being chosen for habitat replacement, many choices are available of sites in the landscape to locate a restored or created wetland. The natural design for a riparian wetland fed primarily by a flooding stream or river (Figure 8.7a) allows for flood events of a river to deposit sediments and chemicals on a seasonal basis in the wetland, and for excess water to drain back to the stream or river. Because there are natural and also often constructed levees along major sections of streams, it is often possible to create such a wetland with minimal construction. The wetland could be designed to capture flooding water and sediments and slowly release the water back to the river after the flood passes, or to receive flooding water and retain it through the use of flap gates.

Wetlands can be designed as in-stream systems by adding control structures to the streams themselves or by impounding a distributary of the stream (Figure 8.7b). Blocking an entire stream is a reasonable alternative only in headwater streams, and it is not generally cost-effective or ecologically advisable. This design is particularly vulnerable during flooding and its stability might be unpredictable, but it has the advantage of potentially "treating" a significant portion of the water that passes that point in the stream. The maintenance of the control structure and the distributary might mean making significant management commitments to this design.

A riparian wetland fed by a pump (Figure 8.7c) creates the most predictable hydrologic conditions for the wetland but at an obvious extensive cost for equipment and maintenance. If it is anticipated that the primary objective of a constructed wetland is the development of a research program to determine design parameters for future wetland construction in the basin, a wetland fed by pumps is a good design. If other objectives are more important, the use of large pumps is usually not appropriate. Small pumps may be necessary to carry a riparian wetland through drought periods. Two examples of wetlands of this type constructed primarily for research and education are the Des Plaines River wetlands in northeastern Illinois (Sanville and Mitsch, 1994) and the Olentangy River wetlands in central Ohio (Mitsch et al., 1998).

The advantages of locating several small wetlands on small streams or intercepting ditches in the upper reaches of a watershed (but not in the streams themselves), rather than creating fewer larger wetlands in the lower reaches, should be considered (Figure 8.7d). The usefulness of wetlands in decreasing flooding increases with the distance the wetland is downstream.

Figure 8.7e shows a design involving the creation of a wetland along a stream to intercept tile drains from agricultural fields. The stream itself is not

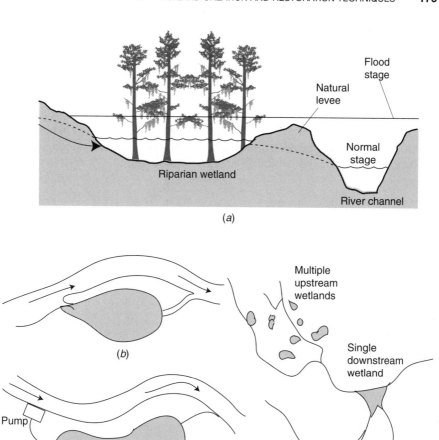

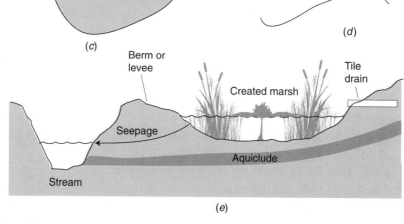

Figure 8.7 Some landscape locations of created and restored wetlands in a riverine setting: (a) riparian wetland that both intercepts groundwater from uplands but also receives an annual flood pulse from adjacent river; (b) riparian wetlands with natural flooding; (c) riparian wetland with pump; (d) multiple upstream wetlands versus single downstream wetland; (e) lateral wetland intercepting groundwater carried by tile drains. (From Mitsch and Gosselink, 2000; copyright 2000; reprinted with permission from John Wiley & Sons, Inc.)

diverted, but the wetlands receive their water, sediments, and nutrients from small tributaries, swales, and especially tile drains that otherwise would empty straight to the stream. If tile drains can be located and broken or blocked upstream to prevent their discharge into tributaries, they can be rerouted to make effective conduits to supply adequate water to constructed wetlands. Because tile drains are often the sources of the highest concentrations of chemicals, such as nitrates from agricultural fields, the lateral wetlands would be an efficient means of controlling certain types of nonpoint-source pollution while creating a needed habitat in an agricultural setting.

Site Selection

Several important factors ultimately determine site selection. When the objective is defined, the appropriate site should allow for the maximum probability that the objective can be met, that construction can be done at a reasonable cost, that the system will perform in a generally predictable way, and that the long-term maintenance costs of the system are not excessive. These factors are:

1. Wetland restoration is generally more feasible than wetland creation.
2. Take into account the surrounding land use and the future plans for the land.
3. Undertake a detailed hydrologic study of the site, including a determination of the potential interaction of groundwater with the proposed wetland.
4. Find a site where natural inundation is frequent.
5. Inspect and characterize the soils in some detail to determine their permeability, texture, and stratigraphy.
6. Determine the chemistry of the soils, groundwater, surface flows, flooding streams and rivers, and tides that may influence the site water quality.
7. Evaluate on-site and nearby seed banks to ascertain their viability and response to hydrologic conditions.
8. Ascertain the availability of necessary fill material, seed, and plant stocks and access to infrastructure (e.g., roads, electricity).
9. Determine the ownership of the land, and hence the price.
10. For wildlife and fisheries enhancement, determine if the wetland site is along ecological corridors such as migratory flyways or spawning runs.
11. Assess site access.
12. Ensure that an adequate amount of land is available to meet the objectives.

Creating and Maintaining the Proper Hydrology

The key to restoring and creating wetlands is to develop appropriate hydrologic conditions. Groundwater inflow is often desired because this offers a more predictable and less seasonal water source. Surface flooding by rivers gives wetlands a seasonal pattern of flooding, but such wetlands can be dry for extended periods in flood-absent periods. Depending on surface runoff and flow from low-ordered streams can be the least predictable. Often, wetlands developed in these conditions are isolated pools and potential mosquito havens for a good part of the growing season; their design should be considered carefully. It is generally considered to be optimum to build wetlands where they used to be and where hydrology is still in place for the wetland to survive. But tile drainage, ditches, and river downcutting have often changed local hydrology from prior conditions. Most biologists have difficulty estimating hydrologic conditions, and engineers often overengineer control structures that need substantial maintenance and are not sustainable.

Wetland basins are constructed either by establishing levees around a basin that may be partially excavated in the landscape or by excavating a depression without constructing a levee. Construction engineers will note that if they use excavated soil for a levee, they can save large sums of money, as excavation is often the largest cost of wetland restoration or creation. This is usually not a good idea, as levees are bound to have problems with leakage and, in many parts of the world, with burrowing animals such as muskrats (*Ondatra zibethicus*).

Some sort of control structure is often needed at the outflow of the basin, whether or not there is a levee. The control devices are the outflow of the wetland. Three such devices are shown in Figure 8.8: (1) drop pipes, (2) flashboard risers, and (3) full-round risers (a combination of drop pipe and flashboard riser). Each has advantages and disadvantages (Massey, 2000). Drop pipes are the least flexible, as they do not allow water-level manipulation. A flashboard riser is more flexible but can easily be vandalized. Full-round risers are a little more secure and can be designed for control of beavers. But they are a little more expensive. Two models of control devices used for holding water level in a created and restored wetlands are shown in Figure 8.9. In both cases the outflow risers include removable *stoplogs* that allow manual changes in water level. This option is desirable when the exact hydrology is not known for the wetland basin; it allows flexibility. But these types of control devices have several disadvantages. They require occasional maintenance, if only for removing accumulation of plant debris and resetting stoplogs. Also, stoplog removal is a favorite pastime of vandals. Control devices such as risers are also favorite locations for nature's ecological engineer—the beaver *Castor canadensis*—to provide its idea of water management, usually creating blockages that can raise the water level by a meter or more, changing the vegetation patterns dramatically.

The best design situation is when the local topography allows the wetland to be naturally flooded without control devices, but this opportunity is rarely

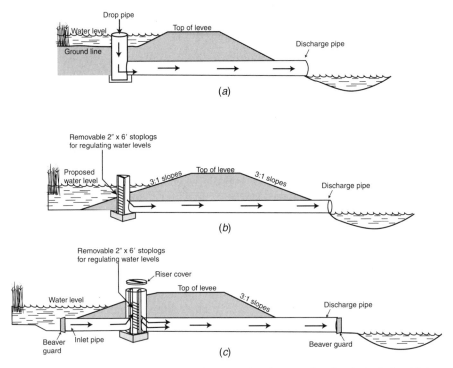

Figure 8.8 Designs for control systems for created and restored wetlands, including (a) drop pipe, (b) flashboard riser, and (c) full-round riser. (Redrawn from Massey, 2000.)

available. For a reliable source of water, groundwater is generally less sensitive than surface water to seasonal highs and lows. Also, a wetland fed by groundwater invariably has better water quality and generally fewer sediments that will eventually fill the wetlands.

Soil Analysis

Often, choice of the site of wetland creation and restoration is limited by property ownership. If a choice exists, a wetland that is restored on former wetland (hydric) soils is much preferred over one constructed on upland soils. *Hydric soils* develop certain color and chemical patterns because they have spent long periods flooded and thus under anaerobic conditions. The soil color is mostly black in mineral hydric soils because iron and manganese minerals have been converted to reduced soluble forms and have leached out of the soil. In most cases, developing wetlands on hydric soils has three advantages:

1. Hydric soils indicate that a site may still have or can be restored to the appropriate hydrology.

(a)

Figure 8.9 Control devices installed in wetlands, including (a) flashboard riser and (b) concrete rectangular weir box with overflow capability for large flows. (Photos by W. J. Mitsch.)

2. Hydric soils may be a *seed bank* of wetland plants still established in the soil.
3. Hydric soils may have the appropriate soil chemistry for enhancing certain wetland processes. For example, mineral hydric soils generally have higher organic carbon than do mineral nonhydric soils. This soil carbon, in turn, stimulates wetland processes such as denitrification and methane production.

(b)

Figure 8.9   (*Continued*)

Otherwise, it is certainly possible to create wetlands on upland soils, and in the long run those soils will develop characteristics typical of hydric soils, such as higher carbon content and seedbanks. It is not well known how long it will take such upland soil wetlands to develop these conditions—it can be over a decade or it may take a century, depending on the soil types and the hydrology. In a study of the wetlands created at the Olentangy River Wetland Research Park in Ohio, hydric soil conditions were shown to develop after only two years of continued flooding (Table 8.3). Before the basins were first flooded, the most prevalent hue was 10YR, and the value/chroma soil color varied between 3/3 and 3/4. The chroma of 3 to 4 indicates nonhydric soils. In 1995, about 18 months after flooding, chromas of 3 or less were common (median = 3/2). The mean value in the surface samples was 3/2; subsurface sample median values were 4/2. Chromas were consistently at 2 or below in 1996, two years after flooding began.

It has also been argued that upland soils often do not allow the development of a major diversity of plant communities but often become *Typha* marshes

Table 8.3   Soil color of the wetland basins created in Olentangy River Wetland Research Park, Ohio, 1992–1999

Year	Color Value/Chroma
1993 (before water was added)	3/3 to 3/4
1994 (flooding began March 1994)	3/3 to 4/3
1995	3/2 to 4/3
1996	3/2
1997	2/2 to 3/3
1998	3/2
1999	3/2

Source: Data from Deshmukh and Mitsch (2000).

[a]Chroma values (second number) below 2 generally indicate hydric soil conditions.

instead because of the absence of seed banks. This fact is as much due to the fact that uplands converted to wetlands have often been used for agriculture for many years and are thus quite eutrophic. The high-nutrient conditions invariably leads to high-productivity, low-diversity systems. Again, the main advantage of using hydric soils in wetland restoration and creation is that they are indicators of the appropriate hydrologic conditions.

Introducing Vegetation

To develop a wetland that will ultimately be a low-maintenance wetland, natural successional processes need to be allowed to proceed. The best strategy is usually to introduce, by seeding and planting, as many native choices as possible to allow natural processes to sort out the species and communities in a timely fashion. Wetlands created or restored by this approach are called *self-design wetlands.* Providing some help to this selection process, for example, selective weeding, may be necessary in the beginning, but ultimately, the system needs to survive with its own successional patterns unless significant labor-intensive management is possible. A somewhat different approach, called *designer wetlands,* occurs when specified plant species are introduced and the success or failure of those plants is used as an indicator of the success or failure of that wetland. The species of vegetation types to be introduced to created and restored wetlands depend on the type of wetland desired, the region, and the climate as well as the design characteristics described above. Common plants used for freshwater marshes include cattails (*Typha* spp.), sedges (*Carex* spp., *Scirpus* spp., and *Schoenoplectus* spp.), and floating-leaved aquatic plants such as white water lilies (*Nymphaea* spp. and spatter-dock (*Nuphar* spp.). Submerged plants are not common in wetland design, and their propagation is often hampered by turbidity and algal growth in the early years of wetland development. Forested wetland restoration and creation usually involve the planting of tree seedlings, but it could take decades to determine if plantings are successful.

An important general consideration of wetland design is whether plant material is going to be allowed to develop naturally from some initial seeding and planting or whether continuous horticultural selection for desired plants will be imposed. W. E. Odum (1987) suggested that "in many freshwater wetland sites it may be an expensive waste of time to plant species which are of high value to wildlife. . . . It may be wiser to simply accept the establishment of disturbance species as a cheaper although somewhat less attractive solution." Reinartz and Warne (1993) found that the way vegetation is established can affect the diversity and value of the mitigation wetland system. Their study showed that early introduction of a diversity of wetland plants may enhance the long-term diversity of vegetation in created wetlands. The study examined the natural colonization of plants in 11 created wetlands in southeastern Wisconsin. The wetlands under study were small, isolated, depressional wetlands. A two-year sampling program was conducted for the created wetlands, aged one to three years. Colonization was compared to five seeded wetlands where 22 species were introduced. The diversity and richness of plants in the colonized wetlands increased with age, size, and proximity to the nearest wetland source. In the colonized sites, *Typha* spp. comprised 15 percent of the vegetation for one-year wetlands, and 55 percent for three-year wetlands, with the possibility of monocultures of *Typha* spp. developing over time in colonized wetlands. The seeded wetlands had a high species diversity and richness after two years. *Typha* cover in these sites was lower than in the colonized sites after two years.

Another study where the effects of planting versus not planting have been observed for several years is at the experimental wetlands at the Olentangy River Wetland Research Park at the Ohio State University (see the following Case Study). That study has completed almost a decade of observations where one full-scale wetland was planted and an identical wetland remained unplanted and showed that there were profound differences in wetland function after several years in planted and unplanted wetland basins that could be traced to the effects of the initial planting.

Identifying Exotic or Undesirable Plant Species

In some cases, certain plants are viewed as desirable or undesirable because of their value to wildlife or their aesthetics. Reed grass (*Phragmites australis*) is often favored in constructed wetlands in Europe, and there is real concern for reed dieback around lakes and ponds in Europe. But reed grass is considered an invasive undesirable plant in much of eastern North America, particularly in coastal freshwater and brackish marshes. Some plants are considered undesirable in wetlands because they are aggressive competitors. In many parts of the tropics and subtropics, the floating aquatic plant water hyacinth (*Eichhornia crassipes*) and alligator weed (*Alternanthera philoxeroides*) are considered undesirable, and in eastern North America, particularly around the Great Lakes, the emergent purple loosestrife (*Lythrum salicaria*) is considered an undesirable alien plant in wetlands. Throughout the United States, cattail (*Typha* spp.) is championed by some and disdained by others, for it is a rapid colonizer but of limited wildlife value. In other parts of the world, *Typha* is considered a perfectly acceptable plant in restored wetlands. In New Zealand, several species of willow (*Salix*) are invading marshes and other wetlands, and programs to eradicate them are common.

CASE STUDY
Planting Experiment, Olengangy River Wetlands

In a multiyear study at the Olentangy River Wetland Research Park (Figure 8.10) in central Ohio, Mitsch et al. (1998) planted a 1-ha constructed wetland receiving nutrient-rich river water with 13 species typical of the midwestern United States. A second hydrologically identical wetland remained as a control. After three years, both wetlands were principally dominated by soft-stem bulrush *Schoenoplectus tabernaemontani* (= *Scirpus validus*) and were thought to be similar. Researchers found that both planted and unplanted wetlands converged in most of the 16 ecological indicators (eight biological measures; eight biophysiochemical measures) in those three years (Figure 8.11). During the second year that the 1-ha planted wetland basin had considerably more vegetation than the 1-ha unplanted wetland basin, only 12 percent of the indices were similar. Continued studies at that site subsequent to those three years showed a persistence of similar cover and vegetation types in the planted and unplanted basins. It appeared that the two wetlands had converged. After six years, however, several communities of vegetation continued to exist in the planted basin, but a monoculture of *Typha* dominated the unplanted basin (Figure 8.12). Residual effects of planting were thus evident six years after planting. The planted wetland had a more spatially diverse macrophyte community, including communities dominated by *Sparganium eurycarpum, Scirpus fluviatilis, Typha* spp., and *Spartina pectinata*.

This study also supported general principles given in Chapter 4 that ecosystems with higher productivity are often less diverse than are low-productivity systems (Figure 8.13). Each of these wetlands had been fed by high-nutrient river water for six years, yet the planted wetland had a more diverse set of plant communities but was considerably lower in productivity. The *Typha* marsh that developed in the unplanted wetland was a monoculture yet almost 50 percent higher in productivity.

(a)

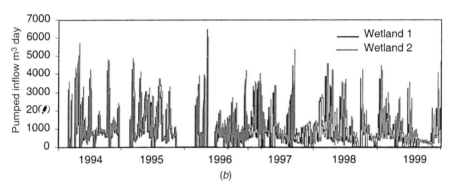

(b)

Figure 8.10 (a) Paired 1-ha experimental wetlands at the Olentangy River Wetland Research Park at the Ohio State University (photo in August 1999, six years after planting). Planted wetland basin (W1) is on the right; unplanted basin (W2) in on the left. (b) Pumped inflow to both basins for the first six years of the experiment. Flow was statistically identical for the six years to both wetland basins. (From Mitsch et al., work in progress.)

Overall, there were functional differences between the two basins after six years that in all probability can be attributed to the initial planting (Figure 8.14). The wetland planted by humans had 43 percent lower aboveground net primary productivity in year 6, mainly because the productive *Typha* dominated the naturally colonizing basin, whereas the planted wetland was dominated by more diverse but less productive, introduced macrophytes. Because there was less aboveground biomass, more sunlight could reach the water column, leading to higher gross primary productivity in the water column and hence higher dissolved oxygen in the planted wetland. This more oxygenated water column favored fish (mostly *Lepomis* sp.) over am-

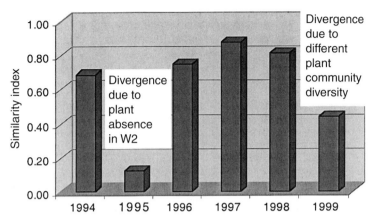

Figure 8.11 Similarity of experimental wetlands (planted and unplanted) at Olentangy River Wetland Research Park for six years as estimated by 16 functional indicators, including plant productivity, water chemistry, nutrient retention, and benthic invertebrate diversity. Note the initial pattern of divergence in the second year (due to the absence of plants in the unplanted basin), convergence, and then divergence again (due to the difference in plant community diversity).

phibians. The "unplanted" wetland, with a higher macrophyte biomass, had higher numbers of muskrats (*Ondatra zibithicus*) (Higgins, 2002), lower aquatic primary productivity, and thus lower dissolved oxygen. That wetland had higher amphibian numbers and biomass (mostly bullfrog, *Rana catesbeiana*), higher snake (*Nerodia sipedon*) populations, but lower fish biomass (Gifford, 2002). Planting does have an effect, but whether to plant depends on the original objective of the wetland. If plant diversity is desired, planting makes sense. If productivity is desired, it may be a waste of effort to plant. But in any regard, there appears to be a major long-term effect on ecosystem function caused by planting.

Planting Techniques

Plants can be introduced to a wetland by transplanting roots, rhizomes, tubers, seedling, or mature plants; by broadcasting seeds obtained commercially or from other sites; by importing substrate and its seed bank from nearby wetlands; or by relying completely on the seed bank of the original and surrounding site. If planting stocks rather than site seed banks are used, it is most desirable to choose plants from wild stock rather than nurseries because the former are generally better adapted to the environmental conditions that they will face in constructed wetlands. The plants should come from nearby, if possible, and should be planted within 36 hours of collection. If nursery plants

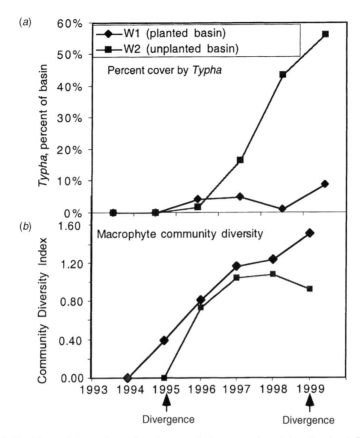

Figure 8.12 Macrophyte patterns for six years in two experimental wetlands at Olentangy River Wetland Research Park: (a) percent cover by *Typha* spp.; (b) community diversity index (CDI) of the two basins. Note the divergence in diversity of the basins in sixth year, 1999. (From Mitsch et al., work in progress.)

are used, they should be from the same general climatic conditions and should be shipped by express service to minimize losses. Brown (1987) suggested that marshes be planted at densities to ensure rapid colonization, adequate seed source, and effective competition with *Typha* spp. Specifically, this could mean introducing from 2000 to 5000 plants per hectare.

For emergent plants, the use of planting materials with at least 20- to 30-cm stems was recommended—and whole plants, rhizomes, or tubers rather than seeds have been most successful. In temperate climates, both fall and spring planting times are possible for certain species, but spring plantings are generally more successful because it is a better time to minimize destructive grazing of plants in the winter by migratory animals and the uprooting of the new plants by ice. Transplanting plugs or cores (8 to 10 cm in diameter) from existing wetlands is another technique that has been used with success, for it

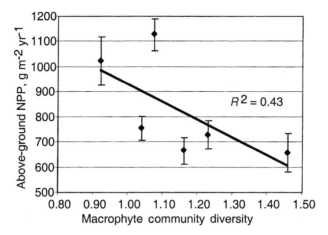

Figure 8.13 Net primary productivity versus community diversity for two experimental wetlands for the period 1997–1999. (From Mitsch et al., work in progress.)

brings seeds, shoots, and roots of a variety of wetland plants to the newly restored or created wetland.

If seeds and seed banks are used for wetland vegetation, several precautions must be taken. The seed bank should be evaluated for seed viability and species present. The use of seed banks from other sites nearby can be an effective way to develop wetland plants in a constructed wetland if the hydrologic conditions in the new wetland are similar. Seed bank transplants have been successful for many different species, including sedges (*Carex* spp.), *Sagittaria* sp., *Scirpus acutus, S. validus,* and *Typha* spp. The disruption of the wetland site where the seed bank is obtained must also be considered.

When seeds are used directly to vegetate a wetland, they must be collected when they are ripe, and stratified if necessary. If commercial stocks are used, the purity of the seed stock should be determined. The seeds can be added with commercial drills or by broadcasting from the ground, watercraft, or aircraft. Seed broadcasting is most effective when there is little to no standing water in the wetland.

8.5 ESTIMATING SUCCESS

There are few satisfactory methods available to determine the "success" of a created or restored wetland or even a mitigation wetland created to replace the functions lost with the original wetland. Figure 3.6 illustrates conceptually how it should be done for replacement wetlands. On the one hand, *legal success* involves a comparison of the lost wetland function and area with that which is gained in the replacement wetland. *Ecological success* should involve a comparison of the replacement wetland with a reference wetland

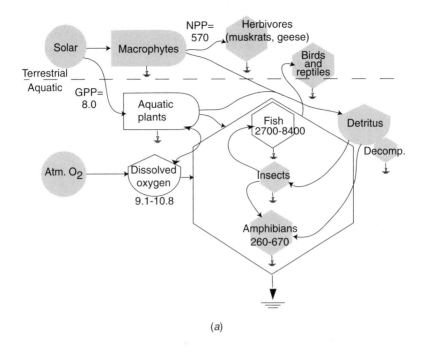

(a)

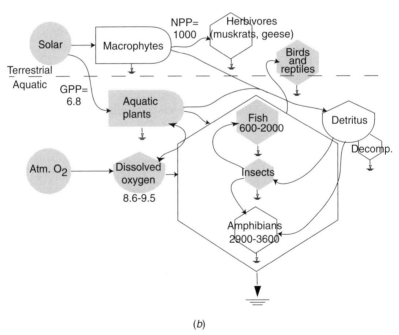

(b)

Figure 8.14 Energy diagrams of (a) planted and (b) unplanted wetland basins after six years. Unshaded symbols illustrate components of each wetland ecosystem that were more pronounced as a result of planting or natural colonization. NPP (net primary productivity) is expressed as g-dry wt m^{-2} yr^{-1}. GPP (gross primary productivity) of the water column is expressed as g-O_2 m^{-2} day^{-1}. Dissolved oxygen is annual average for 1999 and 2000 in mg L^{-1}. Fish and amphibian population estimates for 2000 and 2001 are given as number of individuals ha^{-1}. (Data from Mitsch et al., work in progress; Gifford, 2002; Higgins, 2002.)

(natural wetlands of the same type that may occur in the same setting or generally accepted "standards" of regional wetland function) (Wilson and Mitsch, 1996). Overall success would then be gauged by a combination of the legal and ecological comparisons. Although this model represents an ideal, a comparison involving both standards is rarely made.

Designing for Success

It is clear from the many studies of created and restored wetlands that some cases are successes, while there are still far too many examples of failure of created and restored wetlands to meet expectations. In some cases expectations were unreasonable, as when endangered species habitat was to be established in a heavily urbanized environment (Malakoff, 1998). In such cases, the original wetland should not have been lost to begin with. Where expectations are ecologically reasonable, there is optimism that wetlands can be created and restored and that wetland function can be replaced. The spotty record to date is due, in our opinion, to three factors: (1) little understanding of wetland function by those constructing the wetlands, (2) insufficient time for the wetlands to develop, and (3) a lack or recognition or underestimation of the self-design capacity of nature (Mitsch and Wilson, 1996).

Understanding wetlands enough to be able to create and restore them requires substantial training in plants, soils, wildlife, hydrology, water quality, and engineering. Replacement projects and other restorations involving freshwater marshes need enough time, closer to 15 or 20 years than to five years, before success is apparent. Restoration and creation of forested wetlands, coastal wetlands, or peatlands may require even more time. Peatland restoration could take decades or more; and, of course, forested wetland restoration generally takes a lifetime. Finally, we should recognize that nature remains the chief agent of self-design, ecosystem development, and ecosystem maintenance; humans are not the only participants in these processes. Sometimes we refer to these self-design and time requirements for successful ecosystem restoration and creation as invoking "mother nature and father time" (Mitsch and Wilson, 1996; Mitsch et al., 1998).

Wetland science will continue to make significant contributions to the process of reducing our uncertainty about predicting wetland success. Wetland creation and restoration need to become part of an applied ecological science, not a technique without theoretical underpinnings. Scientists need to make the connections between structures such as vegetation density and diversity; and functions such as productivity, wildlife use, organic sediment accretion, or nutrient retention, in quantitative and carefully designed experiments. Engineers and managers need to recognize that designing systems that emphasize the role of self-design and sustainable structures are more ecologically feasible in the long run than are heavily managed systems.

Summary Principles for Successful Wetland Restoration and Creation

Robin Lewis and Kevin Erwin, two wetland consultants from Florida, have spent about five decades restoring and creating wetlands around the world. Their experience has led to the following 15 recommendations (Lewis et al., 1995), which serve as a fitting summary of the application of principles and practices in effective restoration and creation of wetlands.

1. Wetland restoration and creation proposals must be viewed with great care, particularly when promises are made to restore or recreate a natural system in exchange for a permit.
2. Multidisciplinary expertise in planning and careful project supervision at all project levels is needed.
3. Clear, site-specific measurable goals should be established.
4. A relatively detailed plan concerning all phases of the project should be prepared in advance to help evaluate the probability of success.
5. Site-specific studies should be carried out in the original system prior to wetland alteration if wetlands are being lost in the project.
6. Careful attention to wetland hydrology is needed in design.
7. Wetlands should, in general, be designed to be self-sustaining systems and persistent features of the landscape.
8. Wetland design should consider relationships of the wetland to the watershed, water sources, other wetlands in the watershed, and adjacent upland and deepwater habitat.
9. Buffers, barriers, and other protective measures are often needed.
10. Restoration should be favored over creation.
11. A capability for monitoring and midcourse corrections is needed.
12. A capability for long-term management is needed for some type of systems.
13. Risks inherent in restoration and creation, and the probability of success for restoring or creating particular wetland types and functions, should be reflected in standards and criteria for projects and project design.
14. Restoration for artificial or already altered systems requires special treatment.
15. Emphasis on ecological restoration of watersheds and landscape ecosystem management requires advanced planning.

9

COASTAL RESTORATION

Coastal and nearshore marine ecosystems are being lost and polluted at an alarming rate around the world. Coastlines are being developed at rapid rates, causing a proliferation of high-rise towers, development, and highways that are replacing coastal marshes, mangroves, and even high-energy beaches. In many tropical parts of the world, high-productivity aquaculture producing shrimp for a couple of years and moving on is replacing sustainable mangrove swamps that have supported subsistence cultures for centuries. Estuaries are being dredged and filled for sea-lane transportation by larger and larger ships. The U.S. Army Corps of Engineers annually dredges about 275 million cubic meters of material from U.S. waterways to keep them open, most of that dredging in coastal areas. And now we have found that excessive nutrients from high-yield agriculture, sometimes created hundreds of kilometers inland, are being transported to coastal waters, where they cause extensive coastal eutrophication and resulting "dead zones."

There is a great deal of interest both in upland restoration that will eventually control the sources of pollution reaching the coastline (see other restoration chapters) and in techniques and approaches that can be used to restore disturbed coastal ecosystems themselves. Reviews of what can be done to restore coastal wetlands and other estuarine and coastal ecosystems are described for coastal salt marshes (Broome et al., 1988; Zedler, 1988, 1996a,b, 2001; Broome, 1990; Shisler, 1990) and mangroves (Lewis, 1990b,c). Restoration strategies of a wide range of coastal environments are described by Thayer (1992), Weinstein and Kreeger (2000), and Wilber et al. (2000). The last publication provides a particularly detailed investigation of coastal restoration that resulted from invited papers given at a symposium titled "Goal Setting and Success Criteria for Coastal Habitat Restoration" held in Charleston, South Carolina in January 1998 and published two years later as a special issue of *Ecological Engineering*.

9.1 COASTAL ECOSYSTEMS

Coastal ecosystems can best be categorized by their major forcing functions (e.g., seasonal programming of sunlight and temperature) and stresses (e.g., ice) (Figure 9.1). Salt marshes, found in the type C category of natural temperate ecosystems with seasonal programming, have "light tidal regimes" and "winter cold" as forcing function and stress, respectively. Coral reefs are classified as type B (natural tropical ecosystems) because they have abundant light, show little stress, and reflect little seasonal programming. Three additional classes, type A (naturally stressed systems of wide latitudinal range), type D (natural arctic ecosystems with ice stress), and type E (emerging new systems associated with humans), were included in this classification. The last class, which includes new systems formed by pollution, such as pesticides and oil spills, is still an interesting concept that could be applied to other wetland classifications. Categorizing systems by forcing functions and stresses is a useful way to evaluate ecological engineering approaches to restoring and protecting these systems, as forcing functions need to be intact to have any expectation of coastal system health.

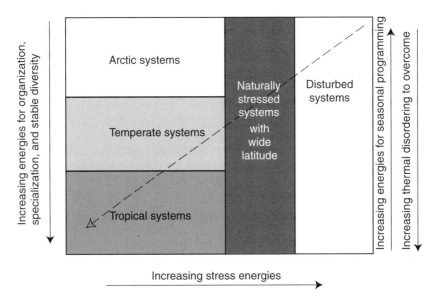

Figure 9.1 Classification of coastal ecosystems. (After H. T. Odum et al., 1974; redrawn by Mitsch and Gosselink, 2000; copyright 2000; reprinted with permission from John Wiley & Sons, Inc.)

The most important ecological part of a coastline is its *estuary* (Figure 9.2). An estuary has been defined as "semi-enclosed coastal body of water which has free connection with the open sea and within which sea water is measurably diluted with fresh water derived from land drainage" (Pritchard, 1967). Day et al. (1989) point out that this definition covers most estuarine-like systems but does not include semiarid coastal waters that receive essentially no fresh water and coastal aquatic systems that can become totally isolated from the sea because of shoreline sediments. Coastal systems vary from full salinity marine systems (30 to 35 ppt) through polyhaline (18 to 30 ppt), mesohaline 5 to 18 ppt), and oligohaline (0.5 to 5 ppt) systems to fresh-water systems such as tidal marshes still influenced by tides. Some of the major types of coastal systems that are frequently restored are described here in more detail.

Salt Marshes

Figure 9.2 illustrates that salt marshes occur when there is both sufficient salinity (>5 ppt), tides, and protection from the open sea. Beeftink (1977) defined a salt marsh as a "natural or semi-natural halophytic grassland and dwarf brushwood on the alluvial sediments bordering saline water bodies whose water level fluctuates either tidally or non-tidally." Salt marshes are found in middle and high latitudes along intertidal shores throughout the

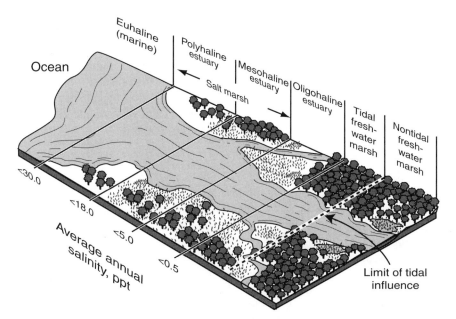

Figure 9.2 Salinity gradient of an estuary. (From Mitsch and Gosselink, 2000; copyright 2000; reprinted with permission from John Wiley & Sons, Inc.)

world and can be narrow fringes on steep shorelines or expanses several kilometers wide. They are found near river mouths, in bays, on protected coastal plains, and around protected lagoons.

The salt marsh ecosystem has diverse biological components that include vegetation and animal and microbe communities in the marsh and plankton, invertebrates, and fish in the tidal creeks, pannes, and estuaries. Characteristic spatial patterns found in coastal salt marshes are shown in Figure 9.3. Salt marshes typical of the U.S. eastern coastal plain (Figure 9.3*a*) include a *low marsh* almost entirely *Spartina alterniflora,* which is tall on the creek bank and shorter behind the natural levee, as elevation gradually increases in an inland direction. The low marsh often contains small vegetated or unvegetated ponds and mud barrens called *pannes.* The *high marsh* is much more diverse, containing short *S. alterniflora* intermixed with associations of *Distichlis spicata, Juncus roemerianus,* and *Salicornia* spp. Along the Mississippi and northwest Florida coasts, there is often a fringe of *S. alterniflora* along the seaward margin, followed, in an inland direction, by large areas of tall and short *J. roemerianus,* mixtures of *S. patens* and *D. spicata,* and *Salicornia* spp. where salt accumulates. In Europe, a totally different salt marsh is found (Figure 9.3*b*), at least compared to eastern U.S. marshes. One of the most notable features is that the intertidal zone between high tide and mean high tide is sparsely covered if it is vegetated at all, so much of what would be called low marsh in Europe is, in fact, a mudflat. Cordgrass found in Europe is generally *S. anglica* or *S. townsendii.*

Tidal marshes have been lost or affected greatly throughout the world with artificial drainage and dikes, changing salinity conditions due to increased runoff, invasion by aggressive species such as *Phragmites* in brackish areas, and conversion to uplands such as hay farms. Often, salt marshes are found where the general population thinks a summer cottage or an airport would be more functional. Many of the coastal cities of the world placed their airports decades ago on fills in coastal marshes. While restoration of marshes is complex, their restoration is generally thought to be relatively simple compared to many other systems if the tidal effects can be restored.

Mangroves

The mangrove swamp is the ecological analog of the salt marsh in tropical and subtropical regions of the world, usually found between 25°N and 25°S latitude. A mangrove swamp (Figure 9.4) is an association of halophytic trees, shrubs, and other plants growing in brackish to saline tidal waters of tropical and subtropical coastlines. This coastal, forested wetland (called a *mangal* by some researchers) is infamous for its impenetrable maze of woody vegetation, its unconsolidated peat that seems to have no bottom, and its many adaptations to the double stresses of flooding and salinity. Like the coastal salt marsh, the mangrove swamp can develop not only where there is salinity but also where there is adequate protection from high-energy wave action. Several

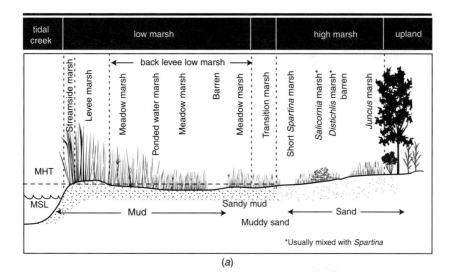

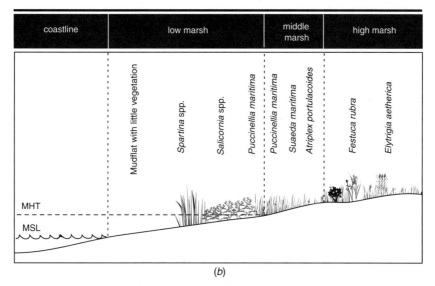

Figure 9.3 Zonation of vegetation in a salt marsh: (*a*) southern Atlantic coast of the United States; (*b*) northern Europe (From Mitsch and Gosselink, 2000; copyright 2000; reprinted with

physiographic settings favor the protection of mangrove swamps, including (1) protected shallow bays, (2) protected estuaries, (3) lagoons, (4) the leeward sides of peninsulas and islands, (5) protected seaways, (6) behind spits, and (7) behind offshore shell or shingle islands. Mangrove swamps are managed throughout the world and are often part of a local culture of low-intensity

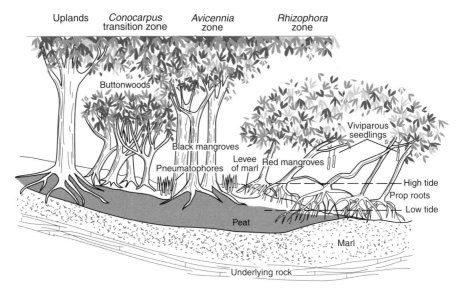

Figure 9.4   Zonation of vegetation in a mangrove swamp. (From Mitsch and Gosselink, 2000; copyright 2000; reprinted with permission from John Wiley & Sons, Inc.)

fishery production. In other locations, mangroves are drained and destroyed to make way for coastal development, much to the dismay of local populations when tropical storms decide to pay a visit. Their protection and restoration in many parts of the world is now a serious government priority. In Florida, major state laws protect mangroves, and their removal or drainage is prohibited.

Sea Grass Beds

Shallow estuarine sediments often develop dense cover of submersed grasses. If the conditions are full salinity, these plants are called sea grasses. The most studied sea grass systems in temperate zones are dominated by eelgrass (*Zostera* spp.) and in tropical regions by turtlegrass (*Thalassia testudinum*) in the eastern United States. Sea grasses in the tropical regions of the Atlantic and Pacific are dominated by a relatively few genera: *Thalassia*, *Cymodocea*, *Halodule*, and *Syringodium*. Sea grass beds are valuable as habitats for a great number of coastal fish and shrimp species.

There was a general decline in *Zostera* eelgrass beds in the 1930s in Europe and North America by a *wasting disease.* Many of these sea grasses have been restored since that time, but sea grass beds are among the most susceptible coastal ecosystems to pollution. High nutrients cause high plankton productivity of water column productivity, reducing light reaching the bottom waters and thus reducing sea grass productivity. High nutrients also cause overabundance of epiphytic algae on sea grasses, inhibiting plant growth. High sediments from coastal runoff or turbid rivers can also have a dramatic effect on sea grass beds. The restoration of sea grass beds, while possible, is one of the most difficult coastal restorations. Generally, the sediment loads or other impacts have to be ameliorated before any substantive restoration, even with significant planting of grasses, can occur.

Coral Reefs

Coral reefs are often described as the aquatic equivalent in diversity and productivity of the tropical rain forest, with high gross primary productivity, high biodiversity, efficient nutrient cycling, several forms of symbiosis, and even layering of strata not unlike the understory and canopy found in forests (Hubbell, 1997). Growing in warm marine waters, corals are found in tropical regions around the world and are characterized by sedentary and rigid coral structures. Individually, the corals grow from 1 to 10 cm yr^{-1} while the reefs themselves grow only 1 to 5 m per 1000 years (0.1 to 0.5 cm yr^{-1}). The corals, in turn, serve as habitat for a great variety of sessile organisms such as microscopic algae (zooxanthellae) living in the corals and other fixed species such as sponges. A symbiotic pattern among producers (algae) and consumers (corals, sponges, anemones) makes these systems highly productive in an otherwise low-nutrient world. The reef structure and food availability then supports a wide variety of mobile organisms such as fish. The diversity of coral reefs is legendary with non-reef-building organisms such as crabs, lobsters, shrimp, sea turtles, manatees, and marine macro algae among the organisms supported. Coral reefs occur as the following types (Maragos, 1992):

1. Fringing reefs grow immediately offshore coastlines and islands on the uppermost slopes of high islands.
2. Barrier reefs are fringing reefs that become separated from an island by a deep lagoon.
3. Atoll reefs are barrier reefs where the original island is gone and a circular reef surrounds a deep lagoon.
4. Bank reefs are offshore systems that have grown in response to rising sea levels in the past 10,000 years but are not true barriers.
5. Table reefs are similar to atolls but do not enclose a lagoon.

The values of coral reefs are many. In addition to all of the habitat support for food production and biodiversity support, corals provide shoreline protection from waves and storms, provide sand for beaches, and are sometimes used as sources of minerals for sand, aggregate, and quarry stone. Coral reefs are exceedingly popular with tourists throughout the tropical world because of the beauty and diversity of coral structures, and especially because of the colorful fish and other nekton. Coral reefs are damaged by hurricanes, cooler-than-normal waters during winter, warmer sea temperatures, channel dredging, and ship anchor damage. Evidence has suggested that climate change and global weather patterns such as El Niño have caused *bleaching* and other losses in coral reefs in many coral reefs around the world. High sediments and nutrients have deleterious effects on coral reefs, in much the same way that they affect sea grass beds.

9.2 COASTAL RESTORATION TECHNIQUES

Salt Marsh Restoration

Early pioneering work on salt marsh restoration was done in Europe (Lambert, 1964; Ranwell, 1967), China (Chung, 1982, 1989), and in the United States on the North Carolina coastline (Woodhouse, 1979; Broome et al., 1988), in the Chesapeake Bay area (Garbisch et al., 1975; Garbisch, 1977), and along the coastlines of Florida, Puerto Rico (Lewis, 1990b,c), and California (Zedler, 1988, 2001; Josselyn et al., 1990). Some of this coastal wetland restoration has been undertaken for habitat development as mitigation for coastal development projects (Zedler, 1988, 1996a).

For coastal salt marshes in the eastern United States, the cordgrass *Spartina alterniflora* is the primary choice for coastal marsh restoration; but the same species is considered an invasive and unwanted plant on the west coast of North America. Both *Spartina townsendii* and *S. anglica* have been used to restore salt marshes in Europe and in China, although the latter species is considered an invasive species in parts of the world, as in New Zealand. Salt marsh grasses tend to distribute easily through seed dispersal, and the spread of these grasses can be quite rapid once reintroduction has begun, as long as the area being revegetated is intertidal; that is, the elevation is between ordinary high and low tides.

The details of successful coastal wetland creation are, of course, site-specific, but several generalizations seem to be valid in most situations:

1. Sediment elevation is the most critical factor determining the successful establishment of vegetation and the plant species that will survive. The site must be intertidal.
2. In general, the upper half of the intertidal zone is more rapidly vegetated than are lower elevations.
3. Sediment composition does not seem to be a critical factor in colonization by plants unless the deposits are almost pure sand that is subject to rapid desiccation at the upper elevations.
4. The site needs to be protected from high wave energy. It is difficult or impossible to establish vegetation at high-energy sites.

5. Most sites revegetate naturally from seeds if the elevation is appropriate and the wave energy is moderate. Sprigging live plants has been accomplished successfully in a number of cases, and seeding has also been successful in the upper half of the intertidal zone.

6. Good stands can be established during the first season of growth, although sediment stabilization does not occur until after two seasons. Within four years, successfully planted sites are superficially indistinguishable from natural marshes.

Several early studies emphasized the importance of restoring tidal conditions, including salinity, to marsh areas that had become more "freshwater" because of isolation from the sea. In cases such as this, the restoration is simple—remove whatever impediment is blocking tidal exchange. In the example shown in Figure 9.5, a 20-ha impounded salt marsh in Connecticut had been impounded and isolated from tidal flushing for many years and had become dominated by the freshwater macrophyte *Typha angustifolia.* In the late 1970s and early 1980s, several major culverts were installed to reintroduce tidal flushing to the marsh, leading to a decline in *T. angustifolia* from 74 percent to 16 percent cover, and a recovery of *Spartina alterniflora* from <1 percent cover to 45 percent cover. *Phragmites australis,* which tolerates brackish conditions, did not decrease as expected but increased from 6 percent to 17 percent cover over that period and was generally found along the edges of the marsh in a stunted (0.3 to 1.5 m tall) condition. A relatively simple alternation of the hydrologic conditions reestablished the salt marsh.

CASE STUDY
Delaware Bay Salt Marsh Restoration

A large coastal wetland restoration project in the eastern United States involves the restoration, enhancement, and preservation of 5000 ha of coastal salt marshes in Delaware Bay in New Jersey and Delaware in northeastern United States (Figure 9.6; Table 9.1). This estuary enhancement, being carried out by Public Service Electric and Gas (PSEG) with advice from a team of scientists and consultants, was undertaken as mitigation for the potential impacts of once-through cooling from a nuclear power plant operated by PSEG on the bay. The reasoning was that the impact of once-through cooling on fin fish, through entrainment and impingement, could be offset by increased fisheries production from restored salt marshes. Because of uncertainties involved in this kind of ecological trading, the area of restoration was estimated as the salt marshes that would be necessary to compensate for the impacts of the power plant on fin fish times a

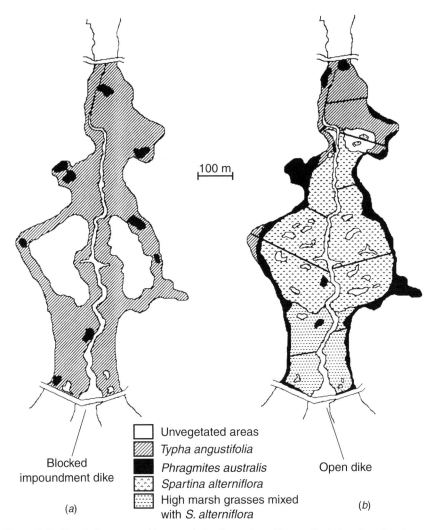

100 m

Unvegetated areas
Typha angustifolia
Phragmites australis
Spartina alterniflora
High marsh grasses mixed
with S. alterniflora

Blocked
impoundment dike

Open dike

(a)

(b)

Figure 9.5 Vegetation maps of impounded salt marsh on Connecticut shoreline showing restoration to salt marsh with reintroduction of tidal flushing: (a) 1976 map prior to opening to tides; (b) 1988 map indicating return of *Spartina alterniflora* after opening to tidal circulation in late 1970s and early 1980s. Lines in the 1988 map indicate vegetation transects. (From Sinicrope et al., 1990; redrawn by Mitsch and Gosselink, 2000, copyright 2000; reprinted with permission from John Wiley & Sons, Inc.)

safety factor of 4. There are three distinct approaches being utilized in this project to restore the Delaware Bay coastline:

1. *Reintroduce flooding.* The most important type of restoration involves the reintroduction of tidal inundation to about 1800 ha of former diked salt hay farms. Many marshes along Delaware

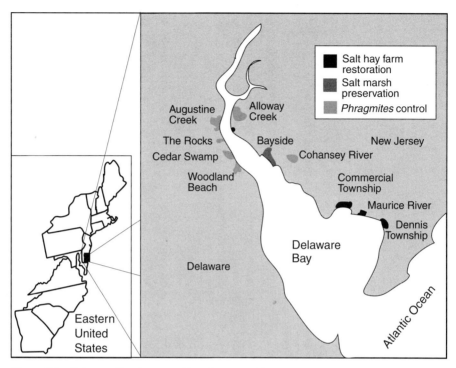

Figure 9.6 Delaware Bay between New Jersey and Delaware, showing locations of a 5800-ha salt marsh restoration. Wetlands are being preserved, restored from salt hay farms by reintroducing flooding, and enhanced by removal of *Phragmites australis*. (From Mitsch and Gosselink, 2000, copyright 2000; reprinted with permission from John Wiley & Sons, Inc.)

Bay have been isolated by dikes from the bay, sometimes for centuries, and put into the commercial production of "salt hay" (*Spartina patens*). Hydrologic restoration was accomplished by excavating breaches in the dikes and, in most cases, connecting these new inlets to a system of recreated tidal creeks and existing canal systems.

2. *Reexcavate tidal marshes.* Additional restoration involves enhancing drainage by reexcavating higher-order tidal creeks in these newly flooded salt marshes, thereby increasing tidal circulation (Figures 1.6*a* and 9.7). This is particularly important in marshes that were formerly diked, as the isolation from the sea has led to the filling of former tidal creeks. After initial tidal creeks were established, it was expected that the system would "self-design" more tidal channels and increase the channel density.

3. *Reduce* Phragmites *domination.* In another set of restoration sites in Delaware and New Jersey, restoration involves the reduction in cover of the aggressive and invasive *Phragmites*

Table 9.1   Location, dominant vegetation, and restoration approach for eight locations along the Delaware Bay as illustrated in Figure 9.6

Site	Type	Restored Area (ha)	Initial Vegetation	Restoration Approach
Dennis Township, New Jersey	Diked salt hay farm	149	*Spartina patens*	Dike breeching, channel excavations
Maurice River, New Jersey	Diked salt hay farm	459	*S. patens*	Dike breeching, channel excavations
Commercial Township, New Jersey	Diked salt hay farm	1171	*S. patens*, *Phragmites australis*	Dike breeching, channel excavations
Cohansey River, New Jersey	*Phragmites*-dominated	368	*P. australis*	Herbicides, prescribed burns
Alloway Creek, New Jersey	*Phragmites*-dominated	1138	*P. australis*	Herbicides, prescribed burns
Cedar Swamp, Delaware	*Phragmites*-dominated	754	*P. australis*	Herbicides, prescribed burns
The Rocks, Delaware	*Phragmites*-dominated	298	*P. australis*	Herbicides, prescribed burns

Source: After Weinstein et al. (2001).

Figure 9.7 Construction activity at a Dennis Township restored salt hay farm on Delaware Bay in 1995. (Courtesy of Ken Strait, PSEG; reprinted with permission from PSEG, Salem, New Jersey.)

australis in 2100 ha of nonimpounded coastal wetlands. Alternatives that were investigated include hydrological modifications such as channel excavation, breaching remnant dikes, microtopographic changes, mowing, planting, and herbicide application.

Early results of this study were reported in many presentations and reports and several early journal articles (e.g., Weinstein et al., 1997, 2001; Teal and Weinstein, 2002). More recent results are reported here.

From a hydrodynamic perspective, in those marshes where tidal exchange was restored, the development of an intricate tidal creek density from the originally constructed tidal creeks has been impressive. Figure 9.8 illustrates the development of a stream network at one of the newly restored marsh sites—Dennis Creek. The "order" of the stream channels increased from 5 or less to 18; that is, there was an increase in the number of small streams and tributaries to small channels. The number of small tributaries increased from "dozens" to "hundreds" at all three salt hay farm sites that were reopened to tidal flushing. For the first three years, there was a rapid increase in the growth of channel orders 3 through 9; in the next three years, there was a rapid increase in the channel orders 10 through 16 (Figure 9.8c). (*Note:* This definition uses channel order opposite to the method described in Chapter 7 on stream order; here the largest channels are designated as channel order 1.) Hydrologic design did occur in self-design fashion after only initial cuts by construction.

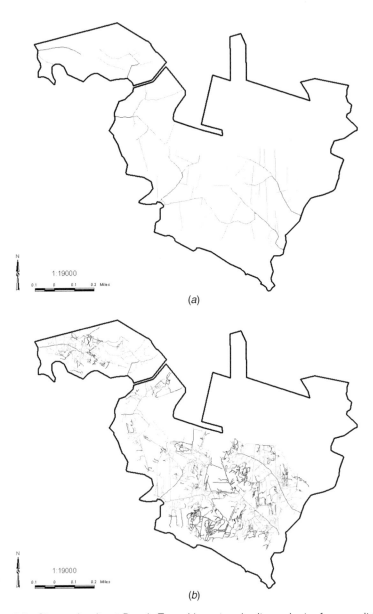

Figure 9.8 Stream density at Dennis Township restored salt marsh at a former salt hay farm in (a) 1996 as restoration begins and (b) 2001. (c) Total number of stream channels by channel class, 1996–2001. [(a) and (b) From URS Corporation, courtesy of Ken Strait; reprinted with permission from PSEG, Salem, New Jersey.)

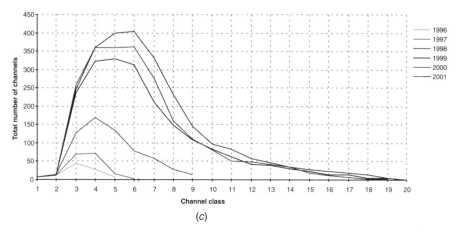

(c)

Figure 9.8 (*Continued*)

For the salt hay farms that were flooded, typical goals include a high percent cover of desirable vegetation such as *Spartina alterniflora*, a relatively low percent of open water, and the absence of the invasive reed grass *Phragmites australis*. The success of this coastal restoration project, subject to a combination of legal, hydrologic, and ecological constraints, is also being estimated through comparison of restored sites to natural reference marshes. Early results of this part of the project are encouraging (Figures 9.9, 9.10; Table 9.2). At the formerly diked salt hay farms (sites in Dennis Township, Maurice

Figure 9.9 Restored Dennis Township salt marsh on Delaware Bay, 2001. (Courtesy of Ken Strait, PSEG; reprinted with permission from PSEG, Salem, New Jersey.)

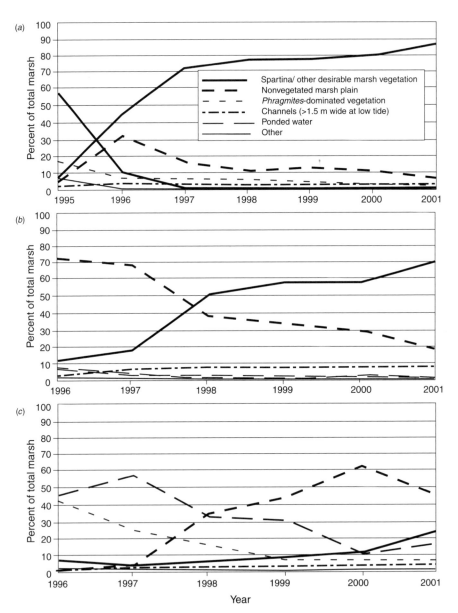

Figure 9.10 Relative area of wetland and other vegetation, 1995–2001, for three salt hay farms being restored to salt marshes on Delaware Bay in New Jersey: (a) Dennis Township (228 ha); (b) Maurice River Township (460 ha); (c) Commercial Township (1172 ha). In all cases, "*Spartina* and other desirable marsh vegetation" is considered an indicator of marsh restoration success. (Data courtesy of Ken Strait, PSEG.)

Table 9.2   Results of salt marsh restoration at Delaware Bay site in 2001, three to five years after construction work

Site	Approach	Year Completed	Restorable Wetland Area (ha)	Desirable Vegetation		Desirable Species
				Percent of Total	Percent of Vegetation	
Dennis Township, New Jersey	Dike breeching, creek excavations	Aug. 1996	228[a]	87	97	*Spartina alterniflora*
Maurice River, New Jersey	Dike breeching, creek excavations	Mar. 1998	460	71	96	*S. alterniflora*
Commercial Township, New Jersey	Dike breeching, creek excavations	Nov. 1997	1172	25	78	*S. alterniflora*

Source: Data courtesy of Ken Strait, PSEG, Salem, NJ.

[a]Initially had 79 ha of *S. alterniflora*.

River, and Commercial Township), reestablishment of *Spartina alterniflora* and other favorable vegetation has been rapid and extensive. At Dennis Township, approximately 64 percent of the site was already dominated by *Spartina alterniflora* after only two growing seasons and 87 percent by the fifth year after construction (Table 9.2). Tidal restoration was completed at the Maurice River site, which is twice the size of Dennis Township, in early 1988 and major revegetation by *Spartina alterniflora* and some *Salicornia* has already occurred, with 71 percent of the site showing desirable vegetation after four growing seasons (Table 9.2). At the third and the largest salt hay farm restoration site, Commercial Township, which is five times larger than Dennis Township site, revegetation is occurring rapidly from the bay side, and 25 percent of the site was in *Spartina* after four growing seasons.

The study has shown that the speed with which restoration takes place depends on three main factors: (1) the degree to which the tidal "circulatory system" works its way through the marsh, (2) the size of the site being restored, and (3) the initial presence of *Spartina* and other desirable species. No planting was necessary on these sites, as *Spartina* seeds arrive by tidal fluxes; but the design of the sites to allow that tidal connectivity (and hence the importance of appropriate site elevations relative to tides) was critical. Self-design works when the proper conditions for propagule disbursement are provided. Extensive ponding in some areas of the marshes, especially in Commercial Township, which has the highest ratio of area to edge, has impeded the reestablishment of *Spartina* in some locations (Teal and Weinstein, 2002). Creating additional streams or waiting for the tidal forces to cause the same effect eventually allows these areas to develop tidal cycles and for *Spartina* to establish itself.

The restoration of 1500 ha of *Phragmites*-dominated marshes (Figures 9.11 and 9.12) in New Jersey and 1050 ha in Delaware is being carried out through spraying of herbicides by ground and by air (Figure 9.13). Burning of the persistent *Phragmites australis* at the Cohansey River and Alloway Creek sites in New Jersey was also carried out in the early years of the restoration. Results are not as promising as those at the hydrologic restoration sites, partially because the restoration methods are much more difficult. Data comparing 1996 to 2001 at the Alloway Creek site (Figure 9.12) suggest an increase in desirable vegetation from 25 percent to 45 percent and a subsequent reduction in *Phragmites* from 58 percent to 22 percent. Burning followed by spraying has not eliminated *Phragmites* at any given site, so repeated treatment is necessary. On these sites, alternative methods to spraying are being investigated for the control of *Phragmites,* including mowing and rhizome cutting, grading of dikes where *Phrag-*

Figure 9.11   Dense growth of *Phragmites* typical of many coastal marshes along the upper Delaware Bay. Marshes have mostly been taken over by *Phragmites* in the past 40 years. (Reprinted with permission from PSEG, Salem, New Jersey.)

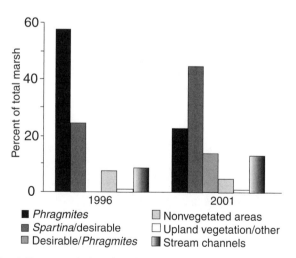

Figure 9.12   Vegetation cover before (1996) and after five-years of treatment (2001) by herbicide and other methods, where *Phragmites* control was attempted at a 650-ha Alloway Creek, New Jersey, site. (Data courtesy of Ken Strait, PSEG.)

(a)

(b)

Figure 9.13 Application of herbicides to control *Phragmites* in Delaware Bay marshes (a) by ground and (b) by air. (Courtesy of Ken Strait, PSEG; reprinted with permission from PSEG, Salem, New Jersey.)

mites tends to dominate, selective planting, and microtopographic modifications of the marsh surface.

Several ecological principles related to the restoration of salt marshes were elucidated by Weinstein et al. (2001):

1. State project goals clearly, as agreed to by the stakeholders; make the goals site-specific and realistic.
2. Restore degraded sites rather than create new wetlands. The factors that favored restoration at the Delaware Bay site included:
   - Appropriate hydrology and topography
   - Establishment of creeks and channels
   - High sediment organic content
   - The presence and proximity of *Spartina* propagules
   - The appropriate salinity
   - A constant supply of sediment to maintain marsh elevation
3. Select sites in a landscape ecology framework.
4. Ecological engineering principles, including self-design, should apply.
5. Restored sites should be self-sustaining but should be guided by *adaptive management*. Adaptive management is an interactive process where management of a wetland restoration is initiated only if expectations are not met.
6. Site monitoring should be planned and implemented and continue until success is assured.
7. Success criteria should include functional as well as structural components (framed by a bound of expectation).
8. A management plan should consider people and property by protecting off-site elements (e.g., upland flooding, salt intrusion into wells, septic systems).
9. Where possible, site should be developed with conservation restrictions to ensure their perpetuity and to protect adjacent property.
10. Site plans should encourage public access for sustainable uses.

Mangrove Restoration

Restoring mangrove swamps in tropical regions of the world has some characteristics similar to those of restoring salt marshes in that the establishment of vegetation in its proper intertidal zone is the key to success. But that is generally where the similarities end. Mangrove restoration is more cosmopolitan, in that it has been attempted throughout the tropical and subtropical world; salt marsh restoration has been attempted primarily on the eastern

North American coastline and to some extent in Europe and the west coast of North America. Salt marsh restoration can often rely on waterborne seeds distributing through an intertidal zone; mangrove restoration often involves the physical planting of trees. In countries such as Viet Nam, mangrove declines have been attributed to the spraying of herbicides during the Viet Nam war and immigration of people to the coastal regions, leading to cutting of lumber for timber, fuel, wood, and charcoal.

More recently in Viet Nam, and currently in many tropical coastlines of the world, mangroves are being cleared for construction of aquaculture ponds at unprecedented rates (Benthem et al., 1999). Most of the edible shrimp sold in the United States and Japan are produced in artificial ponds constructed in mangrove wetlands in Thailand, Indonesia, and Viet Nam. Sold in the United States and Japan at very low prices, these products are the result of massive destruction of mangrove forests. More than 100,000 ha of abandoned ponds located in former mangrove swamps currently exist in these countries (R. Lewis, pers. commun., July 2001). In Viet Nam, mangroves are being restored and protected to provide coastal protection and coastal fisheries support. In the Philippines, despite a presidential proclamation in 1981 prohibiting the cutting of mangroves, is estimated that the country still was losing 3000 ha yr^{-1} in the late 1990s (2.4 percent per year; deLeon and White, 1999). But the insatiable appetite in the United States, Japan, and several other developed countries for shrimp continue to cause mangroves to be destroyed. The shrimp ponds last only about five to six years before they develop toxic levels of sulfur and are then abandoned and more mangroves are destroyed. These abandoned ponds present a challenge for mangrove restoration.

CASE STUDY
A Costly Mangrove Restoration

Lewis (2000) describes a case in southeastern Florida where inattention to the fundamental ecology of how mangrove swamps function led to costly delays and confusion in a mangrove creation/restoration. The project involved the widening of Route 1 between Homestead and Key Largo at the entrance to the Florida Keys. Because of the impact on mangrove wetlands by the highway project, the Florida Department of Transportation was required to mitigate the pending wetland loss by restoring additional wetlands. The project chosen was mangrove restoration on a site that had been filled during previous construction activity but near the site affected. The project was designed and reviewed over a two-year period by consultants, agency personnel, and regulatory agencies and found to be acceptable. In this case, the fill was to be removed with the restoration, depending on natural recruitment of mangrove seedlings. The success criterion was 80 percent total coverage by mangroves two years after excavation, and one of the primary goals was restoring habitat for the American crocodile (*Crocodylus acutus*), a species that required open water. The success criterion was probably doable, but it did not fit the goal of crocodile habitat—tidal streams connecting to deepwater sites. Also, no one thought to determine the exact grade that the land needed to be for mangrove success. With adjacent mangrove communities, that could easily be done. Figure 9.14 illustrates the final revised plan where a specific grade above national geodetic vertical datum is specified and is based on the elevation range in which natural mangroves are found adjacent to the project. Two issues were important here. First, there was so much attention paid to unrealistic goals of 80 percent vegetation cover that caused overdesign on plant introduction techniques that no one thought to review the elevation "design" of the project to see if any mangroves were going to be intertidal. Second, there was a disconnect between biologists who wanted crocodiles and engineers who designed the hydrology. Ecological engineers, with a real knowledge of ecology, might have designed the appropriate system and recognized that 80 percent vegetation cover in two years is neither practical nor necessary. Nature should be trusted when she takes her time.

Delta Restoration

As large rivers connect to the sea, multitributary deltas tend to develop, allowing the river to discharge to the sea in many channels. Many of these rich-soil deltas are among the most important ecological and economic regions of the world, from the ancient Nile delta in Egypt to the modern-day Mississippi River delta in Louisiana. There should be two major ecological resource goals of delta areas: (1) protecting and restoring the functioning of

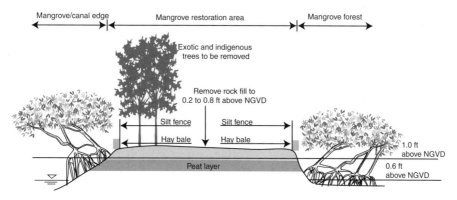

Figure 9.14 Revised mangrove restoration plan for a Harrison Tract wetland mitigation site in southeastern Florida. Elevation numbers are above the national geodetic vertical datum (NGVD). (Redrawn from Lewis, 2000.)

the deltatic ecosystems in the context of a geologically dynamic framework, and (2) controlling pollution from entering the downstream lakes, oceans, gulfs, and bays. Delta restoration should have this dual emphasis where possible—ecosystem enhancement of the delta itself and improvement of coastal water quality downstream. The best strategy for delta restoration when land building is a necessary prerequisite (see the following case study) is to restore the ability of the river to spread out its sediments in deltaic form as wide an area as possible, particularly during flood events and by not discouraging (or encouraging and even creating) river distributaries. When river distributaries are not possible on a large scale due to navigation requirements or population locations, restoring and creating riverine wetlands and constructing river diversions to divert river water to adjacent lands may be the best alternatives to maximize nutrient retention and sediment retention. In some cases this involves the conversion of agricultural lands back to wetlands; in other cases the dikes that "protect" wildlife protection ponds or retain rivers in their channels only need to be carefully breached to allow lateral flow of rivers during flood season.

CASE STUDY
Louisiana Delta Restoration

Louisiana is one of the most wetland-rich regions of the world, with 36,000 km^2 of marshes, swamps, and shallow lakes. Yet Louisiana is suffering a rate of coastal wetland loss of 6600 to 10,000 ha yr^{-1} as it converts to open-water areas on the coastline, due to natural (land subsidence) and human causes such as river levee construction, oil and gas exploration, urban development, sediment diversion, and possibly climate change. Since the early 1990s, there has been a major interest in reversing this rate of loss and even gaining coastal areas, particularly freshwater marshes and salt marshes, the loss of which are the major symptom of this "land loss."

Initially, the solution to the loss of wetlands and uplands in Louisiana was thought to be the passage by the U.S. Congress of the Coastal Wetland Planning, Protection and Restoration Act (CWPPRA; pronounced "quipra") in 1990. This act initiated comprehensive planning in Louisiana aimed at the protection, restoration, and creation of millions of hectares of coastal wetlands. The plans called for diverting the water and sediment of the Mississippi River to build new deltas and smaller *splays* to mimic spring floods and restore subsiding marshes; restoring barrier islands; and instituting measures to protect many smaller wetlands with dikes, plantings, and disposing of dredge spoil (Turner and Boyer, 1997). Several wetland enhancement and creation projects have created *crevasses* in the natural levee of the lower Mississippi River. These crevasses allow river water and sediments to flow into shallow estuaries and create crevasse splays or minideltas that rapidly become vegetated marshes. Their extent and life span can seldom be predicted, and they function in a natural manner because they mimic the natural geomorphic processes of the river. The CWPPRA led to the Louisiana Coastal Wetland Restoration Plan in 1993, and about $40 million annually was being spent on individual projects (Louisiana Coastal Wetlands Conservation and Restoration Task Force, 1998).

A much more ambitious project, Louisiana Coastal Area (LCA), is now being developed in Louisiana to reengineer the coastline to curtail the land loss. If undertaken, this project, estimated to cost $14 billion, would be the largest ecological engineering project in the world. The function is to carry out a suite of dozens of projects that are meant to build land: shoreline protection in critical areas, river diversions, optimizing the functioning of Mississippi River distributaries such as the Atchafalaya River, and restoring barrier islands (Figure 9.15). Because the Mississippi River is maintained for shipping purposes and the navigational channel is now maintained as a channel well downstream of New Orleans and out to the Gulf of Mexico deep waters, simply allowing the river to cut through its artificial levees would cause great economic harm to southern Louisiana. Rather, plans have to be made in a hydrologically complex situation.

River diversions are becoming a large part of the delta restoration in Louisiana. The deltaic wetlands in Louisiana cover more than 20,000 km^2 and are critical to building land and wetlands from open water in the delta and may be critical nutrient removal sites for abatement of Gulf hypoxia (see Chapter 8). The Caernarvon freshwater diversion is one of the largest of several diversions currently in operation on the Mississippi River in Louisiana. The diversion structure is on the east bank of the river below New Orleans 131 km upstream of the Gulf of Mexico (Figure 9.16). The structure is a five-box culvert with vertical lift gates with a maximum flow of 226 m^3 s^{-1}. Freshwater discharge began in August 1991 and discharge from then until December 1993 averaged 21 m^3 s^{-1}; current minimum and maximum flows are 14 and 114 m^3 s^{-1}, respectively, with summer flow rates generally near the minimum and winter flow rates 50 to 80 percent of the maximum (Lane et al., 1999). A typical hydroperiod for the flux of diversion water is illustrated in Figure 9.17. The Caernarvon diversion delivers water to the Breton Sound estuary, a 1100-km^2 area of fresh, brackish, and saline wetlands.

One of the most notable features of the water as it flows through the estuary is the continued decrease in nutrient concentrations as the water passes through the backwaters on its way to the Gulf of Mexico in excess of what would be expected due to dilution from inflowing

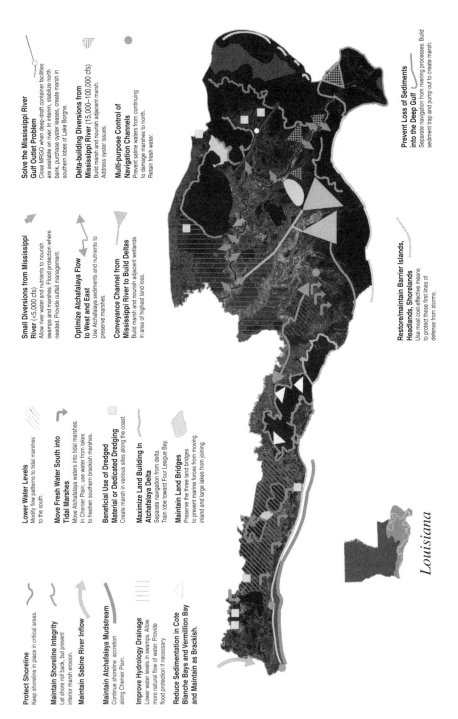

Protect Shoreline
Keep shoreline in place in critical areas.

Maintain Shoreline Integrity
Let shore roll back, but prevent interior marsh erosion.

Maintain Sabine River Inflow

Maintain Atchafalaya Mudstream
Continue shoreline accretion along Chenier Plain.

Improve Hydrology Drainage
Lower water levels in swamps. Allow more natural flow of water. Provide flood protection if necessary.

Reduce Sedimentation in Cote Blanche Bays and Vermillion Bay and Maintain as Brackish.

Lower Water Levels
Modify flow patterns to tidal marshes to the south.

Move Fresh Water South into Tidal Marshes
Move Atchafalaya waters into tidal marshes. In Chenier Plain, use water from lakes to freshen southern brackish marshes.

Beneficial Use of Dredged Material or Dedicated Dredging
Create marsh in various sites along the coast.

Maximize Land Building In Atchafalaya Delta
Separate navigation from delta. Train lobe toward Four League Bay.

Maintain Land Bridges
Preserve the three land bridges to prevent marine forces from moving inland and large lakes from joining.

Small Diversions from Mississippi River (<5,000 cfs)
Allow river water and nutrients to nourish swamps and marshes. Flood protection where needed. Provide outfall management.

Optimize Atchafalaya Flow to West and East
Use Atchafalaya sediments and nutrients to preserve marshes.

Conveyance Channel from Mississippi River to Build Deltas
Build marsh and nourish adjacent wetlands in area of highest land loss.

Solve the Mississippi River Gulf Outlet Problem
Close MRGO when deep-draft container facilities are available on river. In interim, stabilize north bank, purchase oyster leases, create marsh in southern lobes of Lake Borgne.

Delta-building Diversions from Mississippi River (15,000–100,000 cfs)
Build marsh and nourish adjacent marsh. Address oyster issues.

Multi-purpose Control of Navigation Channels
Prevent saline waters from continuing to damage marshes to north. Retain fresh water.

Prevent Loss of Sediments into the Deep Gulf
Separate navigation from rivering processes. Build sediment trap and pump out to create marsh.

Restore/maintain Barrier Islands, Headlands, Shorelands
Use most cost-effective means to protect these first lines of defense from storms.

Louisiana

Figure 9.15 Examples of possible Louisiana Coastal Area ecosystem strategies to stem the loss of land and wetlands in deltaic Louisiana. (Redrawn from Louisiana Coastal Wetlands Conservation and Restoration Task Force, 1998.)

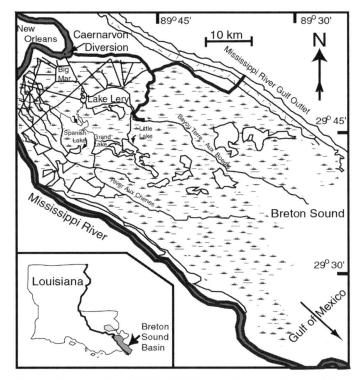

Figure 9.16 Caernarvon diversion structure on Mississippi River and downstream Breton Sound, southeastern Louisiana. (Courtesy of J. W. Day, Jr; reprinted with permission.)

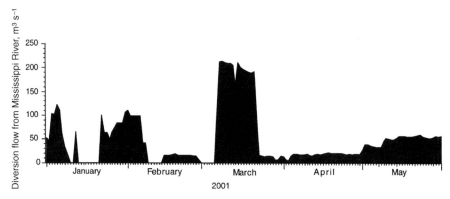

Figure 9.17 Example of pulsing flow through the Caernarvon diversion from the Mississippi River basin to adjacent wetlands in southeastern Louisiana, January–May 2001. (Courtesy of J. W. Day, Jr.; reprinted with permission.)

tidal water. This is illustrated in the plots of nitrate-nitrogen concentrations versus salinity (Figure 9.18); a concave graph illustrates that nutrients are decreasing far more rapidly than dilution due to salt water.

CASE STUDY
Predicting Great Lakes Delta Restoration

Coastal wetlands were once found in great abundance along the Laurentian Great Lakes in North America. Compared to the Atlantic, Gulf, and Pacific shoreline lengths of the United States, the Great Lakes shoreline is similar in length to the Atlantic plus Gulf shoreline. Only a small percentage of the original extent of wetlands remains around the Great Lakes after over 200 years of intensive development and urbanization. As one example, an estimated 95 percent of wetlands were lost around the Western Basin of Lake Erie in Ohio and Michigan, including an area known as the Great Black Swamp, a 4000-km^2 swamp/marsh lowland that was essentially the western-most extension of Lake Erie. Unlike tidal wetlands, which are partially protected because the tide returns twice a day with certainty, Great Lakes wetlands were especially vulnerable to filling during low-lake-level periods, particularly since the water levels of the Great Lakes fluctuate over 1.5 m from a wet year to a dry year. During early European and later American settlement of this coastline, wetland areas were filled and later diked, particularly during low-water years.

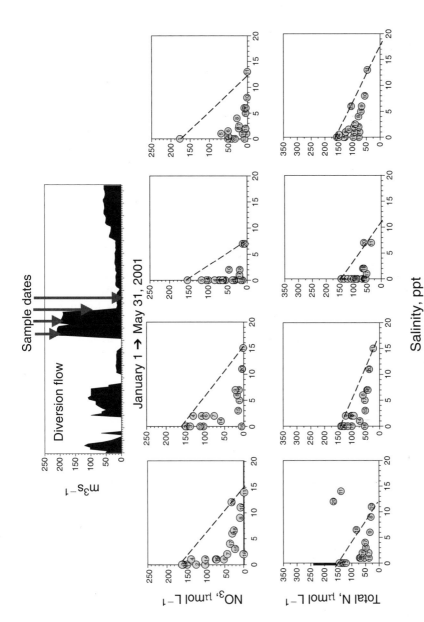

Figure 9.18 Nutrient–salinity diagram for analysis of water through Breton Sound downstream of the Caernarvon diversion on the Mississippi River in southern Louisiana. These nutrient–salinity diagrams indicate that nitrogen is decreasing at a rate more rapid than the dilution due to salt water. Thus, Breton Sound is acting as a nitrogen and nitrate sink as a result of the Mississippi River diversion. (Courtesy of J. W. Day, Jr.; reprinted with permission.)

Wetlands along the Great Lakes are restored with various goals and various criteria for measuring success of the restoration (Table 9.3). Most often, Great Lakes wetlands are restored for the return of vegetation cover and subsequent wildlife enhancement, particularly for waterfowl (Burton and Prince, 1995a,b; Özesmi and Mitsch, 1997). Some recent Great Lakes wetland restorations have focused on restoring wetlands while restricting carp (*Cyprinus carpio*) and on the use of wetlands for spawning and feeding by lake fish. Rarely have wetland restoration projects along rivers that discharge into the Great Lakes been undertaken with a primary goal of enhancing water quality, despite the importance that has been attributed to water quality in the lakes by conventions such as the Great Lakes Water Quality Agreement between the United States and Canada.

In one example of wetland restoration along the Great Lakes, the Michigan Department of Natural Resources has investigated the restoration of wetlands at the delta of the Quanicassee River where it enters Saginaw Bay on Lake Huron, Michigan (Figure 9.19). The river drains into Saginaw Bay through what was former wetland but is now mostly drained hydric soils that support agricultural crops such as potatoes, sugar beets, beans, and corn. The average flow of the Quanicassee River is 11 m^3 s^{-1} with an annual phosphorus loading of 48 metric tons; the river contributes about 3 percent of the total 1544 metric tons of phosphorus loading from the entire 2100-km^2 Saginaw Bay watershed. In the study area, there are presently 131 ha of wetlands along the Quanicassee River and 564 ha of wetlands along the coastline.

The proposed restoration area that served as the basis for initial simulation models ranged up to 3120 ha, roughly 15 percent of the

Table 9.3   Success criteria used for coastal wetland restoration on the Laurentian Great Lakes

Frequently used
Waterfowl production, especially ducks
Vegetation cover
Absence of alien species (e.g., *Lythrum salicaria*)
Sometimes used
Nongame wildlife enhancement
Exclusion of carp (*Cyprinus carpio*)
Fish spawning and feeding
Furbearer harvesting
Rarely used
Detrital export
Water quality enhancement

Source: After Mitsch and Wang (2000).

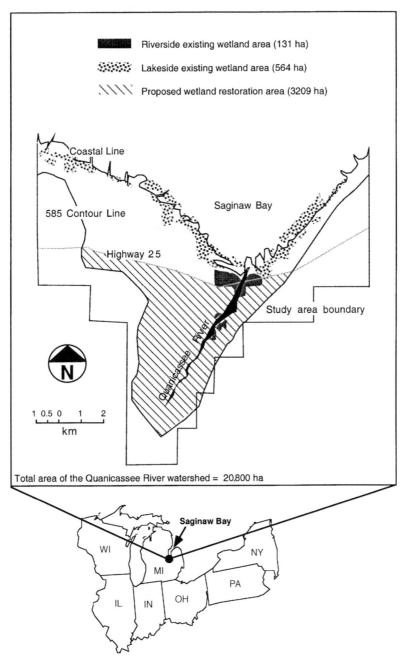

Figure 9.19 Quanicassee River study area on Saginaw Bay, northeastern Michigan. (From Wang and Mitsch, 1998, Fig. 1, p. 71; copyright 1998; reprinted with permission from Kluwer Academic Publishers.)

total Quanicassee River watershed. A simplified model used to estimate the phosphorus retention that would occur with extensive wetland restoration was based on a Vollenweider-type approach (see Chapter 6 for a discussion of the original model used for lakes). The model treated the wetlands as a "black box" and was validated for several wetland years in midwestern U.S. marshes (Mitsch et al., 1995). The model has the form

$$\frac{d\text{TP}}{dt} = P_{\text{in}} - P_{\text{out}} - k\text{TP} + k_s\text{TP}_s \tag{9.1}$$

where P_{in} is the inflow of phosphorus (g day^{-1}), P_{out} the outflow of phosphorus (g day^{-1}), k the phosphorus retention coefficient (day^{-1}), TP the phosphorus in wetland (g), k_s the phosphorus leaching coefficient (day^{-1}), TP$_s$ the excess phosphorus in soil (g). The value for the phosphorus sedimentation rate k is 0.93 day^{-1}, estimated from the validation described above. The last term in equation (9.1), leaching from the soil, was necessary because excessive phosphorus in the soil from years of intensive agriculture would be released to the overlying waters of the restored wetlands for several years.

Model results (Wang and Mitsch, 1998) illustrate the relative importance of inflows, wetland water depth, and antecedent soil conditions (Figure 9.20). The model estimates that the restoration of 15 percent of the Quanicassee River basin to wetlands (3120 ha) would result in a reduction of more than half (53 percent) of the phosphorus that enters Saginaw Bay from that basin, assuming that a proper hydrologic connection between the wetlands and the river could be es-

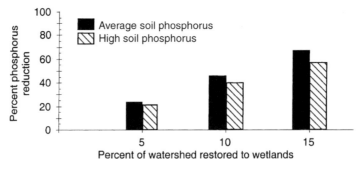

Figure 9.20 Percent of total phosphorus reduction versus percentage of watershed restored to wetlands in the Quanicassee River watershed, Michigan, predicted by simulation modeling. Assumes a 30-cm average depth of wetlands and an inflow of 9.3 m yr^{-1}. Two conditions are for average (10.1 g m^{-2} or 90 lb acre^{-1}) and high (24.9 g m^{-2} or 222 lb acre^{-1}) initial soil phosphorus levels. (From Mitsch and Wang, 2000; copyright 2000; reprinted with permission from Elsevier Science.)

tablished. The model also illustrated that initial phosphorus storage in the soils prior to wetland restoration had a small effect (<15 percent) on phosphorus retention by the wetlands in the first few years but that the long-term effect was not important. Phosphorus loading from the Quanicassee River watershed represents only about 3 percent of the total phosphorus load to Saginaw Bay from all sources. To make a significant impact on the control of phosphorus entering Saginaw Bay and thus to control the symptoms of eutrophication there, a concerted effort of wetland restoration on a similar scale would have to take place throughout all of the river basins that drain into Saginaw Bay.

Creating and Restoring Reefs

Offshore reefs, whether natural or human-made, provide structure for the development of diverse aquatic communities. Although some activity has occurred in temperate-zone aquatic systems to create reefs (such as "disposing of" the old Cleveland, Ohio, football/baseball stadium as a reef structure in Lake Erie to please both fishers and sports fans), most activity related to reef restoration has taken place in tropical regions of the world. Coral reefs, a biologically important coastal system in tropical coastal regions around the world, are concentrated between 22.5°N and 22.5°S, with warm coastal waters extending them farther north on western coasts into subtropical and even temperate climes. Coral reefs depend on growth on the seaward edge of these systems despite low nutrients in seawater and high wave activity. Storms, eutrophic waters caused by excessive nutrients, coral diseases, sedimentation, dredging and filling, overfishing, and overuse by recreationists and visitors all have major effects on coral reefs. Water temperature changes, whether caused by seasonal oscillations such as El Niño or unidirectional changes caused by climate change, can affect reefs.

Coral reefs can be restored in two ways: impact mitigation and active restoration (Table 9.4). In simplest terms, impact mitigation means controlling the nutrients, sediments, fishing, and recreational use that are affecting the reefs.; these are effective techniques for restoring reefs (i.e., controlling the forcing functions), but often the damage is already done and restoration will not occur over a human's lifetime. Active restoration includes the transplantation of corals. These attempts have included the use of nontoxic coated wire on which corals are grown in controlled conditions and the physical transplantation of corals from deeper to shallower waters, "wedging" them in crevices or "cementing" them to hard surfaces (Maragos, 1992). Although it can be effective on small plots, transplantation has two major disadvantages. First, any underwater work of this kind is quite expensive. Second, the transplantation of coral from one location to another does not appear to make sense for the long-term gain of coral overall. It is simply moving it from one

Table 9.4   **Coral reef restoration techniques**

Technique	Hoped-for Effect
Impact mitigation	
Remove of sewage discharges	Reduce anaerobic conditions and competition with benthic algae
Remove of thermal outfalls	Reduce thermal stress on fish and coral
Moor buoy systems	Minimize anchor damage in reefs visited frequently
Establish buffer zones around reefs	Minimize potential damages
Active Restoration	
Transplant coral and sponge	Accelerate coral recolonization and provide shelter for reef life
Construct artificial reefs, reef holes, or breakwaters	Provide fish shelter
Modify current regimes	Promote better flushing through reefs
Replant adjacent mangroves or seagrasses	Stabilize inflow of sediments and nutrients, provide biological recruits
Cement reefs damaged by ships	Stablize hard surfaces to encourage recolonization
Remove disease-bearing organisms and coral predators	Reduce mortality

Source: After Maragos (1992).

location to another. Corals have been attached to dry cement underwater, where a diver swims with a watertight container of cement, builds a mount of cement on a clean reef platform, and attaches coral to the cement. Similarly, marine epoxy is used to attach coral to cleaned reef surfaces. The long-term effectiveness of these transplanting techniques is poorly studied, and the results are quite variable (Japp, 2000).

Reducing the diseases and disease-bearing organisms and predators that affect coral reefs is a bit like treating diseases—it treats the symptoms as much as the cause. Removal or poisoning of the coral-consuming crown-of-thorns starfish (*Acanthaster*) that has infested coral reefs around the Pacific has been effective in some cases but ineffective in others.

The construction of artificial reefs where they did not exist previously is controversial, yet widely practiced. It has occurred accidentally or purposefully throughout the world, almost wherever a ship has sunk in coastal waters. Occasionally, entire "navies" are sunk after wars, such as the shipwrecks on Pacific atolls after World War II. Metal and concrete seem to provide good substrate for attached algae and sometimes coral itself. One of the big difficulties in this approach is the wide range of agendas and objectives that different groups have regarding these reefs (Sheehy and Vik, 1992)—from the demolition engineer who wants to get rid of some material to a fisher who wants the greatest number of sport fish. It is also not well understood if these reefs actually increase the secondary productivity of fish or merely attract it from some other location.

In another application of ecological engineering, artificial habitat can often be designed with a structure that provides optimum habitat or hydraulics for target species. Reflecting on the physical attributes of natural materials will usually lead to the most useful structures, as the genetic information in individual species is attracted to what it knows. Self-design then becomes an important part of these reefs. There is very little "addition" of biological material that occurs to stimulate an artificial reef. The physical structure is provided and nature does the rest. Sheehy and Vik (1992) make an important distinction between traditional fisheries enhancement reefs built from old stadia or bulk concrete waste or sunken ships, and *prefabricated and designed reefs* that take advantage of the particular coastal water depth, temperature, wave exposure, and water quality, to design an ecosystem, not a fish attractor.

10

TREATMENT WETLANDS

Wastewater and polluted water treatment by wetlands is an intriguing concept involving the forging of a partnership between humanity (our wastes) and an ecosystem (wetlands). Therefore, it is a good example of ecological engineering. In Chapters 8 and 9, the ideas of creating and restoring freshwater and coastal wetlands were discussed. In this chapter we discuss the use of wetlands for removing unwanted chemicals from waters, be the waters municipal wastewater, nonpoint-source runoff, or other forms of polluted waters.

Natural wetlands can be sources, sinks, or transformers for a great number of chemicals. A wetland is a *sink* if it has a net retention of an element or a specific form of that element (e.g., organic or inorganic), that is, if the inputs are greater than the outputs (Figure 10.1*a*). If a wetland exports more of an element or material to a downstream or adjacent ecosystem than would occur without that wetland, it is considered a *source* (Figure 10.1*b*). If a wetland transforms a chemical from, say, dissolved to particulate form but does not change the amount going into or out of the wetland, it is considered to be a *transformer* (Figure 10.1*c*).

The idea that wetlands were observed to serve as a sink for inorganic and even organic substances intrigued researchers in the 1960s and 1970s. Researchers in the United States in the 1970s began to investigate the role of natural wetlands, particularly in regions where they are found in abundance, to treat wastewater and thus recycle clean water back to groundwater and surface water. Earlier than this in Europe, German scientists investigated the use of constructed basins with macrophytes (*höhere Pflanzen*) for purification of wastewater. The two different approaches, one utilizing natural wetlands and the other using artificial systems, have converged into the general field of *treatment wetlands*. The field now encompasses the construction and/or

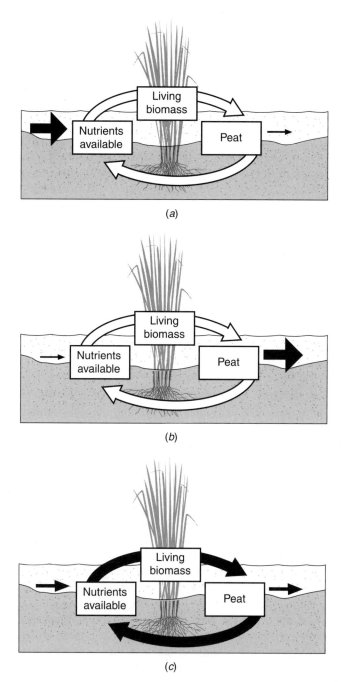

Figure 10.1 Three possible biogeochemical roles of wetlands: (a) sink; (b) source; (c) transformer. (From Mitsch and Gosselink, 2000; copyright 2000; reprinted with permission from John Wiley & Sons, Inc.)

use of wetlands for a myriad of water quality applications, including municipal wastewater, small-scale rural wastewater, acid mine drainage, landfill leachate, and nonpoint-source pollution from both urban and agricultural runoff. Although water quality improvement by treatment wetlands is the primary goal of treatment wetlands, they also provide habitat for a wide diversity of plants and animals and can support many of the other wetland functions described in this book.

Much has been written about treatment wetlands, with a tome by Kadlec and Knight (1996) and the international review report coauthored by Kadlec and colleagues (IWA Specialists Group, 2000) providing the most comprehensive coverage of treatment wetlands. There are also many books and journal special issues covering the general subject of treatment wetlands in general (Godfrey et al., 1985; Reddy and Smith, 1987; Hammer, 1989; Cooper and Findlater, 1990; Knight, 1990; Johnston 1991; Moshiri, 1993; U.S. Environmental Protection Agency, 1993; Reed et al., 1995; Tanner et al., 1999), nonpoint-source wetlands (Olson, 1992; Mitsch et al., 2000), and landfill leachate (Mulamoottil et al., 1999). Papers featuring such systems are frequently published in the journal *Ecological Engineering*. Much of the work given in this chapter is updated from a chapter on treatment wetlands in *Wetlands,* 3rd edition (Mitsch and Gosselink, 2000).

10.1 GENERAL APPROACHES

Three types of wetlands are used to treat wastewater. In the first approach, wastewater is purposefully introduced to existing natural wetlands rather than constructed wetlands (Figure 10.2*a*). In the 1970s, studies involving application of wastewater to natural wetlands were carried out in U.S. locations such as Michigan and Florida where there are abundant wetlands. At that time, legal protection of wetlands had not been institutionalized. These pioneering studies elevated the importance to the general public and governmental agencies of wetlands as "nature's kidneys." This importance was then translated, appropriately, into laws that protected wetlands. But these same laws now generally prohibit the addition of wastewater or polluted water to natural wetlands.

Constructed wetlands are an alternative to using natural wetlands. *Surface-flow constructed wetlands* (Figure 10.2*b*) mimic natural wetlands and can be a better habitat for certain wetland species because of standing water through most, if not all of the year. The second type of constructed wetlands, *subsurface-flow constructed wetlands* (Figure 10.2*c*), more closely resembles wastewater treatment plants than wetlands. In these systems, the water flows horizontally through a porous medium, usually sand or gravel, supporting one or two of a relatively narrow list of macrophytes such as *Phragmites australis*. There is little to no standing water in these systems, as the wastewater passes laterally through the medium. Subsurface treatment wetlands had their

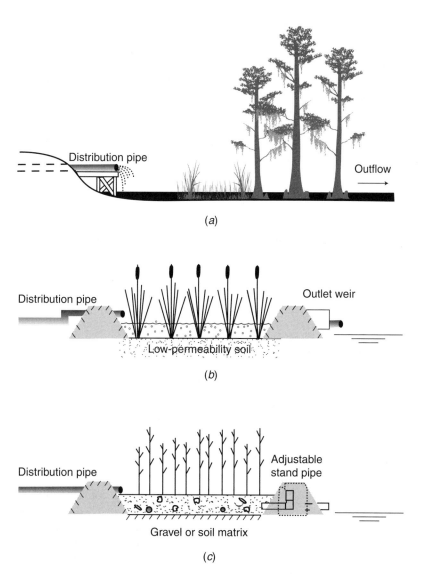

(a)

(b)

(c)

Figure 10.2 Three types of wetland treatment systems: (a) natural wetland; (b) surface-flow wetland; (c) subsurface flow wetland (Based on Kadlec and Knight, 1996; from Mitsch and Gosselink, 2000; copyright 2000; reprinted with permission from John Wiley & Sons, Inc.)

start in the Max-Planck Institute in Germany in the 1950s. Käthe Seidel performed many experiments with emergent macrophytes, *Schoenoplectus lacustris* in particular, and found that the plants contributed to the reduction of bacteria and of organic and inorganic chemicals (Seidel, 1964, 1966). This process was translated into a gravel-bed macrophyte system that became known as the *Max-Planck-Institute process* or *Krefeld system* (Seidel and

Happl, 1981, as cited in Brix, 1994). The development of subsurface wetlands continued in Europe using a system of subsurface-flow basins planted with *Phragmites australis*. These systems were called the *root-zone method* (*Wurzelraumentsorgung*). Subsurface wetland systems continued to be studied and refined through the work of DeJong (1976) in the Netherlands, Brix (1987) in Denmark, and many other scientists in Europe. The appeal of these more "artificial" types of wetlands in Europe (as opposed to North America, where free-water surface wetlands are common) is due to two factors: (1) there are fewer natural wetlands remaining in Europe, and those that are left are protected for nature; and (2) space is at much more of a premium in Europe, and subsurface wetlands require less land area.

Vegetation Classification

Treatment wetlands can also be classified based on the life form of their vegetation. In this case, there are five systems based on herbaceous macrophytes:

1. Free-floating macrophyte systems [e.g., water hyacinth (*Eichhornia crassipes*), duckweed (*Lemna* spp.)]
2. Emergent macrophyte systems [e.g., reed grass (*Phragmites australis*), cattails (*Typha* spp.)]
3. Submerged macrophyte systems
4. Forested wetland systems
5. Multispecies algal systems, particularly algal-scrubber systems

Subsurface-flow constructed wetlands are limited to emergent macrophytes, while surface-flow constructed wetlands often utilize a combination of free-floating, emergent, and submerged macrophytes. Forested wetland treatment systems are generally not constructed wetlands at all, but are natural wetlands to which wastewater is applied. They will often develop extensive communities of all the other vegetation types described in this classification.

10.2 TREATMENT WETLAND TYPES

The general type of wastewater being treated often classifies treatment wetlands. Although many of these systems are used for municipal wastewater, and that is often thought of as the conventional system, there has been much interest in the use of wetlands to treat stormwater from urban areas, acid mine drainage from coal mines, nonpoint-source pollution in rural landscapes, livestock and aquaculture wastewaters, and an array of industrial wastewaters.

Municipal Wastewater Wetlands

In Europe, most of the development of subsurface constructed wetlands was to replace both primary and secondary treatment to remove biological oxygen demand (BOD) and suspended solids as well as inorganic nutrients. Hundreds of subsurface wetland treatment systems for municipal wastewater have been constructed in Europe (Vymazal et al., 1998), particularly in the United Kingdom (Cooper and Findlater, 1990), Denmark (Brix and Schierup, 1989a,b; Brix, 1998), and the Czech Republic (Vymazal, 1995, 1998, 2002). There are also many applications of this technology in Australia (Mitchell et al., 1995) and New Zealand (Cooke, 1992; Tanner, 1996; Nguyen et al., 1997; Nguyen, 2000), some going back several decades.

In North America, most but certainly not all of the wetlands built for treatment of municipal wastewater treatment are surface-water wetland. Locations where wastewater wetlands that have been studied in some detail include Florida (Knight et al., 1987; Jackson, 1989), California (Gerheart et al., 1989; Gerheart, 1992; Sartoris et al., 2000), Louisiana (Boustany et al., 1997), Arizona (Wilhelm et al., 1989), Kentucky (Steiner et al., 1987), Pennsylvania (Conway and Murtha, 1989), Ohio (Spieles and Mitsch, 2000a,b), North Dakota (Litchfield and Schatz, 1989), and Alberta (White et al., 2000). Created wetlands for treating wastewater have been most effective for controlling organic matter, suspended sediments, and nutrients. Their value for controlling trace metals and other toxic materials is more controversial, not because these chemicals are not retained in the wetlands but because of concerns that they might concentrate in wetland substrate and fauna.

CASE STUDY
Houghton Lake, Michigan

Much of the interest in using surface-flow wetlands for water quality management was sparked by two studies begun in the early 1970s. In one of those studies, peatlands in Michigan were investigated by researchers from the University of Michigan for the wetlands' capacity to treat wastewater (Figure 10.3). A pilot operation for disposing of up to 380 m^3 day^{-1} (100,000 gallons per day) of secondarily treated wastewater in a rich fen at Houghton Lake led to significant reductions in ammonia nitrogen and total dissolved phosphorus as the wastewater passed from the point of discharge through the wetlands. Inert materials such as chloride did not change as the wastewater passed through the wetland. In 1978 the flow was increased to approximately 5000 m^3 day^{-1} over a much larger area, essentially all the wastewater from the local treatment plant. Data after more than 22 years of operation at this high flow show that although the area

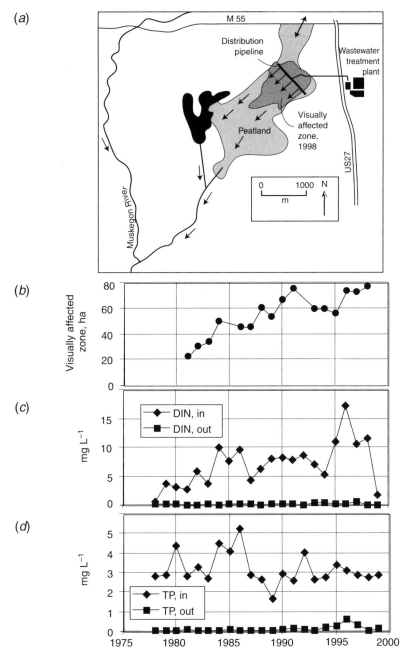

Figure 10.3 Houghton Lake treatment wetland in Michigan, where treated sewage effluent has been applied to a natural peatland since 1978: (a) map of site showing visually affected area in 1998; (b) area of visually affected zone in peatland, primarily where vegetation changes have occurred, 1981–1998; (c) dissolved inorganic nitrogen of influent and outlet stream, 1978–1999; (d) total phosphorus of influent and outlet stream. (Based on Kadlec and Knight, 1996, and R. H. Kadlec, pers. commun., January 2000; from Mitsch and Gosselink, 2000; copyright 2000; reprinted with permission from John Wiley & Sons, Inc.)

of influence of the wastewater on the peatland has grown from 23 ha to 77 ha (Figure 10.3*a,b*), the effectiveness of the wetland in removing both inorganic nitrogen and total phosphorus remained extremely high (Figure 10.3*c,d*).

Mine Drainage Wetlands

Wetlands have frequently been used as downstream treatment systems for mineral mines. Acid mine drainage water, with its low pH and high concentrations of iron, sulfate, aluminum, and trace metals, is a major water pollution problem in many coal mining regions of the world, and constructed wetlands are a feasible treatment option. The use of wetlands for coal mine drainage control was probably considered first when volunteer *Typha* wetlands were observed near acid seeps in a harsh environment where no other vegetation could grow. The application and design of these types of treatment wetlands are discussed more fully in Chapter 12.

Urban Stormwater Treatment Wetlands

The control of stormwater pollution with wetlands is a valid application of ecological engineering of wetlands. Unlike municipal wastewater, stormwater and other nonpoint-source pollution are seasonal, often quite sporadic, and variable in quality, depending on season and recent land use. Wetlands are one of several choices for systems to control urban runoff. More conventional approaches involve either dry detention ponds that fill only during storms and wet detention ponds, which are usually deepwater systems where the edge is usually stabilized with rocks and plant growth is actually discouraged. Wetlands have been designed for capturing stormwater in urban areas in Florida (Johengen and LaRock, 1993), Washington (Reinelt and Horner, 1995), and England (Shutes et al., 1993)

Stormwater from urban areas is particularly rapid as it comes from impervious sources such as roofs, parking lots, and highways. One of the features of stormwater wetland systems is that severe storms have a dramatic effect on treatment efficiency. High flows resulting from high-intensity rainstorms usually result in lower nutrient and other chemical retention as a percentage of inflow, and the storms sometimes cause a net release of nutrients. The very nature of the sudden but short stormwater pulses make management of these systems particularly difficult. The pollutants that are transported to treatment wetlands can also create problems. In addition to nutrients and organic wastes, stormwater runoff can either be very high in sediments if coming from construction sites or relatively low in sediments if coming from roofs. If parking lots and highways are part of the wetland watershed, pollutants result from vehicle exhaust, asphalt erosion, road salts, rubber, oil, grease, metals, and even large rubbish. It is not uncommon to see these wetlands severely stressed with some of these materials, and frequent maintenance can sometimes be necessary, if for no other reason than to remove urban litter from the basins.

Design approaches to these types of wetlands are few despite several studies that have been done on these systems since the early 1980s. Schueler (1992) summarized almost 60 stormwater treatment wetland systems and estimated significant long-term pollution capabilities of these systems (Table 10.1). The sediment retention capability is the strong point of these wetlands, but if any significant construction projects occur upstream, even this capacity can be temporarily or permanently overwhelmed. There is some nutrient and organic matter retention, but it is generally not greater than 50 percent retention. Layout of an ideal stormwater treatment wetland (Figure 10.4) illustrates that a combination of deep ponds and marshes may be most appropriate, with the pond dampening the rapid stormwater pulse, allowing the wetland to "treat" the runoff in a more effective manner. Multiple cells of marshes and a small outflow deepwater pond can contribute to the effectiveness of the system.

Agricultural Runoff Wetlands

One of the most important applications of wetland treatment systems—yet an application that is far behind municipal treatment wetlands in understanding design issues—is the use of nonpoint-source wetlands for treating subsurface and surface runoff from agricultural fields. Research projects illustrating the effects and functioning of these types of wetlands in agricultural watersheds have been carried out in southeastern Australia (Raisin and Mitchell, 1995; Raisin et al., 1997), northeastern Spain (Comin et al., 1997), Illinois (Kadlec and Hey, 1994; Phipps and Crumpton, 1994; Mitsch et al., 1995; Kovacic et al., 2000; Larson et al., 2000; Hoagland et al., 2001), Florida (Moustafa, 1999), Ohio (Nairn and Mitsch, 2000; Spieles and Mitsch, 2000a), and Sweden (Arheimer and Wittgren, 1994; Jacks et al., 1994; Leonardson et al., 1994). Several wetland sites have received the equivalent of nonpoint-

Table 10.1 Average retention of chemicals in stormwater wetlands

Pollutant	Percent Reduction
Suspended solids	75
Total nitrogen	25
Total phosphorus	45
Organic carbon	15
Lead	75
Zinc	50
Bacterial count	10^{-2} decrease

Source: Data from Schueler (1992).

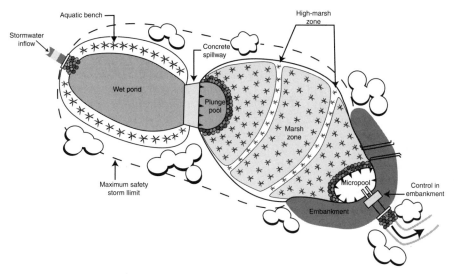

Figure 10.4 General design of a stormwater treatment wetland. (Redrawn From Schueler, 1992.)

source pollution but under somewhat controlled hydrologic conditions (e.g., river overflow to riparian basins) over several years of study. Bony Marsh, a constructed wetland located along the Kissimmee River in southern Florida, was investigated for nutrient retention of river water for nine years (1978–1986) by the South Florida Water Management District (Moustafa et al., 1996) and found to be a consistent sink of nitrogen and phosphorus but at relatively low levels. At constructed riparian wetlands at the Des Plaines River Wetland Research Park in northeastern Illinois and the Olentangy River Wetland Research Park in central Ohio consistent patterns of nutrient and sediment retention have been observed over multiple years of study of these systems, both of which received pumped river water, thus simulating wetlands receiving dilute nonpoint-source pollution. One of the largest wetlands constructed for the control of nutrients in stormwater, the Everglades Nutrient Removal Project, a 1545-ha constructed marsh, removed 82 percent of the phosphorus and 55 percent of the total nitrogen applied to it over three years (Moustafa, 1999).

CASE STUDY
Agricultural Runoff Wetland

An agricultural runoff wetland (Figure 10.5) was constructed in the spring of 1998 in Logan County, Ohio, several kilometers upstream of a popular recreational lake in northwestern Ohio called Indian

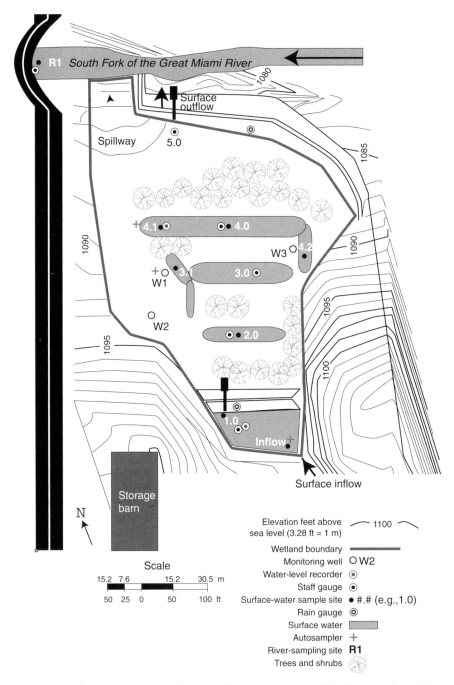

Figure 10.5 Indian Lake agricultural runoff wetland, northwestern Ohio. (Redrawn from Mitsch and Fink, 2001.)

Lake. The multicelled Indian Lake wetland was 1.2 ha and receives drainage from a 17-ha watershed, 14.2 ha of which was used for intensive row-crop agriculture and 2.8 ha of which was forested. Thus the wetland had a watershed ratio of 14:1. It was estimated that surface inflow in 2000 was 664 cm yr^{-1}, and groundwater discharge at multiple locations within the site amounted to almost the same amount of inflow. Surface-water levels of a two-year period of study varied over 40 cm in depth (Figure 10.6), and muskrat activity in one of the cells actually led to a 30-cm water-level decrease in the second year of study. Major storm events led to dramatic but short increases in water level of over 20 cm; these storm events, primarily in the late winter and early spring, led to rapid flow-through and poorer water quality enhancement. Surface inflow had two-year-average concentrations of 0.79 mg-N L^{-1} for nitrate-nitrite, 0.03 mg L^{-1} for soluble reactive phosphorus (SRP), and 0.16 mg L^{-1} for total phosphorus (TP). Groundwater had lower concentrations for SRP and TP, but higher concentrations of nitrate-nitrite (1.97 mg-N L^{-1}). SRP and TP exported from the wetland increased significantly during precipitation events in 2000 compared to dry weather flows, but concentrations of nitrate-nitrite did not increase significantly. Overall, the wetlands retained 59 percent of total phosphorus and 40 percent of nitrate-nitrite (Table 10.2). The overall design of this wetland, with multiple cells

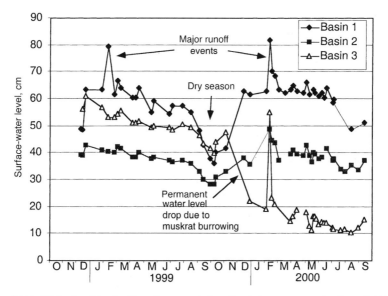

Figure 10.6 Water-level fluctuations in a created wetland downstream of a farm field in the Indian Lake agricultural wetland, northwestern Ohio. Note the short peaks associated with runoff events and seasonal changes in water levels. Also note the effect of muskrats on the water level in one cell.

Table 10.2   Percent reduction of nutrients at each sampling site within the Indian Lake agricultural runoff wetland in Ohio compared to the inflow concentration

Nutrient Type	1.0	2.0	3.0	4.2	4.0	4.1
	\multicolumn{6}{c}{Site[a]}
Soluble reactive phosphorus	15	71.7	71	64	49	56
Total phosphorus	−41	56	65	48	30	59
Nitrate-nitrite	20	34	55	37	43	40

Source: After Fink (2001).
[a]A minus sign indicates a net export of nutrient. Sites are indicated on Figure 10.5.

and a watershed:wetland ratio of 14:1, appeared to be well designed to receive storm pulses of surface runoff coupled with more consistent, yet variable amounts of groundwater inflow.

Landfill Leachate Wetlands

Impermeable liners are used to collect groundwater that has passed through the landfill. This leachate is often quite variable in water quality but generally has very high concentrations of ammonium-nitrogen and chemical oxygen demand (Kadlec, 1999). This wastewater has always presented a problem to landfill operators, and stricter water quality standards are making it necessary for advanced treatment. Wetlands are one of several options for management of leachate; other options include spray irrigation, physical–chemical treatment, biological treatment, and piping to a wastewater treatment plant. Mulamoottil et al. (1999) present a summary of results from several dozen constructed wetlands that are treating landfill leachate in Canada, the United States, and Europe.

Agricultural Wastewater Wetlands

In addition to the nonpoint sources from agriculture that are discussed above, serious water pollution problems occur in many parts of the world due to runoff from confined animals, particularly dairy, cattle, and swine operations (Tanner et al., 1995; Cronk, 1996; CH2M-Hill and Payne Engineering, 1997). As more animals are concentrated per unit area to increase food production, the concentrations and volumes of effluents are becoming more noticeable, both by the public and by water pollution control authorities. Concentrations of organic matter, organic nitrogen, ammonia-nitrogen, phosphorus, and fecal coliforms from animal feedlots far exceed concentrations in most municipal sewer systems (Table 10.3). Two examples from the eastern United States of the effectiveness of wetlands for treating wastewater from dairy milkhouses (Table 10.3) showed significant reductions in most pollutants in the treated

Table 10.3 Hydrology and water quality of two wetlands constructed to deal with heavily polluted dairy milkhouse effluent

	Connecticut[a]		Maryland[b]	
Wetland area (m^2)	400		1,160	
Flow (m^3 wk^{-1})	18.8		—	
Retention time (days)	41		—	
	Inflow	Outflow	Inflow	Outflow
BOD (mg L^{-1})	2,680	611	1,914	59
Total N	103	74	170	13
Ammonia-N	8	52	72	32
Nitrate-N	0.3	0.1	5.5	10.0
Total P (mg L^{-1})	26	14	53	2.2
TSS (mg L^{-1})	1,284	130	1,645	65
Coliform, (no./100 mL)	557,000	13,700	—	—

[a]Data from Newman et al. (2000).
[b]Data from Schaafsma et al. (2000).

water, although ammonia-nitrogen increased substantially in the Connecticut wetland and nitrate-nitrogen increased by 80 percent in the Maryland case.

In addition to livestock waste from land-based agriculture, constructed wetlands have been used to treat effluent from a number of aquaculture operations, including shrimp ponds in Thailand and tilapia fish ponds in the UK.

10.3 WETLAND DESIGN

The need for rigor in designing a wetland varies widely depending on the site and application. In general, a design that uses natural processes to achieve objectives yields a less expensive and more satisfactory solution in the long run. On the other hand, "naturally" designed wetlands may not develop as predictably as more tightly designed systems should. The choice of design is strongly affected by the site and the objectives. In Europe and many parts of North America, subsurface wetlands are designed in rectangular basins to very specific design criteria. In coastal Louisiana, by contrast, there are now several projects where wetlands are being used as tertiary treatment systems for the removal of nutrients from wastewater. In the following sections we focus on rigidly designed wetlands in part because this kind of wetland creation requires much greater ecotechnological sophistication.

Hydrology

Hydrology is an important variable in any wetland design. If the proper hydrologic conditions are developed, chemical and biological conditions will respond accordingly. Several parameters used to describe the hydrologic conditions of treatment wetlands include hydroperiod, depth, seasonal pulses, hydraulic loading rates, and retention time.

Hydroperiod and Depth. In wetlands, *hydroperiod* is the depth of water in a wetland over time (Figures 10.6 and 10.7). In a constructed wastewater wetland with a similar inflow of wastewater every day, water levels often vary little seasonally unless stormwater is part of the treatment inflow (see, e.g., 1995 data in Figure 10.7). Water levels are controlled by a combination of all inflows and outflows, with important variables being the total inflow and the setting of outflow structures:

$$\frac{\Delta(d \times A)}{\Delta t} = S_{\text{in}} + S_{\text{out}} + G_{\text{in/out}} + P - \text{ET} \tag{10.1}$$

where $\Delta(d \times A)/\Delta t$ is the change of water volume in wetland over time (d is the average water depth and A is the wetland surface area), S_{in} is the inflow of wastewater, S_{out} the surface outflow, $G_{\text{in/out}}$ the groundwater exchange (e.g., seepage), P the precipitation, and ET the evapotranspiration. During the start-up period of constructed wetlands, low water levels are needed to avoid flooding newly emerged plants. Startup periods for the establishment of vegetation may involve two to three years of careful attention to water levels.

Although storms and seasonal patterns of floods rarely affect constructed wastewater wetlands built for municipal treatment (except when storm sewers are part of the inflow), they can significantly affect the performance of wetlands designed for the control of nonpoint-source runoff. A variable hydroperiod, exhibiting dry periods interspersed with flooding, is a natural cycle in nonpoint-source wetlands, and fluctuating water levels should be considered a natural feature. A fluctuating water level could provide needed oxidation of

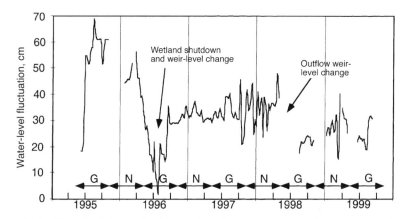

Figure 10.7   Water-level fluctuation in a surface-flow wastewater wetland in Licking County, Ohio. Note that the water levels are generally consistent through a year. From year to year, different levels were maintained by changing outflow weir settings.

organic sediments and can, in some cases, rejuvenate a system to higher levels of chemical retention. A typical water-level variation in a newly constructed multicell wetland built in an agricultural region to control nutrients was illustrated in Figure 10.6. There is a definitive seasonal cycle plus sudden bursts of water levels during winter and spring storms if they occur during nonfrozen conditions. Furthermore, noting the importance of biology in wetland types, the water level in one basin drops almost 30 cm in late fall because of burrowing activity by muskrats (*Ondatra zibethicus*).

Hydraulic Loading Rate. The *hydraulic loading rate* (HLR), one of the most important variables in treatment wetlands, is defined as

$$q = \frac{Q/A}{100} \tag{10.2}$$

where q is the inflowing hydraulic loading rate, the volume per unit time per unit area, which is equivalent to the depth of flooding over the treatment area per unit time (cm day^{-1}), Q is the flow rate (m^3 day^{-1}), and A is the wetland surface area (m^2).

Table 10.4 summarizes hydrologic loading rates for some of the many surface- and subsurface-flow wastewater wetlands in North America and Europe. Loading rates to surface-flow constructed wetlands for wastewater treatment from small municipalities ranged from 1.4 to 22 cm day^{-1} (average = 5.4 cm day^{-1}), while rates to subsurface-flow constructed wetlands varied between 1.3 and 26 cm day^{-1} (average = 7.5 cm day^{-1}). Knight (1990) reviewed several dozen wetlands constructed for wastewater treatment and found loading rates to vary between 0.7 and 50 cm day^{-1}. He recommended a rate of 2.5 to 5 cm day^{-1} for surface-flow constructed wetlands and 6 to 8 cm day^{-1} for subsurface flow wetlands.

Detention Time. *Detention time* of treatment wetlands is calculated as

$$t = \frac{Vp}{Q} \tag{10.3}$$

where t is the theoretical detention time (days), V the volume of wetland

Table 10.4 Hydrologic loading rates for treatment wetlands in North America and Europe

Type of Wetland	Loading Rate (cm day^{-1})
Surface-flow treatment wetlands ($n = 15$)	5.4 ± 1.7
Subsurface flow treatment wetlands ($n = 23$)	7.5 ± 1.0

Source: Data from Mitsch and Gosselink (2000).

basin (m³) or the volume of water for surface-flow wetlands or the volume of medium through which the wastewater flows for subsurface-flow, p is the porosity of the medium (e.g., sand or gravel for subsurface-flow wetlands; about 1.0 for surface-flow wetlands), and Q is the flow rate through wetland (m³ day^{-1}) = $(Q_i + Q_o)/2$, where Q_i is the inflow and Q_o is the outflow.

The optimum detention time (or nominal residence time) has been suggested to be from 5 to 14 days for the treatment of municipal wastewater. Florida regulations on wetlands require that the volume in the permanent pools of the wetland must provide for a residence time of at least 14 days. M. T. Brown (1987) suggested a detention time for a riparian treatment wetland system in Florida of 21 days in the dry season and more than 7 days in the wet season.

Calculation of detention time or nominal residence time with equation (10-3) is not always realistic because of short-circuiting and the ineffective spreading of the waters as they pass through the wetland. Tracer studies of flow through wetlands illustrate the importance of not overrelying on the theoretical detention time to design treatment wetlands. Not all parcels of water that enter at a certain time leave the wetland at the same time. In some instances, water will "short-circuit" through the wetland; in other instances water will remain in backwater locations for considerably more time than the theoretical detention time.

Basin Morphology

Several aspects related to the morphology of constructed wetland basins need to be considered when designing wetlands. For example, Florida regulations for the Orlando area require, for littoral zones, a shelf with a gentle slope of 6:1 or flatter to a point of from 60 to 77 cm below the water surface. Slopes of 10:1 or flatter are even better. A flat littoral zone maximizes the area of appropriate water depth for emergent plants, thus allowing more wetland plants to develop more quickly and allowing wider bands of different plant communities. Plants will also have room to move "uphill" if water levels are raised in the basins due to flows being higher than predicted or to enhanced treatment. Bottom slopes of less than 1 percent are recommended for wetlands built to control runoff, whereas a substrate slope from inlet to outlet of 0.5 percent or less has been recommended for surface-flow wetlands used to treat wastewater.

Flow conditions should be designed so that the entire wetland is effective in nutrient and sediment retention if these are desired objectives. This may necessitate several inflow locations and a wetland configuration to avoid channelization of flows. Steiner and Freeman (1989) suggested a length/width ratio (L/W) (called the *aspect ratio*) of at least 10:1 if water is purposely introduced to the system. A minimum aspect ratio of 2:1 to 3:1 has been recommended for surface-flow wastewater wetlands.

Providing a variety of deep and shallow areas is optimum. Deep areas (>50 cm), although too deep for continuous emergent vegetation, offer habitat for fish, increase the capacity of the wetland to retain sediments, can enhance nitrification as a prelude to later denitrification if nitrogen removal is desired, and can provide low-velocity areas where water flow can be redistributed. Shallow depths (<50 cm) provide maximum soil–water contact for certain chemical reactions such as denitrification and can accommodate a greater variety of emergent vascular plants.

Individual wetland cells, placed in series or parallel, often offer an effective design to create different habitats or establish different functions. Cells can be parallel, so that alternate drawdowns can be accomplished for mosquito control or redox enhancement, or they can be in a series to enhance biological processes.

Chemical Loadings

When water flows into a wetland, it brings chemicals that may be beneficial or possibly detrimental to the functioning of the wetland. In an agricultural watershed, this inflow will include nutrients such as nitrogen and phosphorus as well as sediments and possibly pesticides. Wetlands in urban areas can have all of these chemicals plus other contaminants, such as oils and salts. When added to wetlands, wastewater has high concentrations of nutrients and, with incomplete primary treatment, high concentrations of organic matter and suspended solids. At one time or another, wetlands have been subjected to all of these chemicals, and they often serve as effective sinks.

Design Graphs. The simplest model available to estimate the retention of nutrients or other chemicals by wetlands is to use design graphs that give some measure of chemical retention versus chemical loading, either areal (e.g., g m^{-2} yr^{-1}) or volumetric (e.g., g m^{-3} yr^{-1}). If a wetland were designed to retain nutrients, for example, it would be desirable to know how well that retention would occur for various nutrient inflows. Data compiled from a large number of wetland sites in North America and Europe provide an indication of the nutrient retention of wetlands. For example, Figure 10.8, compiled from some of those data sets, illustrates the percent removal of nitrate-nitrogen versus loading for the midwestern United States in three ways: (1) mass retention per unit area, (2) percent retention by mass, and (3) percent retention by concentration. Each of the data points is based on one year's data at one wetland basin at either the Olentangy River Wetland Research Park in Ohio or the Des Plains River Wetland Demonstration Project in Illinois.

Retention Rates. Another approach to estimating the retention of nutrients is simply to compare a number of studies and estimate the chemical retention that happens consistently in wetlands. Averages from data from many constructed wastewater wetlands are shown in Table 10.5. In general, as suggested by the hydraulic loading rate data in Table 10.4, subsurface wetlands receive more wastewater and thus received greater loadings of chemicals and sediments. The high-average mass retention of nitrate-nitrogen in subsurface

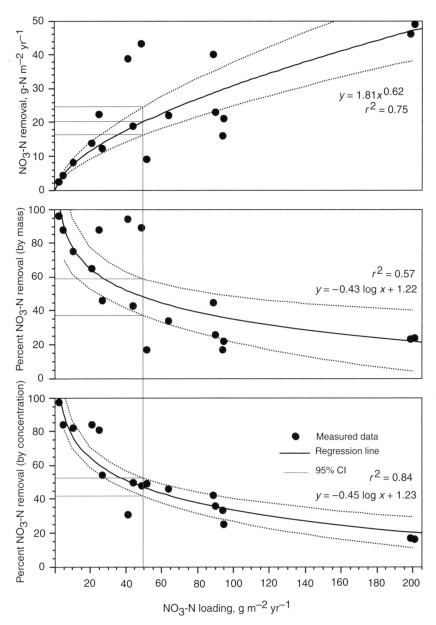

Figure 10.8 Nitrate-nitrogen retention of river-fed constructed wetland basins as a function of nitrate loading. Nitrogen retention is shown as areal-based removal, percent nitrate decrease by mass, and percent nitrate-nitrogen decrease by concentration. Each data point represents one full year of data in one wetland basin at the Des Plaines River wetlands in Illinois or Olentangy River wetlands in Ohio. (From Mitsch and Gosselink, 2000; copyright 2000; reprinted with permission from John Wiley & Sons, Inc.)

Table 10.5 Nutrient and sediment removal rates and efficiency in constructed wastewater wetlands

Wetland Type, Parameter	Loading (g m^{-2} yr^{-1})	Retention (g m^{-2} yr^{-1})	Percent Retention
Surface-flow constructed wetlands			
Nitrate + nitrate-N	29	13	44.4
Total N	277	126	45.6
Total P	4.7–56	2.1–45	46–80
Suspended solids	107–6520	65–5570	61–98
Subsurface-flow constructed wetlands			
Nitrate + nitrate-N	5767	547	9.4
Total N	1058	569	53.8
Total P	131–631	11–540	8–89
Suspended solids	1500–5880	1100–4930	49–89

Source: Data from Knight (1990) and Kadlec and Knight (1996).

wetlands is due more to these high loading rates in subsurface-flow wetlands than it is to the ability of these systems to sequester more nitrate-nitrogen. Note that the percent nitrate-nitrogen retention is much higher in surface-flow than in subsurface-flow wetlands. The retention of phosphorus tends to be more variable in subsurface wetlands than in surface wetlands.

Summaries of retention rates for several wetlands intercepting nonpoint-source pollution are given in Table 10.6. A rule of thumb for this type of wetland is that wetlands can consistently retain phosphorus in amounts of 1 to 2 g-P m^{-2} yr^{-1} and nitrogen in amounts of about 10 to 20 g-N m^{-2} yr^{-1} (Mitsch et al., 2000; Figure 10.8). Nitrate-nitrogen retention capabilities of freshwater marshes receiving nonpoint-source pollution in seasonal to cold climates shows a range of nitrogen retention from 3 to 93 g-N m^{-2} yr^{-1} and a phosphorus retention rate of 0.1 to 6 g-P m^{-2} yr^{-1}. Low retention numbers are generally from wetlands that are "underfed" nutrients. High numbers are usually only periodic and therefore would be inappropriate for design purposes.

Empirical Models. A third method for estimating the ability of wetlands to retain chemicals is to use equations that either have a theoretical base or are determined empirically from large databases of existing wastewater wetlands. One such general model, developed by Kadlec and Knight (1996) and others, is based on a mass balance approach called the *k–C* model* and is given as

$$q \frac{dC}{dy} = k_A(C - C^*) \tag{10.4}$$

where C is the chemical concentration (g m^{-3}), y the fractional distance from

Table 10.6 Nutrient retention in constructed and natural wetlands receiving low-concentration (i.e., nonwastewater) nutrient loading from rivers, overflows, or nonpoint-source pollution

Wetland Location and Type	Wetland Size (ha)	Nitrogen (g-N m⁻² yr⁻¹) $(g\text{-}N\ m^{-2}\ yr^{-1})$	Phosphorus $(g\text{-}P\ m^{-2}\ yr^{-1})$	References
Warm climate				
Everglades marsh, southern Florida	8000	—	0.4–0.6[a]	Richardson and Craft, 1993; Richardson et al., 1997
Boney Marsh, southern Florida	49	4.9	0.36	Moustafa et al., 1996
Everglades Nutrient Removal Project, Southern Florida	1545	10.8	0.94	Moustafa, 1999
Restored marshes, Mediterranean delta, Spain	3.5	69	—	Comin et al., 1997
Constructed rural wetland, Victoria, Australia	0.045	23	2.8	Raisin et al., 1997
Cold climate				
Constructed wetlands, northeastern Illinois River-fed and high-flow	2	11–38[b]	1.4–2.9	Mitsch, 1992; Phipps and Crumpton, 1994; Mitsch et al., 1995
River-fed and low-flow	2–3	3–13[b]	0.4–1.7	
Artificially flooded meadows, southern Sweden	180	43–46	—	Leonardson et al., 1994
Constructed wetland basins, Norway	0.035–0.09	50–285	26–71	Braskerud, 2002a,b
Palustrine freshwater wetlands, northwestern Washington				Reinhelt and Horner, 1995
Urban area	2	—	0.44	
Rural area	15	—	3.0	
Constructed urban in-stream, Ohio	6	—	2.9	Niswander and Mitsch, 1995
Constructed river wetlands, Ohio (2)	1	58–66[b]	5.2–5.6	Mitsch et al., 1998; Nairn and Mitsch, 2000; Spieles and Mitsch, 2000a
Agricultural wetlands, Ohio	1.2	39[b]	6.2	Mitsch and Fink, 2001
Agricultural wetlands, Illinois (3)	0.3–0.8	33[b]	0.1	Kovacic et al., 2000
Natural marsh, Alberta, Canada	360	—	0.43[a]	White et al., 2000

[a]Estimated by phosphorus accumulation in sediment.
[b]Nitrate-Nitrogen only.

inlet to outlet (unitless), k_A the areal removal rate constant (m yr^{-1}), and $C*$ the residual or background chemical concentration (g m^{-3}). This equation is based on an assumption that processes can be described on an aerial basis. Thus, the coefficient k_A has units of velocity and can be recognized as being similar to a settling velocity coefficient used in sedimentation models. $C*$ represents a background concentration of a chemical or constituent below which it is generally agreed that treatment wetlands cannot go. Integrating this equation over the entire length of the wetland, the solution can be expressed as a first-order areal model:

$$\frac{C_o - C*}{C_i - C} = \exp\left(\frac{-k_A}{q}\right) \qquad (10.5)$$

where C_o is the outflow concentration (g m^{-3}), C_i the inflow concentration (g m^{-3}), and q the hydraulic loading rate (m yr^{-1}).

Estimates of the two parameters needed for this model, $C*$ and k_A, are listed in Table 10.7. This equation does not work equally well for all parameters but does provide a way of estimating the area of a wetland necessary for achieving a certain removal. Rearranging equations (10.5) and (10.2) gives the following calculation of wetland area for the results given:

Table 10.7 Parameters for first-order areal model given in equations (10.4) to (10.6) for several constituents of wastewater wetlands[a]

Constituent and Wetland Type	k_A (m yr^{-1})	$C*$ (g m^{-3})
BOD		
Surface flow	34	$3.5 + 0.053C_i$
Subsurface flow	180	$3.5 + 0.053C_i$
Suspended solids, surface flow	1000	$5.1 + 0.16C_i$
Total P, surface and subsurface flow	12	0.02
Total N		
Surface flow	22	1.5
Subsurface flow	27	1.5
Ammonia-N		
Surface flow	18	0
Subsurface flow	34	0
Nitrate-N		
Surface flow	35	0
Subsurface flow	50	0

Source: Data from Kadlec and Knight (1996).
[a]Subsurface-flow constructed wetlands and surface-flow constructed wetlands are given as wetland type where appropriate.

$$A = Q\,\frac{\ln[(C_o - C^*)/(C_i - C^*)]}{k_A} \tag{10.6}$$

where Q is the flow rate through the wetland (m^3 yr^{-1}).

Where this model is insufficiently backed with good data or does not work properly, strictly empirical relationships of the outflow concentration C_o as a function of the inflow concentration C_i and the hydraulic loading rate (q) have been developed. Table 10.8 illustrates several regression equations of this nature that could be used, with other approaches, to estimate outflow concentrations and, in one case, wetland area. For example, the equation estimating total phosphorus efflux in a surface-flow wastewater marsh is

$$C_o = 0.195 C_i^{0.91} q^{0.53} \tag{10.7}$$

Table 10.8   Empirical equations for the estimation of outflow concentrations or wetland area based on inflow concentrations and hydraulic retention time

Constituent	Equation[a]	Correlation Coefficient, r^2	Number of Wetlands Used in Analysis
BOD			
Surface-flow wetlands	$C_o = 4.7 + 0.173 C_i$	0.62	440
Subsurface-flow, soil	$C_o = 1.87 + 0.11 C_i$	0.74	73
Subsurface-flow, gravel	$C_o = 1.4 + 0.33 C_i$	0.48	100
Suspended solids			
Surface-flow wetlands	$C_o = 5.1 + 0.158 C_i$	0.23	1582
Subsurface-flow wetlands	$C_o = 4.7 + 0.09 C_i$	0.67	77
Ammonia-N			
Surface-flow wetlands	$A = 0.01 Q/\exp(1.527 \ln C_o - 1.05 \ln C_i + 1.69)$		
Surface-flow marshes	$C_o = 0.336 C_i^{0.728} q^{0.456}$	0.44	542
Subsurface-flow wetlands	$C_o = 3.3 + 0.46 C_i$	0.63	92
Nitrate-N			
Surface-flow marshes	$C_o = 0.093 C_i^{0.474} q^{0.745}$	0.35	553
Subsurface-flow wetlands	$C_o = 0.62 C_i$	0.80	95
Total nitrogen			
Surface-flow marshes	$C_o = 0.409 C_i + 0.122 q$	0.48	408
Subsurface-flow wetlands	$C_o = 2.6 + 0.46 C_i + 0.124 q$	0.45	135
Total phosphorus			
Surface-flow marshes	$C_o = 0.195 C_i^{0.91} q^{0.53}$	0.77	373
Surface-flow swamps	$C_o = 0.37 C_i^{0.70} q^{0.53}$	0.33	166
Subsurface-flow wetlands	$C_o = 0.51 C_i^{1.10}$	0.64	90

Source: Kadlec and Knight (1996).
[a] C_i, inflow concentration (g m^{-3}); C_o, outflow concentration (g m^{-3}); A, area of wetland (ha); Q, wetland inflow (m^3 day^{-1}); q, hydraulic retention time (cm day^{-1}).

where C_o and C_i are the outflow and inflow concentrations, respectively (g m^{-3}), and q is the hydraulic loading rate (cm day^{-1}).

Other Chemicals. Although most evaluations of the efficiency of wetlands have been concerned with this capacity to remove nutrients, sediments, and organic carbon, there is some literature on other chemicals, such as iron, cadmium, manganese, chromium, copper, lead, mercury, nickel and zinc. Wetland soils or biota or both often easily sequestered metals. But that is the basic problem in using wetlands as sinks for such chemicals. One very notable example stands out. The accumulation of selenium became a problem in marshes in the Kesterson National Wildlife Refuge in California when it began to accumulate in biota there as a result of the inflow of spent irrigation waters for several years. Concentrations of selenium in these agricultural irrigation return flows were as high as 300 μg L^{-1} (Ohlendorf et al., 1986, 1990) compared to an average of 0.1 μg L^{-1} in the world's rivers. As a result, the Kesterson marshes were found to be a threat to fish and wildlife in the region and were dewatered.

Soils

Topsoil is important to the overall function of a constructed wetland (Figure 10.9). It is the primary medium supporting rooted vegetation, and particularly for subsurface wetlands, it is part of the treatment system itself. The sediments retain certain chemicals and provide habitat for micro- and macroflora and fauna that are involved in chemical transformations. Constructed wetland soil texture depends on whether surface flow over the substrate or subsurface flow through the substrate is being considered (Figure 10.9). Surface-flow wetland soils are generally less effective in removing pollutants per unit area but are closer in design to natural wetlands. Their ability to provide structure and nutrition to wetland plants is important. Clay material, although favored as a subsurface liner, limits root and rhizome penetration and may prevent water from reaching plant roots. Silt clay or loam soils are preferable for the overlying soils in constructed wetlands. Sandy soil is less preferred for surface-flow wetlands. For subsurface-flow wetlands, high permeability is preferred. The material needs to be sand, gravel, or some other highly permeable medium.

   The subsoil of constructed wetlands (usually below the root zone and referred to as a *liner*) must have permeability low enough to cause standing water or saturated soils. If clay is not available on site, it may be advisable to add a layer of clay to minimize percolation. Studies have also been undertaken to investigate other materials as liners for constructed wetlands. The most frequently used liners for constructed wetlands are clays, clay–bentonite mixtures, or synthetic materials such as polyvinyl chloride and high-density polyethylene (Kadlec and Knight, 1996). Experiments have been conducted in recycling materials such as coal combustion waste products (Ahn et al., 2001; Ahn and Mitsch, 2001). As it turns out, using calcium-rich sulfur-

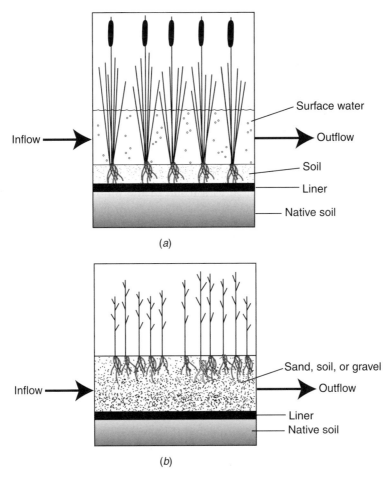

Figure 10.9   Soil cross section of (*a*) surface-flow wetland and (*b*) subsurface-flow wetland. (Redrawn from Knight, 1990.)

scrubber waste material may actually increase the phosphorus-retention capability of the wetlands (Ahn et al., 2001). But care must be taken that the material seals the wetland completely, as leachate from this liner material is highly alkaline.

Subsurface flow through subsurface wetlands can be through soil media (*root-zone method*) or through rocks, gravel, or sand (*rock-reed filters*). Flow in both cases is 15 to 30 cm below the surface. Gravel is sometimes added to the substrate of subsurface-flow wetlands (*gravel bed*) to provide a relatively high permeability that allows water to percolate into the root zone of plants where microbial activity is high. In a survey of several hundreds of wetlands build in Europe for sewage treatment in rural settings, Cooper and Hobson (1989) reported that gravel has been used in combination with soil but that the substrate remains the greatest uncertainty in artificial reed (*Phrag-*

mites) wetlands. Gravel can be silica- or limestone-based; the former has less capacity for phosphorus retention. Another evaluation of the European-design subsurface wetlands indicated that they often decrease in hydrologic conductivity after several years and become clogged, essentially becoming partial surface-flow wetlands (Steiner and Freeman, 1989).

Organic Content. The organic content of soils has some significance for the retention of chemicals in a wetland. Mineral soils generally have lower cation-exchange capacity than organic soils do; the former is dominated by various metal cations, and the latter is dominated by the hydrogen ion. Organic soils can therefore remove some contaminants (e.g., certain metals) through ion exchange and can enhance nitrogen removal by providing an energy source and anaerobic conditions appropriate for denitrification. Organic matter in wetland soils varies between 5 and 75 percent, with higher concentrations in peat-building systems such as bogs and fens and lower concentrations in mineral-soil wetlands such as riparian bottomland wetlands subject to mineral sedimentation or erosion. When wetlands are constructed, especially subsurface-flow wetlands, organic matter such as composted mushrooms, peat, or detritus is often added in one of the layers. For construction of many wetlands, however, organic soils are avoided because they are low in nutrients, can cause low pH, and often provide inadequate support for rooted aquatic plants.

Depth and Layering of Soil. The depth of substrate is an important design consideration for wastewater wetlands, particularly those that use subsurface flow. The depth of suitable topsoil or substrate should be adequate to support and hold vegetation roots. A common substrate depth for constructed wetlands is 60 to 100 cm. In some case, layering more elaborate than that shown in Figure 10.9 is suggested. Meyer (1985) described a layered substrate in wetlands to control stormwater runoff as having the following materials from the base upward from the liner: 60 cm of 1.9-cm limestone, 30 cm of 2-mm crushed limestone to raise pH and aid in the precipitation of dissolved heavy metals and phosphate, 60 cm of coarse to medium sand as filter, and 50-cm organic soil (see "Vegetation" below).

Soil Chemistry. Although exact specifications of nutrient conditions in a wetland soil necessary to support aquatic plants are not well known, low nutrient levels characteristic of organic, clay, or sandy soils can cause problems for initial plant growth. Although fertilization may be necessary in some cases to establish plants and enhance growth, it should be avoided if possible in wetlands that eventually will be used as sinks for the same macronutrients. When fertilization is required to get plants started in constructed wetlands, slow-release granular and tablet fertilizers are often useful.

When soils are submerged and anoxic conditions result, iron is reduced from ferric (Fe^{3+}) to the ferrous (Fe^{2+}) ions, releasing phosphorus that was held previously as insoluble ferric phosphate compounds. The Fe–P compound can be a significant source of phosphorus to overlying and interstitial waters after flooding and anaerobic conditions occur, particularly if the wetland was constructed on land previously agricultural. After an initial pulse of released phosphorus in such constructed wetlands, the iron and aluminum contents of a wetland soil exert significant influences on the ability of that wetland to retain phosphorus. All things being equal, soils with higher aluminum and iron concentrations are more desirable because their affinity for phosphorus is higher.

Vegetation

Just as the question "What plants should be used?" arises for creating and restoring wetlands as discussed in Chapter 8, vegetation choice is also a consideration for treatment wetlands, but there is at least one significant difference. Whereas creation and restoration of wetlands are done principally to develop a diverse vegetation cover and provide habitat, treatment wetlands are constructed with the main goal of improving water quality. The plants in created and restored wetlands are part of the solution; in treatment wetlands they are the partial cause of the solution. Furthermore, treatment wetlands invariably have higher concentrations of chemicals in the water, which by their very nature limit the number of plant species that will survive in those wetlands. Experience has shown that relatively few plants thrive in the high-nutrient, high-BOD (biological oxygen demand) wastewaters that are used in treatment wetlands. Among those plants are cattails (*Typha* spp.), bulrushes (*Schoenoplectus* spp., *Scirpus* spp.), and reed grass (*Phragmites australis*). The last is the preferred plant in subsurface-flow wetlands around the world but is not favored in many parts of North America because of its aggressive behavior in freshwater and brackish marshes (see the Delaware Bay Case Study in Chapter 9).

When water is deeper than 30 cm, emergent plants often have difficulty growing. In these cases, surface-flow wetlands can become covered with duckweed (*Lemna* spp.) in temperate zones and with water hyacinths (*Eichhornia crassipes*) and water lettuce (*Pistia* spp.) in the subtropics and tropics. While rooted floating aquatics such as *Nymphaea*, *Nuphar*, and *Nelumbo* are favored for their aesthetics, they thrive only in rare instances in treatment wetlands, where due to high-nutrient conditions, they are easily overwhelmed by duckweed and filamentous algae.

Tanner (1996) compared relative nutrient uptake and pollutant removal of eight macrophytes in gravel-bed wetland mesocosms fed by dairy wastes in New Zealand. Based on key growth characteristics of the plants in this wastewater, three productive gramminoids (*Zizania latifolia*, *Glyceria maxima*, and *Phragmites australis*) had the highest overall scores. *Baumea articulata*, *Cyperus involucratus*, and *Schoenoplectus validus* had medium scores, and *Scirpus fluviatilis* and *Juncus effusus* had the lowest scores and are least likely to be effective plants in wastewater wetlands.

10.4 SITE MANAGEMENT

Wildlife Control

Although the development of wildlife is a welcomed and often desired aspect of treatment wetlands, managing plant and animal populations often becomes necessary maintenance of constructed wetlands. In North America, beaver (*Castor canadensis*) and muskrat (*Ondatra zibethicus*) will create obstructions to inflows and outflows, destroy vegetation, or burrow into dikes. (This is one reason why dikes should not be built up around constructed wetlands if they can be avoided.) Major vegetation removal, particularly by herbivorous muskrats that use the plant material for both food and shelter can take a fully vegetated marsh and turn it into a plant-devoid pond in a matter of weeks or months. These events are referred to as *eatouts*. There is very little that can be done to prevent eatouts except to trap the animals, a laborious task.

In other cases, animals such as the above and large birds such as Canada geese (*Branta cadadennsis*) and snow geese (*Chen* spp.) that graze on newly planted perennial herbs and seedlings are particularly destructive. The timing of planting is important, especially if migratory animals are involved in destructive grazing in the winter. Using gunshot devices and extract of grape juice as a "hot foot" material on the adjacent landscape have all been suggested but without permanent success. Probably the easiest approach we have noted in many years of observing geese is to have a wide band of emergent vegetation between where they land (on the water) and where they like to graze (upland lawns). But, of course, you will have to get the local muskrats to cooperate and not remove the vegetation.

Trapping of muskrats and beavers can alleviate their impact on constructed wetland basins, but it can also be time consuming and often ineffective. Grazing by geese is more difficult to control, although their impact is most significant soon after vegetation is planted. If vegetation can get established, the grazing effects of geese can be controlled. Similarly, deeper wetlands often become havens for undesirable fish such as carp (*Cyprinus carpio*) that can cause excessive turbidity and uproot vegetation. Carnivores such as northern pike have been discussed for potential use to control carp. Total removal of fish by drawdown is probably necessary if carp begin to degrade outflow water quality excessively. The problem is that this fish removal might affect mosquito control (see below).

Attracting Wildlife

Just as many animals can cause maintenance headaches, the attraction of wildlife to constructed wetlands is one of the reasons why public support for such projects can be high in the first place. So every attempt should be made to have a diverse ecosystem, not just a pond with water flowing through it. Weller (1994) recommends a 50:50 ratio of open water to vegetation cover in marshes to attract water birds, and this kind of ratio, with proper development of the initial bathymetry of the ponds, is quite easy. Also, creating diverse habitats with live and dead vegetation, islands, and floating structures is desirable. In many cases of wetland construction, wildlife enhancement begins soon after construction. At a constructed wetland at Pintail Lake in Arizona, the area's waterfowl population increased dramatically by the second year of use; duck nest density increased 97 percent over the first year (Wilhelm et al., 1989). A considerable increase in avian activity was also noted at the Des Plaines River Wetlands Demonstration Project in northeastern Illinois. Migrating waterfowl increased from 3 to 15 species and from 13 to 617 individuals between 1985 (preconstruction) and 1990 (one year after water was introduced to the wetlands). The number of wetland-dependent breeding birds increased from 8 to 17 species, and two state-designated endangered birds, the least bittern and the yellow-headed blackbird, nested at the site after wetland construction (Hickman and Mosca, 1991).

Mosquito Control

Mosquito control will always be brought up when wetlands are being constructed, particularly when the wetlands receive runoff or wastewater. Mosquitoes can be controlled in constructed wetlands by changing the hydrologic conditions of the wetland to inhibit mosquito larvae development (flow-through conditions discourage mosquitoes) or by using chemical or biological control. Many have proposed mosquito control by fish, especially the air-gulping mosquito fish *(Gambusia affinis)* or similar small fish. One reason to maintain some deeper areas in wetlands in temperate zones is to allow fish such as *Gambusia* and other top minnows and sunfish to survive the winter and feed on mosquito larvae. Little is known about the role that water quality has on encouraging or discouraging mosquitoes directly, but the effect of poor water quality in removing fish can play a dramatic role in causing mosquito increases. Bacterial insecticides (e.g., *Bacillus sphaericus*) and the fungus *Lagenidium giganteum* are known pathogens of mosquito larvae, but they have not been tested extensively. Constructing boxes to encourage nesting by swallows (Hirundinidae), swifts (Apodidae), and bats have also been used to control adult mosquito populations at constructed wetlands.

Pathogens

Since many treatment wetlands are specifically built to deal with human and animal wastewater, proper sanitary engineering techniques should be used to minimize human exposure to pathogens. Treatment wetlands are meant to be biologically rich systems, and microbial activity is a major part of the treatment process. Measurements of indicator organisms such as fecal and total coliforms should be part of the monitoring of municipal wastewater treatment wetlands. Nearby wells should also be sampled, as water seeping from a wastewater wetland near potable water supplies should be monitored carefully. If a wetland is being used as tertiary treatment in a conventional treatment plant, in the design of the treatment plant, consideration has to be given to the disinfection system. Chlorine disinfection and the resulting chlorine residual would cause significant problems in treatment wetlands, so other means of disinfection (ozonation or ultraviolet radiation) should be used if there is disinfection before the wastewater enters the wetland.

Water Level Management

The water level of surface-flow treatment wetlands is the key to both water quality enhancement and vegetation success. Most constructed municipal wastewater wetlands have little control on the overall inflow of wastewater. Flow and hence depth are first controlled by designing a basin large enough to create the proper hydraulic loading rates (HLRs). Most constructed wetlands have a control structure such as a flume or weir to control outflow; these structures should be designed flexibly so that they can be manipulated to control water depth. Too much water stresses macrophytes as much as too little water. Water depths of 30 cm or less are optimum for most herbaceous macrophytes used in treatment wetlands. Water depths greater than 30 cm can lead to vegetation reduction.

Compounding the effect that water level has on vegetation is the effect that it has on wastewater treatment itself. High water levels favor high HLR and sediment and phosphorus retention associated with sedimentation and similar processes; it also leads to less resuspension and longer retention time. Shallow water leads to closer proximity of sediments and overlying water, often causing anaerobic or nearly anaerobic conditions during the growing season. Low water thus favors the reduction of nitrate-nitrogen through denitrification. Providing wastewater treatment with vegetation success is a continual balancing act.

10.5 ECONOMICS AND BENEFITS OF TREATMENT WETLANDS

It is generally believed that treatment wetlands are less expensive to build and maintain than conventional wastewater treatment, and that is the appeal of these systems to many. However, cost comparisons should be made carefully before investing in these systems. Any estimate of the cost of a new wetland's development should include the following items: (1) engineering plan, (2) preconstruction site preparation, (3) construction costs (labor, equipment, materials, supervision, indirect and overhead charges), and (4) cost of land.

An equation estimating the cost of constructing wastewater wetlands (not including the cost of land) was developed by Mitsch and Gosselink (2000):

$$C_A = \$196,336A^{-0.511} \tag{10.8}$$

where C_A is the capital cost of wetland construction per unit area ($ ha^{-1}) and A is the wetland area (ha). This relationship suggests that a 1-ha wetland would cost almost $200,000, a 10-ha wetland would cost $60,000 per hectare, and a 100-ha wetland would cost $19,000 per hectare. The data clearly suggest that there is an economy of scale involved in wetland construction.

Operating and maintenance costs vary according to the wetland's use and to the amount and complexity of mechanical parts and plumbing that the wetland contains. Fewer data on operational costs are available. Kadlec and Knight (1996) estimate the operation and maintenance costs for one wastewater wetland to be about $85,500 per year. That estimate included $50,000 per year for personnel to be in charge of the 175-ha wetland. A wide range of $5000 to $50,000 per year of operating and maintenance costs was estimated by Kadlec and Knight (1996) from smaller wetlands. Gravity-fed wetlands are far less expensive to maintain than are highly mechanized wetlands that need significant plumbing and pumps.

Subsurface treatment wetlands provide little additional benefit beyond the water quality improvement they were designed to provide. But surface-flow treatment wetlands have a variety of additional benefits. The watery habitat that is created can be a major ancillary benefit of these systems. In addition to providing habitat for mammals such as nutria, beavers, muskrats, amphibians, fish, and voles, surface-flow treatment wetlands are often a haven for waterfowl and wading birds. Human uses such as trapping and hunting are not incompatible with some wastewater wetlands. When designed properly in an urban area, wetlands are locations where the public can visit and learn about their important water quality role. This message is a powerful one to the uninitiated, and they often become ardent wetland conservationists as a result of seeing "wetlands at work." Another benefit of using both natural and constructed wetlands for water quality improvement relates to areas where land building is needed. In the subsiding environment of Louisiana's Gulf coast, nutrients are permanently retained in the peat of wetlands, which receives high-nutrient wastewater as the wetland aggrades to match subsidence. In this case, wastewater discharge into a wetland can occur without saturating the system and, simultaneously, helps counteract the deleterious effects of land subsidence.

10.6 SUMMARY CONSIDERATIONS

Wastewater treatment wetlands are not the solution to all water quality problems and should not be viewed as such. Many pollution problems, such as excessive BOD or metal contamination, may require more conventional approaches. Yet the fact that thousands of wetlands have been constructed around the world for pollution control attests to their importance and value.

There are many considerations, both technical and institutional, that must be considered as treatment wetlands are designed and built.

Technical Considerations

1. Values of the wetlands such as wildlife habitat should be considered in any treatment wetland development.
2. Acceptable pollutant and hydrologic loadings must be determined for the use of wetlands in wastewater management. Appropriate loadings, in turn, determine the size of the wetland to be constructed. Overloading a constructed wetland can be worse than not building it at all.
3. All existing characteristics of local natural wetlands, including vegetation, geomorphology, hydrology, and water quality, should be well understood so that natural wetlands can be "copied" in the construction of treatment wetlands.
4. Particular care should be taken in wetland design to address public health, including mosquito control and protection of groundwater resources.

Institutional Considerations

1. Potential conflicts over the protection and use of wetlands can arise among state agencies, federal agencies, and local groups. Some agencies view treatment wetlands as a beneficial use. Others view it as polluting a wetland habitat. Still others bring up questions of disease vectors and potential groundwater contamination.
2. Wastewater treatment by wetlands can often serve the dual purposes of both wetland habitat development and wastewater treatment and recycling. The creation of treatment wetlands as mitigation for lost wetlands is still generally unacceptable because of the lack of sustainability and the high level of pollution of treatment wetlands compared to restored wetlands.
3. Many permit processes in governments do not recognize treatment wetland systems as alternatives for wastewater treatment. In these cases, experimental systems should first be established for a given region. Modification of requirements for granting permits for pilot wetlands is needed to make effective progress in developing approaches.

It is useful to remember that wetland design is an inexact science and that perturbation and biological change are the only things we can be sure of in these created ecosystems. Traditional engineering approaches to wastewater wetlands, without an appreciation of self-design in ecosystems, are sure to cause disappointment. If a treatment wetland continues to function according to its main goal, improving water quality, changes in species and forms are not that significant.

11

BIOREMEDIATION: RESTORATION OF CONTAMINATED SOILS

The numbers of registered contaminated sites are alarming and continue to grow. Environmental remediation is thus an admittedly rapid growth area ripe for technological application and innovation. The high cost of remediation has driven interest in the direction of ecological engineering applications of bioremediation technologies. These technologies apply biological processes to remove contaminants, mainly through the use of microorganisms or plants.

Bioremediation is an attractive alternative in the cleanup of polluted sites because it is often possible to solve the pollution problem satisfactorily by this ecotechnology-based methodology without the hazard and expense involved in removing polluted materials for treatment elsewhere by traditional environmental technology methods. Bioremediation may be applied for both organic wastes and heavy metals, although the methods applied may differ. The practical examples of bioremediation are therefore treated below in two different sections, one for organic compounds and another for heavy metals. The success of any bioremediation technology depends on a number of factors, including site characteristics, environmental factors such as temperature, pH, redox potential, concentrations of nutrients, the contaminant, the presence of microorganisms, and bioavailability.

11.1 BIOAVAILABILITY

Bioavailability is a crucial factor for the application of bioremediation. It is defined as the amount of contaminant present that can be taken up readily by organisms. This section is therefore devoted to a discussion of the factors that determine bioavailability. Bioavailability controls the biodegradation rates for organic contaminants because microbial cells must expend energy to induce the catabolic processes used in biodegradation. If the perceived contaminant concentration is too low, induction will not occur. Ingenious soil microbial populations are typically slow-growing organisms, often exposed to nutrient-poor environments (Koszak and Colwell, 1987). Bioavailability also determines, however, the toxicity for both organic and inorganic contaminants to organisms other than the organisms used for bioremediation. There is therefore an increased need for bioremediation when the bioavailability is high, making bioremediation more attractive.

Three cases can be envisioned that would result from the differing bioavailability of contaminants (Maier, 2000):

1. Biodegradation will not occur because the amount of bioavailable contaminant is insufficient and/or the biodegradation rate for the focal contaminant is too low to justify the energy expenditure to induce biodegradation.
2. Microbial cells may degrade the contaminant at low bioavailable concentrations and/or low biodegradation rates but in a resting or maintenance stage rather than in a growing stage.
3. At a sufficient bioavailability and biodegradation rate, there is enough bioavailable contaminant to induce biodegradation in a growing stage, thus allowing for optimal rates of remediation.

The biodegradability of organic contaminants is highly dependent on chemical structure (Jørgensen et al., 1997). In Section 11.2 we discuss this relationship in more detail and attempt to establish some rough but applicable rules for a very first estimation of the biodegradability of organic compounds.

The bioavailability of heavy metals is also a significant factor for the applicability of bioremediation. Heavy metals are of course not degraded but are removed by bioremediation. The uptake by organisms of heavy metals that determines the overall removal efficiency is controlled entirely by the bioavailable amount of heavy metals. The ecological model presented in Section 11.3 illustrates this strong dependence of the bioavailability of heavy metals.

Most of the factors influencing bioremediation rates have a direct effect on bioavailability. It is therefore not surprising that a research committee of the U.S. Environmental Protection Agency appointed to identify the factors of most critical importance to bioremediation came to the conclusion that bioavailability was the most important factor determining bioremediation (US EPA, 1991). In 1993, a second panel, the National Research Council, released a report entitled *In Situ Bioremediation: When Does It Work?* (NRC, 1993). In this report bioavailability was identified as a major limitation to widespread use of in situ bioremediation, bioremediation based on ecotechnology and ecological engineering principles.

Factors That Affect Bioavailability

Bioavailability is influenced by a number of factors:

1. *Low water solubility* can limit the availability of the substrate to bacterial cells and hence constrain biodegradation (Fogel et al., 1981; Zhang and Miller, 1992). Microbial cells are 70 to 90 percent water, and the food they utilize comes from the water surrounding the cells. Plants take up water for the evapotranspiration needed in the maintenance of their life functions. Therefore, uptake and transport are feasible only for water-soluble material. If first-order biodegradation kinetics is presumed, the biodegradation rate becomes proportional to the concentration in the water phase. This means that components with low water solubility biodegrade very slowly.

There is a clear relationship between the water solubility of an organic compound and the chemical structure that can be utilized to estimate the water solubility (Jørgensen et al., 1997). There is also a relationship between the water solubility and the octanol–water distribution coefficient (Figure 11.1). A high octanol–water distribution coefficient indicates that an organic compound is lipophile and therefore soluble in the organic fraction of soil rather than in water.

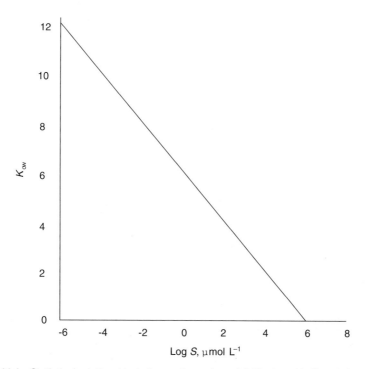

Figure 11.1 Statistical relationship between the water solubility (μmol L^{-1}) and the octanol–water distribution coefficient.

Side reactions may change the water solubility. This is of particular interest for heavy metal ions, which can increase the solubility by formation of complexes with either organic or inorganic compounds. The formation of complexes with humic acid and fluvic acid plays a major role in the solubility of metal ions in soil water. Hydrocarbons, which are frequently found as soil contaminants, have a low water solubility: 2 to 6 μg L^{-1} for pentacyclic aromatic hydrocarbons and n-alkanes of chain length 18 to 30. The solubility decreases with increasing molecular weight (Jørgensen et al., 1997).

The state of the contaminant in combination with water solubility is also of importance. There is evidence that liquid-phase hydrocarbons are more bioavailable than are solid-phase hydrocarbons (Miller, 1995). In practical terms this means that the maximum growth rate occurs in different solubility ranges for liquid-phase (0.01 to 1 mg L^{-1}) and solid-phase components (1 to 10 mg L^{-1}). The degradation is described by the Michaelis–Menten expression discussed in Chapter 4. These limitations are easy to observe because instead of Monod kinetics, linear or first-order biodegradation occurs. This is demonstrated in Figure 11.2, which compares biodegradation of 400 mg L^{-1} octadecane at 25 and 34°C. Usually, the water solubility increases with increasing temperature, but this is not sufficient to explain the very significant differences observed. The more than fivefold increase in biodegradation rates at a temperature difference of only 9°C is explained simply by a change in state from solid phase to liquid phase.

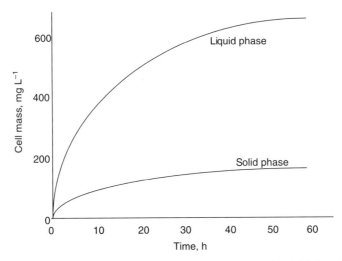

Figure 11.2 Biodegradation of octadecane at 25°C (solid form) and at 34°C (liquid form). The latter biodegradation rate is about five times faster than at a temperature only 9°C lower. The temperature difference could at the most explain a two fold-faster biodegradation rate. The difference in biodegradation rate is therefore explained primarily by the difference in physical state.

Many authors have found that surfactants increase the mineralization rate due to increasing dissolution (see, e.g., Volkering et al., 1995) or decreased interfacial tension (Aronstein et al., 1991). Also, surfactants may provide an additional carbon source that is preferentially utilized by the bacteria. There may, however, also be a negative effect by surfactants due to their toxicity to the bacterial population.

Some organic solvents (e.g., methanol) are miscible with water—they are *co-solvents.* A co-solvent system can increase the solubility and enhance desorption of contaminants (Knox et al., 1993), because the presence of an organic solvent in water makes the resulting solution less polar than water alone.

2. *Sorption on the solid phase of soil* may be a limiting factor for biodegradation of microorganisms and uptake by plants. Several reports suggest that organic chemicals are not mineralized while associated with solid phases (Miller and Alexander, 1991; Greer and Shelton, 1992). Experiments by Robinson et al. (1990) show that sorbed-phase substrate was not degraded and that long-term biodegradation was limited by the slowly desorbing fraction of substrate. These results suggest that rate-limited mass transfer processes (primary desorption) may significantly affect the rate at which a compound is degraded in the presence of a solid phase.

The model presented in Section 11.3 uses the fraction soluble in the soil water of heavy metal ions to determine the uptake (Jørgensen and Bendoricchio, 2001). The sorption is dependent of pH, redox potential, and humus, clay, and sand fractions in the soil. The relationship between these factors and the sorption is included in the model. If the sorption of organic compounds to soil is not known, the soil–water partition coefficient, K_{oc}, can be estimated from the octanol–water partition coefficient:

$$\log K_{oc} = \begin{cases} -0.006 + 0.937 \log K_{ow} & \text{(Brown and Flagg, 1981)} \quad (11.1) \\ -0.35 + 0.99 \log K_{ow} & \text{(Leeuwen and Hermens, 1995)} \quad (11.2) \end{cases}$$

Where the carbon fraction of organic carbon in soil is f, the distribution coefficient, K_D, for the ratio of the concentration in soil and in water can be found as $K_D = K_{oc}f$.

It has been suggested that there are different stages of sorption processes and that newly sorbed material is more labile and therefore more bioavailable than aged sorbed material. Numerous experiments have demonstrated that aging affects bioavailability in soil due to changes in the soil structure, resulting in slower desorption processes. The sorption can frequently be described by either Freundlich or Langmuir adsorption isotherms, expressed, respectively, by

$$a = \begin{cases} kc^b & (11.3) \\ \dfrac{k'c}{c + b'} & (11.4) \end{cases}$$

where a is the concentration in soil, c is the concentration in water, and k, k', b, and b' are constants. Equation (11.3), corresponding to Freundlich adsorption isotherms, is a straight line with slope b in a log-log diagram, since $\log a = \log k + b \log c$ (see Figure 11.3).

The Langmuir adsorption isotherm is an expression similar to the Michaelis–Menten equation. If $1/a$ is plotted versus $1/c$, we obtain a straight line (Figure 11.4), the Lineweaver–Burke plot, as $1/a = 1/k' + b'/k'c$. When $1/a = 0$, $1/c = -1/b'$, and when $1/c = 0$, $1/a = 1/k'$. This plot can be applied to assess whether the expression of the type used in the Michaelis–Menten equation and in Langmuir's adsorption isotherm can explain the available observations. b is often close to 1 and c is small for most environmental problems. This implies that the two adsorption isotherms approach $a/c = k$, and k becomes a distribution coefficient. k for 100 percent organic carbon is usually denoted K_{oc}.

The sorption determines the uptake of organic contaminants by plants as it is expressed in the equation

$$\mathrm{BCF} = \frac{f_{\mathrm{lipid}} K_{\mathrm{ow}}^b}{h f K_{\mathrm{ow}}^a} \qquad (11.5)$$

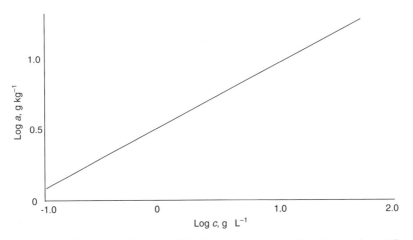

Figure 11.3 Log-log plot of the Freundlich isotherm for adsorption of phenol on soil: $a =$ concentration of phenol in soil; $c =$ concentration in water. The slope, which is $1.15/3 = 0.383$, represents b in equation (11.3) and $\log k = 0.48$, which means that $k = 3.1$. The equation for the plot shown is therefore $a = 3.1c^{0.383}$.

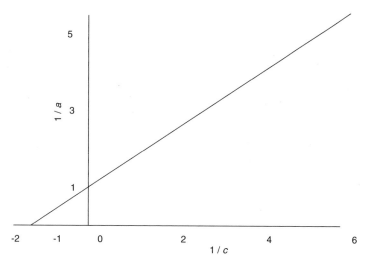

Figure 11.4 Lineweaver–Burke plot where $1/a$ is plotted versus $1/c$. From the plot it is possible to read that $1/b' = -(-1.5)$, or $b' = 2$ and that $k' = 1$ ($1/k' = 1$). This means that it is a Langmuir adsorption isotherm: $a = c/(c + 2)$.

where BCF, the biological concentration factor, expresses the ratio between the concentration in the plant and in the soil; f_{lipid} is the lipid fraction in the plant; f is, as shown above, the fraction of organic carbon in the soil; and a, b, and h are constants. The nominator expresses the fraction of the organic matter that is dissolved in the soil water. h is therefore the constant determined by equations (11.1) and (11.2). If we use equation (11.1), x becomes the antilog to -0.006, or $h = 0.99$. a is 0.935 in equation (11.1), and b is usually close to 1.0. Therefore, equation (11.5) may be reformulated as

$$BCF = \frac{1.01 f_{lipid} K_{ow}^{0.063}}{f} \qquad (11.6)$$

As seen, BCF becomes almost independent of K_{ow} and mainly dependent on the ratio between f_{lipid} and f.

3. *Physical makeup of the soil (pore size distribution)* is important for bioavailability. Bacteria may be excluded from the microporous domain since most bacteria range from 0.5 to 2 μm. If such an exclusion occurs, biodegradation cannot take place in the microporous domain. The degradation rate is therefore limited by the diffusion of solute from the microporous to the macroporous domain. This is obviously of particular importance for organic contaminants with great molecular weight.

In a field situation it is difficult to separate the effects of sorption and micropore exclusion, as some residues are protected from biodegradation by both mechanisms. Steinberg et al. (1987) reported that 1,2-dibromoethane was persistent in field studies for as long as 19 years after application, although 1,2-dibromoethane is readily degraded.

4. *Microbial adaptations* are strategies developed by microorganisms to increase the bioavailability of organic contaminant. One strategy is the development of increased cell affinity for hydrophobic surfaces. This allows the microorganisms to attach to the hydrophobic substrate and adsorb it directly. A second strategy is the production and release of surface-active agents or biosurfactants (Rosenberg, 1986; Fiechter, 1992). These strategies explain the biochemical adaptation of microorganisms to enhance bioavailability to cope with contaminants of a certain toxicity.

Biological adaptation of the microbial population is a generation–by-generation change in properties through a selection of the microorganisms that are best fitted to survive and grow under the prevailing conditions, determined by the environment that included the contaminant. A biological adaptation is widely used to prepare a microorganism population for bioremediation.

A criterion for determining when biodegradation is limited by intraparticle diffusion and sorption and when it is limited by the rate of biodegradation was established under simplifying assumptions by Y. P. Chung et al. (1993). The limiting mechanism is defined by the magnitude of the dimensionless number q:

$$q = R \left[\frac{p + K(1 - p)/k}{D} \right]^{0.5} \tag{11.7}$$

where R is the radius of the soil particles, p the porosity of the soil, K the adsorption partition coefficient $= dK_d/(1 - p)$ [where d is the bulk density of soil (g cm^{-3}) and K_d is the partition coefficient for soil water; see equation (11.2)], k is the first-order reaction coefficient for biodegradation (1/h), and D is the effective diffusion coefficient in the soil pores. If q is small, the diffusional resistance is not important compared with the rate of biodegradation; a large value of q indicates that strong adsorption or slow diffusion is an important determinant.

11.2 BIODEGRADATION

It is possible to describe biodegradation quantitatively by relating growth to the substrate concentration by the Michaelis–Menten expression:

$$\frac{dc}{dt} = - \frac{dB}{Y \, dt} = - \frac{\mu_{max} B c}{Y(K_m + c)} \tag{11.8}$$

where c is the concentration of the compound considered, Y the yield of biomass B per unit of c, B the biomass concentration, μ_{max} the maximum specific growth rate, and K_m the half-saturation constant. If $c \ll K_m$, the expression is reduced to a first-order reaction scheme:

$$\frac{dc}{dt} = -K'Bc \qquad (11.9)$$

where $K' = \mu_{max}/YK_m$. B is determined in nature by the environmental conditions. In aquatic ecosystems, for example, B is highly dependent on the presence of suspended matter. Therefore, under certain conditions, B may be considered a constant that reduces the rate expression to

$$\frac{dc}{dt} = -kc \qquad (11.10)$$

An indication of k in units of $time^{-1}$ can therefore be used to describe the rate of biodegradation. If the biological half-life time is denoted $t_{1/2}$, we get the following relation:

$$\ln 2 = 0.693 = kt_{1/2} \qquad (11.11)$$

This implies that the biological half-life time can also be used to indicate the biodegradation rate, presuming a first-order reaction scheme.

Biodegradation in waste treatment plants is often of particular interest, in which case the percent ThOD (theoretical oxygen demand) may be used. It is defined as the 5-day BOD (the biological oxygen demand expressed as mg oxygen per liter) as a percentage of the theoretical BOD. It may also be indicated as the BOD_5 fraction. For instance, a BOD_5 fraction of 0.7 means that BOD_5 corresponds to 70 percent of the ThOD. It is, however, also possible to find an indication of percentage removal in an activated sludge plant.

In some cases, however, the biodegradation is very dependent on the concentration of microorganisms as expressed in equations (11.8) and (11.9). Therefore, K' indicated in mg g dry wt^{-1} day^{-1} in equation (11.9) will in many cases be more informative and correct.

In the microbiological decomposition of xenobiotic compounds, an acclimatization period from a few days to one to two months should be foreseen before the optimum biodegradation rate can be achieved. We distinguish between primary and ultimate biodegradation. Primary biodegradation is any biologically induced transformation that changes the molecular integrity. Ultimate biodegradation is the biologically mediated conversion of an organic compound to inorganic compounds and products associated with complete and normal metabolic decomposition. The biodegradation rate is expressed by application of a wide range of units:

1. As a first-order rate constant (day^{-1})
2. As a half-life time (days or hours)

3. In milligrams per gram of sludge per day (mg g^{-1} day^{-1})
4. In milligrams per gram of bacteria per day (mg g^{-1} day^{-1})
5. In milliliters of substrate per bacterial cell per day (mL cell^{-1} day^{-1})
6. In milligrams COD (chemical oxygen demand, i.e., the oxygen demand measured by means of a dichromate as oxidator) per gram of biomass per day (mg g^{-1} day^{-1})
7. In milliliters of substrate per gram of volatile solids including microorganisms (mL g^{-1} day^{-1})
8. As BOD_x/BOD_∞ [i.e., the biological oxygen demand in x days compared with complete degradation (unitless), termed the BOD_x coefficient].
9. As BOD_x/COD [i.e., the biological oxygen demand in x days compared with complete degradation, expressed by means of COD (unitless)]

The biodegradation rate in water or soil is difficult to estimate, because the number of microorganisms varies by several orders of magnitude from one type of aquatic ecosystem to the next and from one type of soil to the next. Artificial intelligence has been used as a promising tool to estimate this important parameter (Kompare, 1995). However, a (very) rough first estimation can be made on the basis of the molecular structure and biodegradability. The following rules can be used to establish these estimations:

1. Polymer compounds are generally less biodegradable than monomer compounds: 1 point for a molecular weight >500 and ≤1000, 2 points for a molecular weight >1000.
2. Aliphatic compounds are more biodegradable than aromatic compounds: 1 point for each aromatic ring.
3. Substitutions, especially with halogens and nitro groups, will decrease the biodegradability: 0.5 point for each substitution, although 1 point if it is a halogen or a nitro group.
4. Introduction of a double or triple bond will generally mean an increase in the biodegradability (double bonds in aromatic rings are of course not included in this rule): −1 point for each double or triple bond.
5. Oxygen and nitrogen bridges (—O—and—N—(or =)] in a molecule will decrease the biodegradability: 1 point for each oxygen or nitrogen bridge.
6. Branches (secondary or tertiary compounds) are generally less biodegradable than the corresponding primary compounds: 0.5 point for each branch.

Find the number of points and use the following classification:

≤1.5 *points:* the compound is readily biodegraded. More than 90 percent will be biodegraded in a biological treatment plant.

2.0–3.0 *points:* the compound is biodegradable. Probably about 10 to 90 percent will be removed in a biological treatment plant. BOD_5 is 0.1 to 0.9 of the theoretical oxygen demand.

3.5–4.5 *points:* the compound is slowly biodegradable. Less than 10 percent will be removed in a biological treatment plant. BOD_{10} is ≤ 0.1 of the theoretical oxygen demand.

5.0–5.5 *points:* the compound is very slowly biodegradable. It will hardly be removed in a biological treatment plant and a 90 percent biodegradation in water or soil will take six months or more.

≥ 6.0 *points:* the compound is refractory. The half-life time in soil or water is counted in years.

The estimation obtained by the method must be considered as a very first, very coarse estimation, which gives a relative but not an absolute value. The rate will, furthermore, correspond to the rate after an adaptation phase, while application of well-adapted microorganism strains will probably give a higher rate.

EXAMPLE

A soil with 5 percent organic carbon is contaminated by a pentacyclic aromatic hydrocarbon of molecular weight 260 with a water solubility of 4 μg L^{-1}. The total concentration in the contaminated soil is 5 mg L^{-1}.

(a) Estimate the octanol–water distribution coefficient.

(b) Estimate the distribution soil water of the pentacyclic hydrocarbon.

(c) What is the concentration of the pentacyclic aromatic hydrocarbon in the soil water?

(d) Estimate the biodegradability of the compound.

(e) The half-life time of the compound has been found to be 2000 hours. Find the first-rder rate coefficient for biodegradation. Is it in accordance with the estimated biodegradability?

(f) It is presumed that only the pentacyclic aromatic hydrocarbon in the soil water is biodegraded, but that desorption takes place rapidly. Find the time for the decomposition of 99 percent of the total amount of pentacyclic aromatic hydrocarbon (i.e., the amount in the soil water and the amount adsorbed to soil).

(g) Which concentration should be expected in plants growing in the described pentacyclic aromatic hydrocarbon–contaminated soil if it is presumed that the plants contains 8 percent lipids?

Solution: (a) Octanol–water distribution coefficient: log K_{ow} = 7.5 (from Figure 11.1 for water solubility S = 4 μg L^{-1}/260 μg μmol^{-1} = 0.0154 μmol L^{-1}).

(b) log K_{oc} = $-0.006 + (0.937 \times 7.5)$
= 7.02 [from equation (11.1)]
$K_D = K_{oc} f = 10^{7.02} \times 0.05 = 5 \times 10^5$

(c) The amount in the soil water is very small. We can therefore find the concentration in soil water directly as 5000 μg L^{-1}/(5 $\times$ 10^5) = 0.01 μg L^{-1}.

(d) Five rings give 5 points: 90 percent decomposition will require at least six months.

(e) ln 2 = $kt_{1/2}$ = $k \times 20,000$; k = 3.45 $\times$ 10^{-5} h^{-1} [from equation (11.11)], reasonably good accordance between the two values.

(f) 10^{-5} mg L^{-1} $\times$ 3.45 $\times$ 10^{-5} h^{-1} = 3.45 $\times$ 10^{-10} mg L^{-1} h^{-1} or 3 $\times$ 10^{-6} mg L^{-1} yr^{-1}. The concentration in soil water is constant. The decomposition rate will therefore also be constant: 5 mg L^{-1}/ (3 $\times$ 10^{-6} mg L^{-1} yr^{-1}) $\approx$ 1,660,000 years.

(g) The biological concentration factor BCF = $1.01 f_{lipid}[K_{ow}]^{0.063}$/ f = $(1.01 \times 8/5)(10^{7.5})^{0.063} \approx 5$ [from equation (11.6)]. Therefore, c_{plant} = BCF $\times$ c_{soil} = 5 $\times$ 5 mg L^{-1}. Assuming the specific gravity of plants, wet weight, is approximately 1 kg L^{-1}, the plant concentration of pentacyclic aromatic carbon is 25 mg kg^{-1} wet weight.

11.3   UPTAKE OF HEAVY METALS BY PLANTS

Plants are contaminated by heavy metals originating from deposition of heavy metals (waste sites), air pollution, the application of sludge from municipal wastewater plant as a soil conditioner, and the use of fertilizers. Bioremediation by plants is frequently used to decontaminate harbor sludge before it is reused as soil (Miljøstyrelsen, 2002). Harbor sludge often contains very high concentrations of heavy metals, particularly TBT (3-butyltin), which is used extensively as an algicide.

The uptake of heavy metals from municipal sludge by plants has been modeled previously (see Jørgensen, 1993). The model can be described briefly as follows: Depending on the soil composition, it is possible to find a distribution coefficient (i.e., the fraction of the heavy metal that is dissolved in the soil water relative to the total amount) for various heavy metal ions. The distribution coefficient was found by examination of the dissolved heavy metals relative to the total amount for several different types of soil. Correlation between pH, the concentration of humic substances, clay, and sand in the soil, on the one hand, and the distribution coefficient, on the other, was also determined. The uptake of heavy metals was considered a first-order reaction of the dissolved heavy metal.

Wood and Shelley (1999) use acid-volatile sulfide and organic carbon to describe the metal-binding capacity of sediments in constructed wetlands. This will give approximately the same ratio "bound" to "bioavailable" heavy metals as the correlation above. The basic idea is the same, namely, to find easily measurable soil properties that determine the metal-binding capacity, which is crucial for the uptake of heavy metals by plants. In addition to the uptake from soil water, the model presented below considers (1) direct uptake from atmospheric fallout onto the plants, and (2) other sources of contamination, such as fertilizers and the long-term release of heavy metal bound to the soil and the unharvested parts of the plants.

CASE STUDY
Recycling Heavy Metals in Agricultural Cropland

Published data on lead and cadmium contamination in agriculture are used to calibrate and validate a model that is intended to be used for (1) a more generally applicable risk assessment for the use of fertilizers and sludge that contain heavy metals as contaminants, (2) a risk assessment for the use of plants harvested from a waste site, and (3) a determination of the possibilities of removal of heavy metals by plants that have a particular ability to take up heavy metals. This last intended application of the model makes it useful for the determination of the result of bioremediation.

Figure 11.5 shows a conceptual diagram of the Cd version of the model. As can be seen, it has four state variables: Cd-total, Cd-soil, Cd-detritus, and Cd-plant. An attempt was made to use one or two state variables for cadmium in the soil, but to get acceptable accordance between data and model output, three state variables were needed. This can be explained by the presence of several soil components that bind the heavy metal differently (see EPA Denmark, 1979; Christensen, 1981, 1984; Chubin and Street, 1981; Hansen and Tjell, 1981; Jensen and Tjell, 1981). Cd-total covers the cadmium bound to minerals and to more-or-less refractory material; Cd-soil covers the cadmium bound by adsorption and ion exchange; Cd-detritus is the cadmium bound to organic material with a wide range of biodegradability.

The forcing functions are airpoll, Cd-air, and Cd-input. The atmospheric fallout is known, and the allocation of this source to the soil (airpoll) and to the plants (Cd-air) is according to Hansen and Tjell (1981) and Jensen and Tjell (1981). Cd-input covers the heavy metal in the fertilizer, sludge, compost, and other added sources, and as seen from the equations (Table 11.1), it comes as a pulse at day 1 and afterward with a frequency of every 180 days. The yield corresponds to the part of the plants that is harvested. This is also a

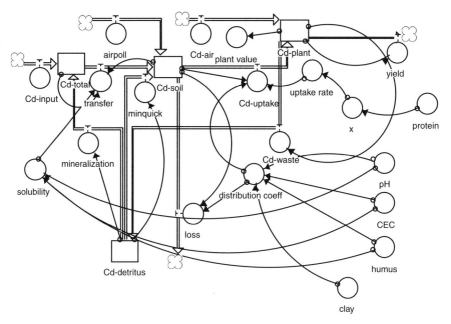

Figure 11.5 Conceptual diagram of the model used to evaluate the fate of metals in agricultural land. Boxes show state variables, double-line arrows give flows, circles give functions, and single-line arrows show feedback mechanisms. Model based on STELLA language. (From Jørgensen and Bendorrichio, 2001; copyright 2001; reprinted with permission from Elsevier Science.)

pulse function at day 180, and afterward with an occurrence every 360 days. As seen from Table 11.1, it is in this case 40 percent of the plant biomass.

The loss covers transfer to the soil and groundwater below the root zone. It is expressed as a first-order reaction with a rate coefficient dependent on the distribution coefficient that is found from the soil composition and pH, according to the correlation found by Jørgensen (1975). Furthermore, the rate constant is dependent on the hydraulic conductivity of the soil. The constant 0.01 in Table 11.1 reflects the dependence of the hydraulic conductivity.

The transfer from Cd-total to Cd-soil indicates the slow release of cadmium due to a slow decomposition of the more-or-less refractory material to which cadmium is bound. The cadmium uptake by plants is expressed as a first-order reaction, where the rate is dependent on the distribution coefficient, as only dissolved cadmium can be taken up. It also depends on the plant species. As will be seen, the uptake is a step function that here (grass) is 0.0005 during the growing season, and, of course, zero after the harvest and until the next growing

Table 11.1   Model equations (STELLA format) for metal fate model shown in Figure 11.5

Cd-detritus = Cd-detritus + dt * (Cd-waste − mineralization − minquick)
INIT(Cd-detritus) = 0.27
Cd-plant = Cd-plant + dt * (Cd-uptake − yield − Cd-waste + Cd-air)
INIT(Cd-plant) = 0.0002
Cd-soil = Cd-soil + dt * (−Cd-uptake − loss + transfer + minquick + airpoll)
INIT(Cd-soil) = 0.08
Cd-total = Cd-total + dt * (Cd-input − transfer + mineralization)
INIT(Cd-total) = 0.19
airpoll = 0.0000014
Cd-air = 0.0000028 + STEP(−0.0000028,180) + STEP(+0.0000028,360) +
    STEP(−0.0000028,540) + STEP(+0.0000028,720) + STEP(−0.0000028,900)
Cd-input = PULSE(0.0014,1,180)
Cd-uptake = distributioncoeff * Cd-soil * uptake rate
Cd-waste = PULSE(0.6 * Cd-plant,180,360) + PULSE(0.6 * Cd-plant,181,360)
CEC = 33
clay = 34.4
distributioncoeff = 0.0001 * (80.01− 6.135 * pH − 0.2603 * clay − 0.5189 *
    humus − 0.93 * CEC)
humus = 2.1
loss = 0.01 * Cd-soil * distributioncoeff
mineralization = 0.012 * Cd-detritus
minquick = IF TIME_180 THEN 0.01 * Cd-detritus ELSE 0.0001 * Cd-detritus
pH = 7.5
plantvalue = 3000 * Cd-plant/14
protein = 47
solubility = 10^(+6.273 − 1.505 * pH + 0.00212 * humus + 0.002414 * CEC) *
    112.4 * 350
transfer = IF Cd-soil<solubility THEN 0.00001 * Cd-total ELSE 0.000001 *
    Cd-total
uptake rate = x + STEP(−x,180) + STEP(x,360) + STEP(−x,540) + STEP(x,720)
    + STEP(−x,900)
x = 0.002157 * (−0.3771 + 0.04544 * protein)
yield = PULSE(0.4 * Cd-plant,180,360) + PULSE(0.4 * Cd-plant,181,360)

season starts. Cd-waste covers the transfer of plant residues to detritus after harvest. It is therefore a pulse function, which here is 60 percent of the plant biomass, as the remaining 40 percent has been harvested.

Cd-detritus covers a wide range of biodegradable matter, and the mineralization is therefore accounted for in the model by the use of two mineralization processes: one to Cd-soil and one to Cd-total. The first one is rapid and is given a higher rate for the first 180 days, as the addition of municipal sludge in this case is at day 0. The second one occurs at about the same rate, but as the cadmium is transferred to the Cd-total, the slow release rate is considered by the very slow transfer from Cd-total to Cd-soil.

Data from Hansen and Tjell (1981) and Jensen and Tjell (1981) were used for calibration and validation of the model. It was in this phase of the modeling procedure that it was revealed that three state variables for heavy metals in soil were needed to get acceptable results. It was particularly difficult to obtain the right values for heavy metal concentrations in the plants the second and third year after municipal sludge had been used as a soil conditioner. This use of models may be called *experimental mathematics* or *modeling*, where simulations with different models are used to deduce which model structure should be preferred. The results of experimental mathematics must, of course, be explained by examination of the processes involved and here can be referred to the references given above.

The results of the validation phase are shown in Figure 11.6. Accordance between observations and model predictions is reasonably good. It is apparent from the validation that the developed model can explain the observations. A wider use of the model would require that still more data from experiments with many plant species be used to test the model, including plant species with a particular ability to take up heavy metals with the aim of removing heavy metals from contaminated soil by bioremediation, using plants.

It can be concluded from these results that the model structure must account for at least three state variables for the heavy metal in soil to cover the ability of different soil components to bind the heavy metal by various processes. The model has been validated on the basis of three years of experiments and measurements, and it was clear from the model exercises that the atmospheric fallout and heavy metal in the plant residues were significant. Translocation of the heavy metal to various parts of the plant was not considered in the model, and this would be a natural next step to include in the model, as it is important to distinguish heavy metal concentrations in various parts of the plants, and when removal of heavy metals by plants is used for bioremediation.

The problem modeled is very complex, and many processes are involved. On the other hand, an ecotoxicological management model should be somewhat simple and not involve too many parameters. The model can obviously be improved, but it gives at least a first rough picture of the important factors determining the contamination of plants and the possibilities for using bioremediation. For the most part it is not possible to get very accurate results with toxic-substance models, but on the other hand, as we want to use somewhat large safety factors, the need for high accuracy is not pressing.

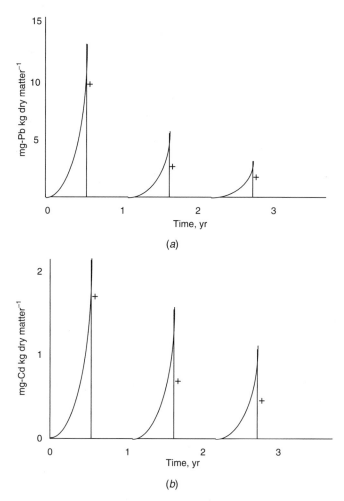

Figure 11.6 Simulation results for plant metal concentrations from the metal fate model in Figure 11.5. The model was validated by use of (*a*) the lead concentration as a function of time for salad plants and (*b*) cadmium concentration as a function of time for red clover at third and fourth harvests. A plus sign indicates actual observations, and the curve gives corresponding model predictions. (After Jørgensen and Bendorrichio, 2001.)

11.4 REMOVAL OF ORGANIC TOXIC COMPOUNDS BY BIOREMEDIATION

Prior to the twentieth century, naturally occurring biodegradation processes were largely adequate in recycling organic materials on the surface of Earth. A number of different biochemical processes were able to cope with the organic substances that rarely accumulated to cause environmental pollution problems. However, in the twentieth century, humankind achieved the ability to synthesize and disseminate large quantities of industrial chemicals. The per capita use of chemicals and the human population increased simultaneously, causing a dramatic increase in the overall production of industrial chemicals. Environmental pollution problems have become particularly prominent due to concomitant improvements in analytical chemistry, epidemiology, and toxicological expertise.

Thus, the rates of production and dispersal of industrial and other toxic substances have completely outpaced naturally occurring biodegradation processes. An unsteady state has resulted, threatening human health and ecosystem function. Development of bioremediation becomes a pivotal ecotechnological issue because failure of ecosystems may jeopardize human health and ecosystem integrity.

The usual procedure for bioremediation is covered by the following six steps:

1. A map showing the geographical distribution of the concentration of the contaminant is developed by intensive use of analytical chemistry. Analyses of pore water are often applied to establish an environmental risk assessment.
2. Laboratory tests are conducted to verify the applicability of bioremediation.
3. Calculations (often by development of a model) are performed to assess the feasibility of the method in situ.
4. An adapted strain of the microorganisms is produced in a sufficient amount.
5. This is applied in situ. If groundwater is high, it is usually lowered. Injection pits are introduced into the soil and air is blown in to reinforce the decomposition of organic matter. When the pollutants are chlorinated compounds, a mixture of methane and air may be used.
6. The results are followed by use of a wide spectrum of analytical methods, including radioactive tracers, detection of intermediary metabolites, and respiration rate.

Ecotechnological Approaches with Microorganisms

At the beginning of the twenty-first century, experience from many case studies of bioremediation using adapted microorganisms has shown that this method can be used successfully to remove organic contaminants. A few successful ecotechnological applications are listed below to illustrate the spectrum of possibilities: (1) dechlorination of chlorinated aliphatic compounds (McCarty, 1997), (2) removal of various organic compounds, including toluene and phenol from aerobic aquifers (Steffan et al., 1999), (3) use of herotrophic microorganisms to remove petroleum components from aerobic soil (Wilson and Jawson, 1995), and (4) removal of aromatic fuel components (toluene and others) under anaerobic conditions (Chapelle et al., 1996).

Biobarriers

A particularly important application of bioremediation is protection of groundwater against toxic compounds. Contamination of groundwater is a serious concern for public health and environmental quality. The problem is commonly manifested as a contaminant plume migrating in the direction of groundwater flow from a point source. Containment of the contaminant plume is important in preventing further migration. Current containment methods include sheet piling and grout curtains. These abiotic barriers require extensive physical manipulation of the site, excavation, and backfilling, and are expensive to construct. Biobarriers (Figure 11.7) are an alternative approach that involves the use of microbial biomass produced in situ to manipulate groundwater flow by biodegradation of the contaminant at a suitable rate (Cunningham et al., 1997).

When bioremediation is applied, an environmental risk assessment is often set up for contamination of the adjacent groundwater sources. The concentration in the pore water is applied in the environmental risk assessment (Miljøstyrelsen, 2000). Generally, it is strongly recommended that all consequences of bioremediation be considered before implementation of the method (Miljøstyrelsen, 2002).

Phytoremediation: Plants for Removing Toxics from Soil

Plants can in some cases be used to remove organic toxic matter. They can efficiently take up organic substances that are moderately hydrophobic, with a K_{ow} value from about 0.5 to 3.0. Hydrophobic substances exceeding a K_{ow}

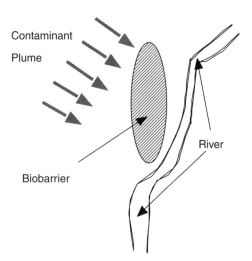

Figure 11.7 Use of a biobarrier to prevent a groundwater contaminant plume from polluting an adjacent river.

of 3.0 bind so strongly to the soil and the roots that they are not easily taken up within the plants. Plants are also able to enhance microbiological decomposition rates of organic matter because the roots pump oxygen to the root zone and thereby create a good environment for the microbiological activity. It is often important to add fertilizers to increase the plant biomass and thereby the biological activity in the root zone. Lin and Mendelssohn (1998) describe the successful removal of crude oil in a marsh by the use of fertilization of *Spartina alterniflora* and *S. patens*. The oil degradation rate in the soil was enhanced significantly by the application of fertilizer in conjunction with the presence of plants. Miljøstyrelsen (2002) reports that polycyclic aromatic hydrocarbons (PAHs) can be effectively removed provided that the roots reach the contaminated layers. Forty percent removal during two years was reported for soil contaminated at about 10 ppm by PAHs.

Plant productivity may also be increased by the addition of ion-exchange substrate (1 percent is sufficient) to completely depleted soil and barren sand. Soldatov et al. (1997) conclude that the addition of ion exchange substrates can be an efficient means for remediation of destroyed soils and fruitless rocks.

Soils contaminated simultaneously by heavy metals and toxic organic components are very difficult to remedy because the presence of heavy metals causes a distinct decrease in organic matter decomposition rate (see Zwolinski, 1994). It implies that it is first necessary to remove the heavy metals by the methods presented in the next section, followed by the removal of toxic organic components.

11.5 REMOVAL OF HEAVY METALS BY BIOREMEDIATION

Three different variations of bioremediation used for the removal of heavy metals are available: use of algae, use of microbial biotraps, and use of metal-tolerant plants as bioaccumulators. All three possibilities may be used to recover contaminated soil or to treat polluted water that may also originate from contaminated soil. This chapter is complemented by a discussion of the restoration of lakes contaminated by heavy metals in Chapter 6 and restoration of disturbed land in Chapter 12. A few methods that are closer related to the bioremediation of soil are mentioned in this section.

All three possibilities are attractive alternatives to environmental technological methods based on physicochemical processes, such as ion exchange, liquid extraction, precipitation, and crystallization. Bioremediation processes are in most cases considerably less expensive than physicochemical processes, particularly if a high concentration of heavy metals can be achieved in the biosorbent.

Recovery and reuse of metals is possible by bioremediation methods, such as acid extraction, but the cost of the recovery process often exceeds the value of the metals. All processes removing heavy metals from water or soil, as well as processes based on environmental technology, only transfer the heavy metals into a more concentrated and more accessible form, which may facilitate deposition or recovery. Consequently, the problem is not solved fully before the problem of recovery or deposition is solved in a satisfactory manner.

Use of Algae

Marine algae possess large quantities of biopolymers, including polysaccharides, uronic acids, and sulfated polysaccharides, which can bind heavy metals. Their metal uptake capacity is therefore high. The marine alga *Sargassum* can take gold up to as much as 40 percent of the algal dry weight (Kuyucak and Volesky, 1989). It is a further advantage that seaweeds are large enough to facilitate immobilization. They may, for instance, be applied in packed columns without pretreatment.

Biosorption in algae has been attributed primarily to the cell wall, where both electrostatic attraction and formation of complexes play a role. It is therefore important to know the cell-wall characteristics of different groups of algae in order to select promising species for biosorption (Schiewer and Volesky, 1995, 1997). The ion-exchange properties of the sulfated polysaccharides, which form a major component in seaweeds, are also of importance (Schiewer and Volesky, 1995) and may often explain a significant part of the biosorption observed.

Langmuir adsorption isotherms can be used to describe the uptake (Holan et al., 1993):

$$M = \frac{BK(M)}{1 + K(M)} \tag{11.12}$$

where M is the concentration of the heavy metal in the algal material expressed, for instance, in mEq g^{-1}, B is the total number of binding sites (dependent on the alga selected), and $K(M)$ expresses the affinity of the metal to the alga, dependent on the alga and the metal.

Soil and wastewater may contain more than one type of heavy metal competing over the same binding sites. A multi-Langmuir adsorption isotherm is applied in that case. If two-component competition is considered (metal M_1 and metal M_2), the multi-Langmuir adsorption isotherm takes the form

$$M = \frac{BK(M_1)}{1 + K(M_1) + K(M_2)} \tag{11.13}$$

The capacity at low pH is usually lower than at high pH, as demonstrated in Figure 11.8. This may be utilized to recover heavy metals removed by marine algae by application of an acid extraction. Notice that a rather high

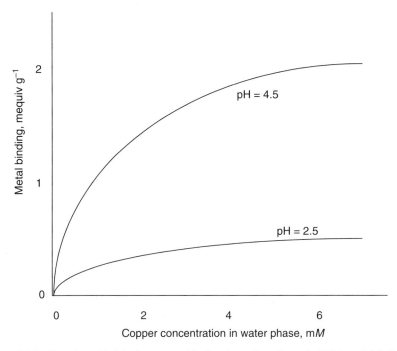

Figure 11.8 Experimental data for copper binding to marine algae at pH 2.5 and 4.5. Notice that a rather high capacity is observed at pH 4.5: 2 mequiv g^{-1} or 64 mg-Cu g^{-1}.

capacity is observed at pH 4.5: 2 milliequivalents (mEq) g^{-1} (equivalent to 64 mg g^{-1} for copper). The metal binding to cell surfaces is a rapid process. A large percentage of the binding is reached in a few minutes, while complete equilibrium may be attained in a few hours. The rate is generally proportional to the displacement from equilibrium and to the external surface area. A more detailed model of the kinetics of metal binding includes film diffusion from the solution to the surface, pore diffusion through the particles, and chemical binding processes.

The practical application of marine algae for metal removal may be realized in situ by either a batch process or a packed-bed column. In both cases it is advantageous to pretreat (reinforce and immobilize) the biosorbent particles. Reinforcement can be achieved by chemical cross-linking with formaldehyde or gluralaldehyde. This technique successfully reduces the swelling and leaching of biomolecules (Holan et al., 1993). Immobilization is possible by treatment of the biosorbent by alginate, silica, and polyacrylamide (Holbein, 1990).

Microbial Biotraps

Heavy metals can also be removed by *microbial biotraps* using mainly adsorption and ion-exchange processes by dead and living biomass (White et al., 1995). This is also observed in wastewater treatment systems, where more than 50 percent removal of heavy metals may occur. The *meander system* passes drain water containing heavy metals through various channels containing bacteria, including cyanobacteria and algae. Metals may be removed by an efficiency greater than 99 percent (Erlich and Brierley, 1990) This process will probably utilize precipitation and entrapment of particulate matter in addition to adsorption and ion exchange. The metal is concentrated in the sediment.

Many microorganisms produce extracellular polymers composed mostly of polysaccharides and sulfated polysaccharides in the form of a slime layer. This layer binds metal ions tightly by an ion exchange–like process. Hydrogen sulfide is produced by sulfate-reducing bacteria. As solubility products of most metal sulfides are extremely low, anaerobic biomass decay causing sulfate reduction can be utilized to remove toxic metals in form of sulfides (Brierley, 1995). Other precipitation processes that may be utilized by this technology are the formation of insoluble uranium phosphate complexes (Macaskie, 1991) and the formation of copper phosphate and copper oxalate (Crusberg et al., 1994).

Plants as Bioaccumulators

Recently, it has been shown that a few plant species (e.g., the alpine penny-cress) can be used as *bioaccumulators* (hyperaccumulators). When growing in contaminated soil, they are able to achieve a heavy metal concentration many times higher (10 to 100 times) than normal plant species. If the plants are harvested, the corresponding amount of heavy metal is removed from the soil. If the soil is only slightly contaminated with heavy metal, this method may be able to remediate the soil in a few years; thus, the method is very attractive because of its low costs. For very contaminated soil, the number of years needed for complete recovery of the soil will probably be too high. By watering the plants carefully with a diluted ethylenediaminetetraacetic acid (EDTA) solution, it is possible to enhance the uptake of heavy metals in the plants. Jørgensen (1993) reports a removal of 11.5 percent lead per harvest of plants watered by 0.02 M EDTA solution at an initial concentration of 380 mg Pb per kilogram of soil (dry matter). Several plant species release chelating ligands and enzymes into the soil and thereby accelerate the removal rate. Chelating agents will also reduce the toxicity of heavy metal ions.

The economy of the various alternatives offered by this method determines the practical applicability of the method. Hyperaccumulators are usually plants that are slow to grow and therefore slow to remove heavy metals as mg m^{-2} yr^{-1}. Fast-growing poplar trees are not as good hyperaccumulators as many other species, but they are promising phytoremediators due to their fast growth.

The aquatic plants *Salvinia* and *Spriodela* have been used to remove chromium and nickel from wastewater. A significant removal efficiency in the concentration range 1 to 8 ppm, although fluctuations have been reported.

Fungal biomass has recently been proposed for removal of heavy metals. The amounts of metal adsorbed per unit weight are several mg g^{-1} for cadmium, copper, lead, and nickel (Kapoor and Viraraghavan, 1995). Sharma and Gaur (1995) have been able to use duckweed, a small vascular plant, for removal of zinc, lead, and nickel.

11.6 CONCLUSIONS

From this brief review of methods to remove in situ toxic substances from water and soil, it can be concluded that ecotechnology offers many possibilities for restoration of soil and water that have been contaminated with toxic substances. Bioremediation can be an alternative for cleaning up polluted soils contaminated by both organic materials and metals in many situations. The prime advantage is that it eliminates the necessity of removing polluted materials from the site, which in essence is simply a shell game—taking pollutants from one location and moving them to another location. Plants and microbes solve the problem on the site. The experience today appears sufficient to utilize these methods in large-scale projects.

12

MINE AND DISTURBED LAND RESTORATION

There is a great deal of our planet, both in industrialized and developing countries, where we have essentially converted the landscape to barren landscape or nearly so—strip mine land for mineral extraction, pits and mounds of waste from previous mines or industrial activity, abandoned industrial sites (often referred to as *brownfields*), and even lands so affected by air and soil pollution that the landscape is barren. There are hundreds of thousands of square kilometers of stripped mine land in the world alone. In total, including agriculture, it has been estimated that humans have transformed between one-third and one-half of Earth's surface (Vitousek et al., 1997). To be able to restore these lands to functional ecosystems is one of the greatest challenges to ecological engineering. As stated by Bradshaw and Hütttl (2001): "What is clear is that a narrow engineering approach is not enough" [for disturbed land restoration]. In the future a more biological approach to restoration will be necessary. If this leads to a more complex programme of work, that is the price that must be paid for the creation of ecosystems that function properly in all respects and that are fully self-sustaining. The ecological engineering that is required is much more subtle and complex than that previously carried out by reclamation teams, which have to broaden their approach and indeed their expertise."

Compiled volumes that emphasize the science and ecological restoration of disturbed lands in general and mined lands specifically are given for forest recovery after mining, grazing, and severe acid deposition in central Europe (Fanta, 1994; Seip et al., 1994; Kilian and Fanta, 1998), remediation of damaged ecosystems in central and eastern Europe (Mitsch and Mander, 1997), terrestrial rehabilitation in general (Wali, 1992), and post–coal mine ecology and restoration (Hüttl and Bradshaw, 2001). Pioneering works on this subject were on bioengineering for land reclamation published in Germany in the 1960s and republished in English 20 years later (Schiechtl, 1980), a classic paper by Bradshaw (1983), and several papers in books edited by Cairns (1980, 1988b) and Wali (1992). The general fields of restoration/rehabilitation, most of which concerns terrestrial ecosystem restoration, was treated in well over 150 books and many more reports in the 1970s through 1990s (Wali, 1999).

12.1 TERMINOLOGY

Terminology in disturbed land recovery is both precise and misused. Bradshaw (1992, 1997) offers two competing definitions of the word *restoration* of terrestrial systems: (1) the act of restoring to a former state or position or to an unimpaired or perfect condition, and (2) reinstating the original functions of the soil in full measure. These should be contrasted with *rehabilitation,* which means partial restoration of ecosystem structure and function or development of ecosystems that are close approximations to what was originally present, or *reclamation,* which implies making the land fit for cultivation. If the restoration involves reinstating the soil function of a disturbed landscape, the restoration of soil from a degraded site to the original soil is illustrated in Figure 12.1. The diagram shows a recovery of both the function and structure of the soil system as true system restoration. Partial restoration to some intermediate point is rehabilitation, while reclamation takes the sys-

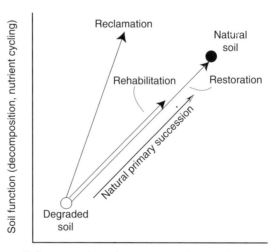

Figure 12.1 Contrasting approaches to the restoration of terrestrial soils in land degraded by mining. (Redrawn from Bradshaw, 1992, 1997.)

tems in an entirely different direction, with more function restored but not necessarily full structure.

12.2 ECOLOGY OF TERRESTRIAL RESTORATION

Restoration of terrestrial soil systems, and hence terrestrial ecosystem restoration of disturbed lands, involves restoration of the following in disturbed soils (Bradshaw, 1997): (1) soil organic matter, (2) soil nitrogen capital, (3) other available nutrients, and (4) nutrient cycling properties. Nature, of course, can overcome these problems on its own, given enough time. Soil organic matter and nutrient storage in the soil, of course, will accumulate with time, due partially to the symbiosis between microbes and vegetation, at least whenever vegetation gets established. Vegetation, in turn, depends on suitable soil moisture, nutrients, organic matter, and soil physics. The ecological way in which a barren landscape restores itself should be the model for the ecological engineer; it is with the time over which it happens that the ecological engineer can be of some assistance.

One of the key nutrients that is in poor supply in many surface mines is nitrogen. There is often essentially none in subsoils and minerals that are left on the surface of excavations. At first, some nitrogen comes from atmospheric sources, particularly with increased nitrate-nitrogen concentrations, due to anthropogenic emissions to the atmosphere. This amount could be on the order of 10 to 30 kg ha^{-1} yr^{-1}. Soil development would continue with nitrogen fixers such as *Rhizobium* and other actinomycetes associated with plant roots, fixing nitrogen from atmospheric gas and storing it directly in plants such as legumes, which in turn release it to the soil upon decay. Nitrogen-fixing species can accumulate nitrogen at rates around 50 to 150 kg ha^{-1} yr^{-1}. Transpiration, shading, and litter accumulation will all change the physics and hydrology of the soil, including soil moisture. Normally, it would take nature decades, if not a century or two, to restore a viable ecosystem on mined or severely disturbed lands.

In addition to the changes in soil biogeochemistry and the development of proper soil microbiota, the most important component to terrestrial restoration is the availability of vegetation seeds and other propagules (Wali, 1999). These seeds and other propagules can, of course, come from natural pathways through the atmosphere, biota, or hydrology; but they can also be introduced by humans. The latter pathway is the one with the most controversy because there are so many cases of restoration failure where ecosystem function was not restored even though major amounts of biomass were through introduces species. Wali (1999) points out properly that understanding ecological succession should be the underpinning for planning and managing restoration and rehabilitation of disturbed ecosystems. How do you know where an ecosystem is going if you don't know where its prototype has been and how long it took to get there?

Enhancing the Natural Restoration Process

The natural restoration processes can, of course, be enhanced by the ecological engineer so that it takes years rather than decades. Bradshaw (1997) suggests that there are critical extreme conditions of mined landscapes that almost necessitate intervention by humans; "otherwise the whole restoration process may either not begin, or fail after a few years" (Bradshaw, 1997). Table 12.1 summarizes a number of problems that can be overcome by human intervention. These problems in the soil are (1) physical, (2) nutritional, or (3) toxic. Some solutions are mechanical or chemical approaches that have been around for decades that change the initial conditions; by themselves these processes are not ecological and are short-term solutions. Examples are fertilizing or liming the soil to change the nutritional or pH situation or irrigating or draining to change the salinity or hydrology. What is more desirable in most cases is adding appropriate native plants to mollify physical extremes, such as compact or unstable soils, adding legumes to increase soil nitrogen, or adding tolerant or phytoremediation plants so as eventually to mitigate the toxic conditions (see Chapter 11).

Generally, microbes have little problem getting to restoration sites. Notable exceptions to this are nitrogen fixers that could take decades to get to a restoration site. For plants to be established, seeds and propagules have to get to the site by one of several vectors: hydrology, atmosphere, animal transport, or human introduction. In large-scale mined land restoration, the last is often the only solution. Plants can be seeded or planted, usually to get a ground cover started, which in turn will start to accumulate organic matter, nitrogen, and soil moisture. The generally harsh environment of exposed subsoils can also be ameliorated by accumulating and then reapplying topsoil, often a very expensive procedure.

Proper restoration includes one more important ingredient already emphasized in the introductory chapters—*time*. Particularly with terrestrial systems that require long-term soil development (can be decades or centuries) and the development of canopy vegetation of trees (can be hundreds of years), time in restoration is not on the same clock as the working career of an average human or the time it takes to do a Ph.D. dissertation. Thus one of the problems with restoring degraded sites is the fact that there are no such sites with sufficient long-term ecological data to use as successional models of what will happen when these disturbed sites are restored (Wali, 1999).

12.3 MINELAND AND DEGRADED LAND RESTORATION

Treating Coal Mine Drainage Ecologically

Restoration of coal mines is a worldwide problem. Coal is the most abundant fossil fuel on Earth, and its use is expected to continue far after society uses up its last supply of natural gas and oil. There are an estimated 11×10^{12}

Table 12.1 Major problems in restoring mined land soil with short- and long-term solutions

Problem	Short-Term Treatment[a]	Long-Term Treatment[a]
Physical conditions		
Soil too compact	Rip and scarify	*Plant native vegetation*
Soil too open	Compact and cover with fine material	*Plant native vegetation*
Unstable soil	*Apply stabilizer, mulch, or nurse*	Regrade, *Plant native vegetation*
Too much soil moisture	Drain	*Plant native wetland species*
Too little soil moisture	*Apply organic mulch or nurse*	*Plant native xeric species*
Nutrition		
Nitrogen deficiency	Apply fertilizer	*Introduce legumes and other nitrogen fixers*
Other macro/micro nutrient deficiency	Apply fertilizer	*Plant tolerant vegetation*
Toxic conditions		
Acidic conditions (low pH)	Apply lime	Apply lime, *plant tolerant vegetation*
Alkaline conditions (high pH)	*Apply pyritic or other low-pH waste or organic matter*	*Weathering, plant tolerant vegetation*
Heavy metal accumulation	*Add organic matter, plant tolerant cultivars*	Add inert covering, *phytoremediation, plant tolerant cultivars*
High salinity	*Apply gypsum, irrigate, plant tolerant vegetation*	*Weathering, plant tolerant vegetation*

Source: After Bradshaw (1983, 1997).

[a]Approaches that would be considered ecological engineering are identified in italic.

tons of coal as a geologic resource in the world, with about 0.66×10^{12} tons of recoverable reserves (currently economical) available. About 80 percent of those resources are found in Russia and the former Soviet republics, the United States, and China. But coal is also considered a "dirty" fossil fuel in that it causes environmental problems when it is extracted as well as when it is combusted. Even when coal exhaust in cleaned, a major environmental issue of disposing of ash and slurry from air pollution control devices remains. When surface mining is employed for the extraction of coal, great expanses of landscape are disrupted. Regulations in most developed countries require full restoration of the landscape after surface mining. The problems of restoring surface coal mines are threefold: (1) establishment of vegetation on an inhospitable landscape from which topsoil has been removed, (2) reclaiming highly toxic locations where coal cleaning and coal waste were stored, and (3) controlling mine drainage that emanates from exposed coal seams. Restoration of stripped areas involves potentially all of the problems described in Table 12.1. Returning topsoil to the mined site is now required in most parts of the world to give vegetation a chance to establish.

One of the most difficult environmental problems with coal mines, both surface and subsurface, is that coal seams leak to the surface and acid mine drainage (AMD) results. The major pollutants in AMD are dissolved iron, manganese, and aluminum, the hydrogen ion, and the production of ferric hydroxide ($FeOH_3$) precipitate, which coats stream bottoms emanating from coal mine areas, killing most benthic invertebrates and fish (Letterman and Mitsch, 1978).

One of the ecological engineering methods for controlling AMD is with mine drainage treatment wetlands. Several hundred of these wetlands have been constructed in U.S. Appalachia alone (see, e.g., Fennessy and Mitsch, 1989; Wieder, 1989, Weider et al., 1990; Flanagan et al., 1994; Stark and Williams, 1995; Manyin et al., 1997; Mitsch and Wise, 1998; Tarutis et al., 1999). The use of wetlands for coal mine drainage control was probably considered first when volunteer *Typha* wetlands were observed near acid seeps in a harsh environment where no other vegetation could grow. By the end of the 1980s, more than 400 wetlands were constructed in the eastern United States alone to treat mine drainage water. The most common goal of these systems is usually the removal of iron from the water column to avoid its discharge downstream, but sulfate reduction and the alleviation of extremely acidic conditions are also appropriate goals.

Design criteria for these wetlands have been developed, but they are neither consistent from site to site nor generally accepted. Hydraulic loading rates as high as 29 cm day^{-1} have been suggested for wetlands designed for acid mine drainage, although Fennessy and Mitsch (1989) recommended 5 cm day^{-1} as a conservative loading rate for this type of wetland and a minimum detention time of 1 day, with much longer periods for more effective iron removal (Table 12.2). Loading rates of 2 to 10 g-Fe m^{-2} day^{-1} were suggested for circumneutral mine drainage (Fennessy and Mitsch, 1989) and 0.72 g-Fe m^{-2}

Table 12.2 Suggested design parameters for constructed wetlands used for controlling coal mine drainage

Parameter	Design	Reference
Hydrologic loading rate (cm day^{-1})	5	Fennessy and Mitsch, 1989
Retention time (days)	>1	Fennessy and Mitsch, 1989
Iron loading (g-Fe m^{-2} day^{-1})		
pH < 5.5	0.72	Brodie et al., 1988
pH > 5.5	1.29	Brodie et al., 1988
For 90% removal, pH 6	2–10	Fennessy and Mitsch, 1989
For 50% removal, pH 6	20–40	Fennessy and Mitsch, 1989
pH 3.5, outflow < 3.5 mg-Fe L^{-1}	2.55	Manyin et al., 1997
Basin characteristics		
Depth (m)	<0.3	
Number of cells	>3	
Plant material	*Typha* spp.	
Substrate material	Organic peat over clay seal, spent mushroom material	

day^{-1} by Brodie et al. (1988) for mine drainage with pH < 5.5. Manyin et al. (1997) ran a series of mesocosm experiments and found that a rate less than 2.5 g-Fe m^{-2} day^{-1} would be necessary to achieve a consistent iron concentration of less than the water quality standard of 3.5 mg-Fe L^{-1}.

Stark and Williams (1995) found that design features that enhanced iron removal and decreased acidity included broad drainage basins, nonchannelized flow patterns, high plant diversity, southern exposure, low flow rates and loadings, and shallow depths. It is not always cost-effective to construct wetlands when extremely high (>85 to 90 percent) iron removal efficiencies are necessary or when the pH of the mine drainage water is less than 4. Furthermore, effluent from constructed wetlands does not always meet strict regulatory requirements. Nevertheless, where no other alternative is feasible, the use of wetlands to reduce this type of water pollution is viewed as a low-cost alternative to costly chemical treatment or to downstream water pollution (Baker et al., 1991).

Vegetation in acid mine drainage wetlands is generally limited and is usually *Typha* because of that plant's hearty nature and resistance to pollution. Other vegetation species have been attempted in mine drainage wetlands, but in almost all cases, only *Typha* survives. The plant can withstand both the flooding and anoxia of AMD wetlands and is extremely tolerant of high concentrations of iron, sulfur, and other elements. Because it leaks oxygen be-

Table 12.3   Comparison of *Typha Typha* spp. aboveground peak biomass (g dry wt m^{-2}) in mine drainage wetlands versus freshwater marshes

Species	Location	Peak Biomass[a] (g dry wt m^{-2})	Reference
Mine drainage wetlands			
Typha latifolia	Coshocton County, Ohio	447 (350–540)	Fenessey, 1988
	Lick Run wetland, Ohio	502 (128–1135)	Mitsch and Wise, 1998
Freshwater marshes			
Typha glauca	Prarie pothole, Iowa	2297	van der Valk and Davis, 1978
Typha spp.	Constructed wetland, Illinois	634 + 56 (1990) 714 + 65 (1991)	Fennessey et al. 1994

[a]Range is shown in parentheses and standard error is given when available.

yond its metabolic needs to the soil surrounding the root hairs, it can withstand high concentrations of sulfides, which would oxidize to sulfates in this rhizosphere. From the limited data on mine drainage wetlands, it also appears that primary productivity, as estimated by peak biomass, is about 50 percent of that of a productive *Typha* marsh (Table 12.3). In essence, *Typha* survives in these mine drainage marshes but does not usually flourish.

CASE STUDY
Predicting the Effectiveness of a Mine Drainage Wetland

Models can be used to predict the effectiveness of AMD wetlands. A simulation model (Figure 12.2) was developed that predicted the retention of iron and aluminum in an AMD constructed wetland in southeastern Ohio before the wetland was built (Flanagan et al., 1994). After the wetland was constructed, field results were compared to model predictions (Mitsch and Wise, 1998). The model, which emphasized the processes of advection, diffusion, and precipitation/ sedimentation in both surface and subsurface state variables, was calibrated and verified using data from five constructed wetlands in Pennsylvania and Ohio. Predicted retention ranges from 0 to 93 percent for aluminum and from 50 to 99 percent for iron for the proposed wetland, depending on season and whether surface or subsurface flow is used. Diffusion of metals from water to sediments limits metal retention in low-pH wetlands, while metal precipitation rates from

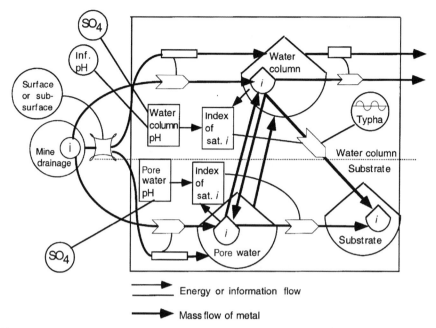

Figure 12.2 Simulation model of metal retention in an acid mine drainage (AMD) wetland. (From Flanagan et al., 1994; copyright 1994; reprinted with permission from Elsevier Science.)

the water column limit metal retention in circumneutral wetlands. Simulations published prior to construction of the wetland predicted iron removal retentions of 50 to 98 percent and 6.1 g Fe m^{-2} day^{-1}. There was a good match between model prediction and system performance. On a monthly basis, excepting the January 1993 flood flow through the wetland when there was a net release of iron, monthly iron retention ranged from 52 to 98 percent based on concentration and 62 to 98 percent based on mass. Overall, 80 percent of the iron mass was retained in the wetland, very close to the overall model prediction. When influent pH, water flow rates, and iron data from 1993 were entered into the simulation model (Mitsch and Wise, 1998), model results were compared to field data and to the preconstruction predictions (Figure 12.3). The new model simulation, using 1993 data, estimated a removal rate for iron of 3.60 g Fe m^{-2} day^{-1}. Field data from 1993 showed a removal rate of 4.4 g Fe m^{-2} day^{-1}. Thus, the model as published by Flanagan et al. (1994) overestimated the actual iron retention of this wetland by 39 percent. But when the model was rerun with actual 1993 hydrologic and chemical data, the model underestimated that actual iron retention by only 18 percent. Greater accuracy in model prediction is possible with better inflow (hydrology and iron concentration) data, but the model gave a rea-

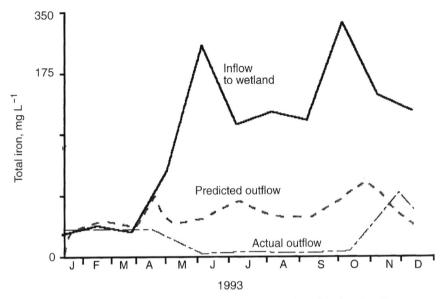

Figure 12.3 Comparison of actual inflow, actual outflow, and predicted outflow iron concentrations for the Lick Run wetland in Ohio for 1993. Iron data predicted were obtained from simulation of the preconstruction mine drainage model shown in Figure 12.2. (From Mitsch and Wise, 1998.)

sonable estimate of the performance of this wetland well before it was built.

The long-term suitability of wetland treatment systems is poorly understood, although it appears that *Typha*-dominated systems can survive decades in a mine drainage system. For surface mining restoration, the restoration of the larger site, probably through terrestrial succession either imposed or natural, means that the postmining hydrology needs to be considered in the project design as much as the hydrology when the wetland is built (Kalin, 2001). With mine drainage systems, the accumulation of iron hydroxides can eventually cause the system to begin to export materials unless the design and management includes adequate storage capacity and/or material removal. There are some who suggest that these wetlands, after several decades, can become mineral mines in their own sense, effectively recycling minerals that otherwise would be lost to downstream watersheds back to the economy.

Restoring Degraded Forests

The degradation of forests worldwide is caused by a number of factors, including biotic impacts and disease outbreaks, atmospheric pollution and subsequent soil acidification, poor forestry practices and harvesting regimes, and more recently, changes in climate (Hüttl and Schneider, 1998). In the past, practices such as soil liming or the planting of seemingly "better" trees has been tried to slow forest degradation. In central Europe it has been determined that there has probably not been 1 ha of virgin forest in 7000 years; deforestation in this region reached its first peak in the twelfth century followed by another in the seventeenth century, continuing into the beginnings of the industrial period. Early techniques of forest management included grazing by domestic animals (cattle, pigs, goats, sheep) and a process called *litter raking* that removed both organic substrate and some topsoil. Fire cultivation was also used, with clear-cutting, until the mid-twentieth century. Supposed "reforestation" in the nineteenth century included replacing broadleaved species with conifer plantations.

Over these centuries of degradation, soils were acidified and left devoid of nutrients, organic matter, and water, and the soil microbiota were less active and considerably less diverse. Overall, hundreds of years of poor practices have left forests in this part of the world significantly degraded (Kilian, 1998). Although not as acute in North America as in Europe, forest degradation is fast becoming a major issue. The degradation of forests in North America is of a different nature; although many beautiful deciduous and coniferous forests are still found throughout the continent, issues such as invasive species and wildfires, the latter caused by excessive fire prevention and human presence, are becoming major issues.

Ecological practices to restore or rehabilitate forests are many, but ways to improve forest conditions instantly do not exist. Some of the techniques that have been attempted include developing mixed stands of original canopy vegetation (e.g., broadleaved trees to replace conifers). Liu et al. (1998) point out that mixed stands of several different tree species at different strata would be an effective substitute for monospecific stands of larch (*Larix gmelinii, L. olgensis,* and *L. leptolepis*) that produce poor leaf litter that does not decompose. This accumulating litter inhibits thermal exchange between the atmosphere and the soil, thereby inhibiting soil microbiota. Hüttl and Schneider (1998) report that German scientists found two to three times more fine-root biomass for European beech than for Norway spruce and a higher acid-neutralizing capacity for beech leaves than for spruce needles.

Sometimes, trees should not automatically be put back on degraded land. This may sound like heresy to foresters, but reforesting degraded land may not always be the ecological solution. Hunter et al. (1998) point out that foresters are frequently the ones given money to "put back trees" when that may not be the most appropriate method. This concept is illustrated in Figure 12.4. Although most recovery of a degraded ecosystem to the original system would occur along the line E to D, foresters often plant monospecific plantations that quickly take the ecosystem from E to A or even A+ (more productivity than the original ecosystem). Native understory species will enter the stand (B) to add a little diversity. Sometimes foresters will add canopy

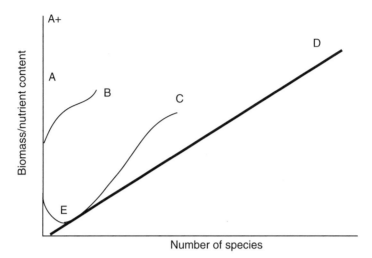

Figure 12.4 Model of landscape degradation and restoration by forest introduction. E indicates a degraded ecosystem and D the original ecosystem. Often, degraded land is restored automatically to monospecific stands of exotic species (A) to achieve rapid biomass but little biodiversity. Sometimes the productivity is even higher than that of the original ecosystem (A+). Natural vegetation comes into the stand later (B). Sometimes a combination or exotics and native species are planted (C), but the system never achieves either natural diversity or productivity. (After Hunter et al., 1998, and Lamb, 1994.)

mixtures of native and exotic trees (E to C), but this still stops far from achieving the ultimate goal desired for the original sustainable ecosystem D.

CASE STUDY
Restoring the Black Triangle Forests in Central Europe

The degraded Black Triangle is an area along the German–Czech–Polish border in central Europe where the forests have been affected heavily by industrial pollution since World War II and by poor forestry practices since the first decades of the eighteenth century (Figure 12.5). Although technically not a primary succession restoration, there is a need to restore the soils of the region so that deciduous forests can once again flourish in the region. The wholesale introduction of coniferous trees starting 300 years ago, coupled with acid deposition, has led to the loss of 80,000 to 100,000 ha of mountain forest. Even as late as the 1950s, foresters limed, drained, ploughed, removed topsoil, and replanted with Norway spruce (*Picea abies*)

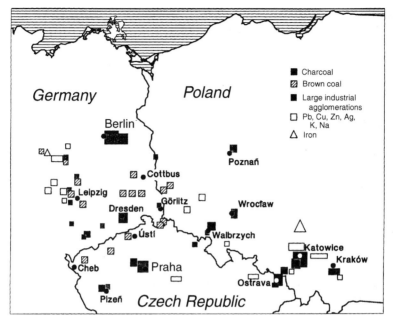

Figure 12.5 Black Triangle region of Saxony, Bohemia, and Silesia in central Europe, where there is extensive forest decline due to past practices and chemical degradation of soils because of acid deposition. (From Fanta, 1994; copyright 1994; reprinted with permission from Elsevier Science.)

plantations (Fanta, 1997). Forest decline in the region has been particularly noted in the period 1980–1990 (Fanta, 1994, 1997).

The ecological solution to this kind of degradation is first to try to remediate the extreme soil conditions that will otherwise thwart any restoration effort. Before that can happen, atmospheric sources of air pollution must be reduced, a situation that is now partially occurring with the changes in political systems in central and eastern Europe. But after the pollution sources are cut off, the damage remains.

An ecological engineering solution to the forest decline is to reintroduce native broadleaved vegetation to the region and to allow a long-term reclamation of the region that may take centuries. Moravcik (1994) describes some studies in the Krušnéhory Mountains between Germany and the Czech Republic. These mountains are acidic and naturally poor in nutrients, yet have been determined to have a potential natural vegetation of European beech, silver fir, Norway spruce, and other species. This poor environment has been affected additionally by reforestation of monocultures of Norway spruce (*Picea abies*) and by subsequent air pollution damage. In the mid-1970s, forest regeneration was oriented toward "emergency" forest plantings of tolerant species such as *Betula* sp. and *Picea pungens* (Figure 12.6), done to stimulate fast reforestation. A review of advanced stands of four stands (*Picea pungens, Betula verrucosa, Picea abies,* and *Sorbus aucuparia* by Moravcik (1994) found that both *Picea* stands showed notable symptoms of air pollution damage: needle loss and crown thinning. Their findings were an impressive support of the idea of self-design. The *Sorbus aucuparia* stand, which was not planted but which resulted from spontaneous regeneration at a declining spruce stand, had fast growth and high productivity and biomass almost as high as the twice-as-old *P. abies* stand. The birch (*B. verrucosa*) stands also grew well, leading Moravcik (1994) to conclude that broadleaved trees, one of which developed spontaneously on degraded sites, could "effectively contribute to the development of new forest ecosystems in the heavily polluted regions in the Krušnéhory." This is another example where self-designing systems have been shown to be superior to planted systems in restoring heavily polluted landscapes.

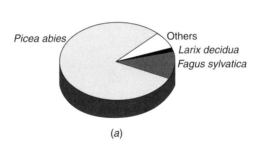

(a)

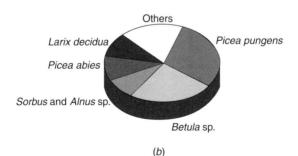

(b)

Figure 12.6   Changes in three species composition of forest stands in the Krušnéhory region of current German–Czech Republic border for (a) 1957 (all stands) and (b) 1991 (young stands, age 1–30 yr). (Redrawn from Moravcik, 1994.)

Tropical Mine Restoration

Restoring surface mining sites in tropical forests is a different proposition than temperate forest restoration on mine lands. The mining generally affects smaller areas, yet the downstream pollution effects are quite significant. The soils are shallow and fragile, and stockpiling and returning them to the site is almost mandatory. The restoration of such tropical mines, which are often for the production of bauxite, involves planting of either native or exotic species to achieve plant cover rapidly. The use of exotic species is now generally not favored. One such restoration site that has been studied since the 1980s is the Trombetas bauxite mine site in the evergreen equatorial rain forest near the Amazon River in Pará, Brazil (Parrotta et al., 1997; Parrotta and Knowles, 1999, 2001). The general approach to restoration in these areas includes leveling of the clay overburden, the replacement of about 15 cm of topsoil and woody debris that had been stockpiled prior to mining, and planting of native tree species at 2-m centers. In a series of plantings in the mid-1980s, the following treatments were attempted: (1) mixed native tree species, (2) mixed native tree species with insufficient topsoil application, (3) mixed commercial tree species (mostly *Eucalyptus* spp. and *Acadia* sp.), (4) direct seeding combined with mowing to promote sprouting, and (5) natural generation from topsoil seed bank.

After 9 to 13 years, the basal area on the foregoing treatments ranged from 1.5 m^2 ha^{-1} at the site with insufficient topsoil application to 24.9 m^2 ha^{-1} at the mix commercial species site (Figure 12.7a). If plant richness of the woody species is compared, the direct seeding plots had an average of 35 species per plot, while the commercial tree and insufficient topsoil sites had the lowest number of species per plot: 17 and 15 species per plot, respectively (Figure 12.7b). Parrotta and Knowles (2001) concluded a number of points about restoring tropical forests from mining.

1. Mining companies, with a modest investment in research, can develop cost-effective tropical forest restoration plans.
2. Site preparation, particularly with topsoil handling and reapplication prior to planting, is essential for establishment of forest cover.
3. In addition to the mixed native species planting, reliance of the seed bank in the reapplied topsoil was an effective means of restoration.
4. Although commercial planting, direct seeding, and natural regeneration from seed banks led to more basal area development, those treatments were generally less diverse in species. The risk of early canopy mortality and subsequent reinvasion by fire-prone grasses is higher in the fast-growing alternatives than in the mixed native species treatments.
5. The enrichment of all treatments with naturally colonizing species (i.e., those not in the seed bank or directly planted) depends on seed-dispersing wildlife such as bats, birds, and terrestrial mammals. Con-

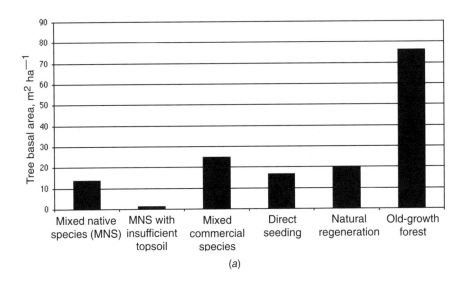

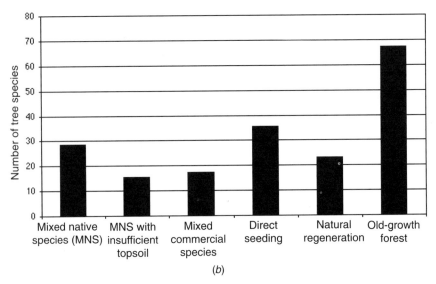

Figure 12.7 Comparison of (*a*) tree basal area and (*b*) woody species richness for five forest restoration approaches at the Trombetas bauxite mine site near the Amazon River in Pará, Brazil. Results are compared with similar data from a nearby old-growth forest. (Data from Parrotta and Knowles, 2001.)

servation of surrounding forests, including bans on hunting, greatly accelerate the restoration process.

6. Density and diversity of mixed native species planted sites is affected greatly by proximity to seed sources. For large-seed tree species, it is important to introduce these species after a canopy has developed to ensure their introduction.

7. Highly diverse and complex Amazonian forests can be restored after their destruction by mining through replanting techniques as described above.

Phosphate Mine Restoration

Phosphate mines are common in many parts of the world but are especially prevalent where limestone, rich in the phosphorus-rich mineral apatite, is abundant. It is estimated that 128,000 ha of land had been drastically altered by phosphate mining in central Florida alone (M. Brown, pers. comm., May 2003). If not reclaimed, these lands generally end up as clay settling pits, areas of mine cuts and overburden, and deposits of sand tailings. Brown et al. (1992) argue for the restoration of these mine lands because it involves recontouring the land into usable and ecologically sustainable form, restoring the groundwater and surface-water dynamics, and enhancing ecological succession. Investigators have argued that successful succession to a functioning landscape could take as long as 75 years (Kangas, 1981; Best et al., 1988; Odum et al., 1990) with seed sources available, and much longer in cases where there are no readily available seed sources, such as tree islands in the mined areas. Among the problems most acute in this mining are the altered groundwater and surface-water hydrology and the drastic changes in soil structure and chemistry with the topsoil removed.

CASE STUDY
Importance of Refugia and Seed Dispersal in Phosphate Mine Restoration

Brown et al. (1992) demonstrate the restoration of one such small 13-ha phosphate mine area in central Florida where created wetlands and sinuous streams, and tree and shrub revegetation, were used to restore part of a phosphate mine land (Figure 12.8). They found that natural seed dispersal from an adjacent natural forest by wind decreased exponentially from the edge of the forest and generally had a range of influence of 45 to 60 m (Figure 12.9). Bird dispersal of seeds was also seen beneath snags, and artificial perches were important, particularly for shrub species. The presence of perches and

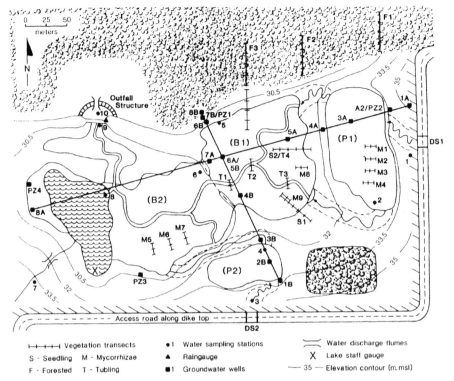

Figure 12.8 Hydrology and vegetation monitoring network of the Gardinier phosphate mine restoration site near Ft. Meade in central Florida. Outflow structure (station 10) flows to an adjacent creek. P, perched wetlands; B, lake border wetlands, PZ, piezometers. Solid lines are locations of vegetation and hydrology transects. (From Brown et al., 1992; copyright 1992; reprinted with permission from Elsevier Science.)

snags and perches was much more important than distance from the seed source, This research project also found that water dispersal could be quite important if there was a seed source upstream and the downstream floodplains received flooding.

The Florida case study findings included the following recommendations for these types of forest/hydrology restoration projects. These recommendations are general enough that they could be followed on any mine land or disturbed-land restoration project (Brown et al., 1992).

1. Remnant forest islands serve as important refugia for wildlife and seed sources and should be preserved during mining to accelerate restoration.
2. Water displersal of seeds can be important if headwater wetlands are left intact and surface-water flows are restored after mining.

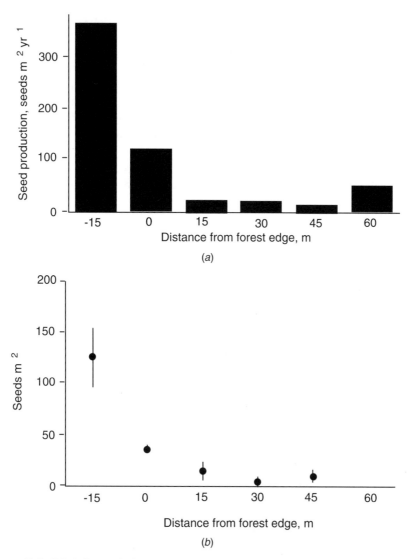

Figure 12.9 Wind dispersal of seeds as distance from adjacent floodplain forest to Gardinier restoration site: (a) seed deposition rate, 1985; (b) total seeds deposited, July–December 1983. (Data from Brown et al., 1992.)

3. Bird and wind dispersal of seeds can be important to the self-design of restoration sites. Wind dispersal is most rapid closest to the seed source and drops exponentially from the source.

4. Bird dispersal of seeds can be enhanced by bird perches and shags, but if the right soil moisture conditions are not present at these locations, survival of seeds is low.

5. Clay settling areas, which often make up over 60 percent of these phosphate mines often act more as sources of groundwater than of surface water. This suggests that any wetlands restored in these areas might be designed to be groundwater rather than surface-water wetlands.

6. Land contour, groundwater elevations, and subsequent soil moisture conditions are probably among the most important factors controlling revegetation success. Therefore, after recountouring, the site should be left alone for about one year before planting to better determine soil moisture conditions.

7. Planting of riparian vegetation should also be delayed until surface-water patterns and flooding frequency along streams are determined and streams are somewhat stabilized.

Reconstruction of Forests in Urban Environments: The Miyawaki Approach

Restoring natural systems in urban or crowded environments is challenging. Contaminated soil, air and water pollution, and confined spaces make any attempt at restoration quite difficult. Akira Miyawaki and his colleagues and students have used ecological engineering approaches for years to restoring ecosystems in the Japanese archipelago. The approach is contrasted with more conventional forest or terrestrial restoration by the fact that restoration would try to "direct and manage natural processes of recovery to obtain vegetation patterns that fit the expected regional patterns for the site," while the ecological engineering approach needed in many cases "directly and intentionally manipulating natural materials for specific purposes within human-made designs" (Miyawaki and Golley, 1993). The latter approach is based on (1) knowledge of the potential natural vegetation, (2) an understanding of germination and establishment biology of dominant species, and (3) knowing methods of planting large number of seedlings in prepared seedbeds. Knowledge of natural vegetation of a region allows reconstruction of the vegetation community with considerable confidence and avoids the need to use exotic vegetation. If a forest is being developed, it may start with planting of the canopy trees, letting understory plants and animals come in later (Miyawaki and Golley, 1993).

The first step in this process is the establishment of a given region. Miyawaki and his colleagues have identified over 900 maps for Japan that indicate potential native vegetation (Miyawaki and Fujiwara, 1988) based on field studies around mature vegetation sites found around Shinto shrines, Buddhist temples, and villages (Miyawaki and Golley, 1993), and specific climate, soil characteristics, and geology are noted for all species. This results in a complete ecological understanding of each potential plant so that the proper species can be recommended for any site reconstruction.

The second step, understanding the germination and establishment of each species, is vital to ensure strong growth of seedlings in containers for one to

Figure 12.10 Urban forest restoration in Japan. (*a*) Twelve hundred people including school-children, planting 15,000 seedlings along the Kashihara Bypass in Nara Prefecture in March 1977; (*b*) same site after 15 years. A 12-m environmental protection forest belt has formed along the highway. (Courtesy of A. Miyawaki; reprinted with permission.)

two years before planting. Because of a long history of horticultural experience in Japan, techniques to allow propagation of most species are well known.

The third step, the actual planting of a site, is actually a social approach where, not experts, but the local community is involved (Figure 12.10). After proper soil replenishment in urban areas or the use of bamboo revetments for planting on steep slopes, planting is scheduled as a planting festival that includes local authorities and hundreds of volunteers from schools and other groups. The groups are educated on the importance of these restored forests. There is little attempt to plant according to an exact plan of location for each plant, as the large number of planters and plants already causes a complex pattern of tree distribution, almost an invitation to self-design (Miyawaki and Golley, 1993). Planting is usually restricted to what is anticipated as the canopy vegetation, with the idea that understory and soil fauna will come in naturally with time. After mulching, little further management is required and the saying is "no management is good management"—an important point for the establishment of sustainable systems. The species planted take over in self-design, compete for resources (with some species being eliminated), and a closed canopy usually results in five years. The overall costs of this approach are high initial costs for research, plant propagation, and training but almost no maintenance costs afterward.

13

ECOLOGICAL ENGINEERING
IN CHINA

It is said of the great endeavors of the world,
that the fragmented must eventually be united,
and the united must at length be fragmented.

Chinese saying on order–disorder.
Opening words in Lo Kuanchung,
The Romance of the Three Kingdoms,
c. 1500 (quoted by van Slyke, 1988)

13.1 CHINESE BASIS FOR ECOLOGICAL ENGINEERING

The Western world may have much to learn from the Chinese about ways to live with the landscape. The design of the landscape with human society has been a part of Chinese culture for thousands of years. Taoism, an ancient philosophy–religion that still influences Chinese thought, has as its basic principle that nature, not humankind, is the center of its teachings. One has only to visit the exquisitely designed gardens of Suzhou (which appeared in the Han Dynasty, 206 B.C. to A.D. 220) to appreciate the Chinese ability to design nature and art in a harmonious whole. The endless expanses of rice paddies and seemingly infinite number of fishponds in China have not impoverished the Earth in the thousands of years that they have been in place, partly because for everything the Chinese take from Earth, something is always returned. In a country that must now support over 1 billion people, the Chinese have perfected the art of maximizing production with approaches that embody recycling and natural subsidies. On the other hand, the landscape is filled with both the odor and presence of natural organic materials, and stream and river pollution is the rule, not the exception. Westerners are quick to view the Chinese landscape as unorganized, overused, and depleted of all its resources; such a view would be much too simple and often wrong.

Ecological engineering is a concept that has been used informally in China for centuries and formally since the 1970s, the latter particularly by Ma Shijun, who has been described as the father of ecological engineering in China (see Yan, 1992). Ecological engineering in China was described as "the systems analysis, design, planning, and regulating of the structure, process, feedback, and engine of the artificial ecosystem according to natural ecosystem principles of holism, symbiosis, circulation, and self-regulation in order to gain as much benefits as possible in the long-term and large scale for humans" (Ma et al., 1988). Later, Ma (1988) described ecological engineering as "a specially designed system of production process in which the principles of the species symbiosis and the cycling and regeneration of substances in an ecological system are applied."

Specific applications of what is called ecological engineering in China include agroecological engineering (over 2000 sites in China), integrated fishponds, water pollution control systems, salt marsh restoration, and wetland, lacustrine, and forest management. Differences between Western and Chinese systems relate to design principles, objectives, human manipulation of ecosystem structure, and recognized values and economics. Good descriptions of the application of ecological engineering in China are contained in two special issues of *Ecological Engineering* (Mitsch et al., 1993; Wang et al., 1998b) and in many other papers from China in the journal that have passed the peer-review process.

Part of the Chinese special approach to ecological engineering is based on China's historical background, some of which is compatible with modern ecological theory (Yan et al., 1993). The most influential theory in Chinese philosophy is based on *yin and yang*. The symbol resembles two fish, one eating the tail of the other (Figure 13.1). Dating from the Zhou Dynasty (c. 1100–250 B.C.), the two parts represent earth (*yin*) and heaven (*yang*) with humans located between the two, their existence depending on both (Jiang et al., 1992). Heaven could be interpreted as weather, climate, season, and so on, with Earth as geomorphology, resources, and so on. The rotation implied by the symbol suggests a process of ceaseless cyclic motion and recycling. A related philosophy is that of *five elements* in which there is mutual restraint of the elements: fire, water, wood, metal, and soil (Figure 13.2). This also suggests a balance of promotion and restraint, building and decay, anabolism and catabolism.

With these philosophies, a concept close to ecological engineering developed in China and the East probably centuries ago. Whereas the emphasis of more recent definitions of ecological engineering in the West has been on a partnership with nature and research has been carried out primarily in experimental ecosystems with some full-scale applications in aquatic systems, particularly shallow ponds and wetlands, ecological engineering as pioneered by

Figure 13.1 Ancient Chinese symbol of yin and yang, which together suggest the basis of why ecological engineering is implicit in Chinese philosophy.

Ma and others in China has been applied to a wide variety of natural resource and environmental problems, ranging from fisheries and agriculture to wastewater control and coastline protection. Yan et al. (1993) describe the early development of ecological engineering practices in China, long before they were referred to as ecological engineering, as "models . . . that have existed for thousands of years [that] may be considered naïve and spontaneous ecological engineering."

13.2 COMPARISON WITH WESTERN ECOLOGICAL ENGINEERING

An overview of the contrasts between Western and Chinese ecological engineering is shown in Table 13.1. Contrasting ecological engineering in the theory-based West with the experience-strong China will help us come to

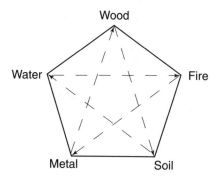

Figure 13.2 Five elements in which there is mutual promotion and restraint of five elements: fire, water, wood, metal, and soil. (From Yan et al., 1993; copyright 1993; reprinted with permission from Elsevier Science.)

Table 13.1   Contrast between ecological engineering concept in the United States and China

	United States and West	China
Basis	Principles of ecology	
Design principles	Self-design with some human work	Heavy intervention by humans to maintain system
Emphasis on:	Changing forcing functions to produce outcome	Changing ecosystem structure to produce outcome
People (no./ha)	<1	>10
Subsidy type	Fossil-fuel economy supporting ecologists	Human energy of resource managers
Subsidy size	Should be small	Sometime surprisingly high, especially humans
Recycling	Acceptable	Absolute requirement
Commercial yield	Not usual	Always food production
Species diversity	Often monospecific	Multilayered with many niches filled
Values considered	Aesthetics, conservation of natural resources, nonmarket values (e.g., water pollution control)	Commercial products
Economic values	Willingness to pay, contributory, or replacement	Utilitarian
Experience	40 years (since early 1960s)	3000 years

some common ground regarding the ecological theories that are truly "global" bases for ecological engineering.

Design Principles

As often practiced in the West, ecological engineering relies on the self-design or self-organization capacity of ecosystems (see Chapter 2). Chinese systems do not often rely on self-design; rather, they tend to involve significant manipulation of the ecosystem structure to achieve certain ends (i.e., fishponds or agroecological systems). Human intervention in terms of both the number of people involved and ecosystem manipulation, is much more important in Chinese ecological engineering systems. Some of this is probably due to the lower cost of labor, and hence system maintenance, in China compared to the West. As China's economy develops, its ecologically engineered systems may tend closer toward self-design.

Goals

In the West, the goal of ecological engineering projects is usually environmental protection, with the development of resources introduced only occasionally. The goal of ecological engineering in China is not only environmental protection, but also economic and social benefits. Ma and Wang (1984, 1989) and Mitsch et al. (1993) describe this multiobjective approach as a social–economic–natural complex ecosystem (SENCE). The structure of SENCE includes a three-tiered relationship (Figure 13.3). The core is *human society*, including organization, culture, and technology; this is the controlling part of SENCE and is called the eco-core. The second layer within SENCE is the *direct environment* of human activities; it includes geographical, artificial, and biological environments that are the fundamental media for human activities (eco-base). The third layer is the external environment or *surroundings* of SENCE, also called the *eco-pool,* which includes sources, sinks, and stores of the elements that exchange with our direct environment. Each layer relates to the others, and the multiple objectives of ecological engineering projects in China are a product of this relationship. Given this model, ecological engineering in China is described as trying to find "an environmentally sound, economically productive, and systematically responsible way of constructing and maintaining a sustainable ecosystem" (Wang and Yan, 1998). Wang and Yan (1998) argue that this can take place only when *hardware* (technology, innovation) is integrated with *software* (institutional reform, system optimization) and *mindware* (human behavior, human ecology) (Figure 13.4). In the end, the goals of ecological engineering in China as described by Wang and Yan can be summarized as:

1. Comprehensive wealth (money assets, natural assets, human resources, etc.)
2. Health (functional state of humans)

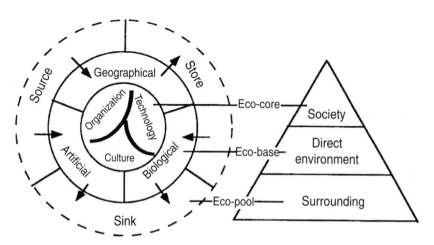

Figure 13.3 Social–economic–natural complex ecosystem (SENCE) model that describes ecological engineering in China. (From Mitsch et al., 1993; copyright 1993; reprinted with permission from Elsevier Science.)

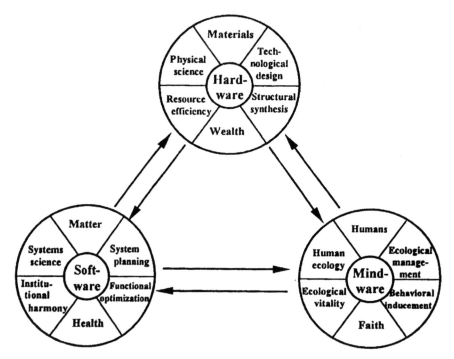

Figure 13.4 Framework of ecological engineering in China as an integration of hardware, software, and mindware. (From Wang and Yan, 1998; copyright 1998; reprinted with permission from Elsevier Science.)

3. Faith (values, material attitudes, and perceptions, concepts, and beliefs by humans)

It is no coincidence that the three systems in Figures 13.3 and 13.4 and the three goals of wealth, health, and faith all related to the yin and yang symbolism of the Earth, the heavens, and the human beings in between.

Ecosystem Structure

Constant manipulation and management of ecosystem structure are the methods of ecological engineering in China. Every niche is filled and ecological diversity is often high; the opposite is usually the case in the West. Even in systems used for the control of pH 2.0 mine runoff, Tang (1993) used a diversity of plants in his experiments on mine drainage control, in contrast to the *Typha* monocultures used in the West for mine drainage (see Chapter 12). Yan and Zhang (1992) classified ecotechniques (i.e., their approaches to ecological engineering) into five fundamental approaches: (1) adding new food chain components, (2) promoting parallel connections of symbiosis, (3) multilayer use of wastes as resources in the ecosystem food chain, (4) promoting beneficial recycling of substances, and (5) the restoration of ecosystems.

Recycling of Wastes

Recycling of materials in Chinese systems is not an option; it is a must. It leads to a reduction of external pollution and an increased yield of commercial products. Agroecological engineering systems depend on the waste products from one subsystem feeding another system. Similar feedbacks are illustrated by Yao (1993) for fishponds. Chung (1993) for *Spartina* marshes and Yuan et al. (1993) for crop–hog–fish agroecosystems. As summarized by Yan and Zhang (1992): "Natural ecosystems have inexhaustible resources due to the circulation and regeneration of materials within the system. Another implication is one of diversion. A cycle is an interruption or eddy in the straight-line progression towards entropy. It is the special provenance of life to cycle materials."

Values and Economics

There is a big difference in the way the West and East view the economics of ecologically engineered systems. First, the emphasis in the Chinese systems has been on applications that generate wealth. The West first emphasizes experimentation with mesocosms and pilot-scale projects. Second, the systems in China value the production of food and fiber more than environmental protection or habitat creation. In the West, improvements in water quality or ecological restoration are ends in themselves. In China, ecological engineering projects usually have a *utilitarian* value, that is, a commercial product such as food or fodder for other agricultural systems as well as environmental protection. In the Western world these values are often estimated from *willingness to pay*, *replacement value*, or *contributory value* approaches. In other words, a direct economic benefit does not accrue to individuals or organizations but to society as a whole.

Experience

The Chinese have almost 3000 years of experience in ecological engineering systems, although that experience is primarily empirical. The West has a stronger position in applying ecological theory toward ecosystem construction and management but with only a few decades of experience in formal ecological engineering. The world's environment is now being recognized as one system, and there will be a gradual joining together of the two approaches. We have already begun to see that integration. The West is now looking at multispecies systems and more complex food webs in our ecological systems as a result of seeing the experience in China. China, on the other hand, is beginning to recognize the intrinsic value of natural systems such as wetlands and is beginning to report on research in systems such as wetlands for controlling mine drainage without the necessity of developing commercial products from the system. As stated in a previous paper (Mitsch, 1991):

> While it is not possible to coordinate all terminology and concepts, at least identification of common theories (presently under different names) and establishment of new theories in a common scientific language will facilitate the future application of ecological engineering projects. The Western scientific community could benefit from the scholarly review of Chinese ecological engineering theories, some now published only in Chinese, and from seeing how an overpopulated country has developed many ecological engineering approaches out of the necessity of conserving resources and maximizing their use of the landscape. The Western world could also benefit from this most detailed glimpse into an ecotechnological approach that could be useful if we are ever confronted with the necessity to develop a sustainable economy in period of low energy. If one difference can be stated, Chinese use space (solar-based systems) for ecological engineering systems with time at less of a premium (labor is relatively inexpensive and information has a long residence time); Westerners maximize the use of time (labor costs are high and information is abundant and decays rapidly), with space at less of a premium. Our theories on ecological engineering will be truly general if they span this wide cultural difference.

13.3   EXAMPLES OF ECOLOGICAL ENGINEERING IN CHINA

Ecological engineering, as pioneered by Ma (1985) and others, has been applied to a wide variety of natural resource and environmental problems in China, ranging from fisheries and agriculture to wastewater control and coastline protection. The emphasis in the Chinese systems has been on applications rather than experimentation and more toward the production of food and fiber than on environmental protection. Some case studies are presented here to give the range of projects that could be considered ecological engineering. Locations of some of these projects are illustrated in Figure 13.5.

Fish Production in Wetlands

The Lake Go Reed Wetland and Fish Farm, located in Yixing County, China, integrates a *Phragmites* wetland with fisheries production. Water levels are manipulated in such a way as to maximize production of herbivorous grass carp (*Ctenopharyngodon idella*) and Wuchang fish (*Megalobrama amblycephala*) as well as the harvesting of reed grass (*Phragmites* sp.) for various uses, including fuel (Figure 13.6). This application of ecological engineering is accomplished primarily through water-level manipulation that is synchronized with fish growing seasons and harvesting schedules and a series of deeper channels for fish overwintering extending through wetland areas that are much shallower. Water-level manipulation maintains a dry wetland in

Figure 13.5 Map of China indicating some locations of ecological engineering projects discussed in this chapter.

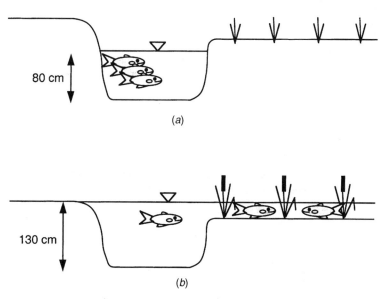

Figure 13.6 Lake Go Reed Wetland and Fish Farm in Yixing County, Jiangsu Province, China, illustrating the role of water-level manipulation used for the feeding and harvesting of fish: (*a*) winter season, low water levels; (*b*) summer season, high water levels.

January through March with about 80 cm of water in the channel. In April, pumped water from adjacent Lake Go raises the water level in the channel, allowing some water to spill over into the wetland. The water level is maintained at about 130 cm (50 cm in the wetland in May) and water is continually pumped through October until there is up to 1 m of water in the wetland. From October to December, the water level is decreased until the wetland is dry. Sometimes the channels are drained completely to harvest benthic fish.

Culture ponds for fish fry surround the wetlands and channels. When water is pumped from the lake, it passes first through the surrounding fishponds before entering the wetland–channel system. When wetlands are drained, the water is pumped directly to the lake. Pumped water is passed through screens to remove possible contamination of the system by lake fish. Reed yield is reported to be 75 metric tons (mt) fresh weight per hectare per year (with a dry/wet ratio of 30 to 50 percent). Fish yield is approximately 9 mt ha^{-1} with 60 percent as grass carp and Wuchang fish, 30 percent as silver carp and bighead carp, and 10 percent as others, including crucian and common carp. Table 13.2 summarizes the areas of this particular wetland–fishpond system. There are five fish channels for fish harvesting and culture. The system employs 104 people (with fewer of that number full time). Some artificial chemicals are added from time to time, and supplementary food is occasionally used for grass carp in the reeds. Pesticides are sometimes employed in the open channels and ponds but not in the wetlands.

A similar set of approaches for integrating fish production in wetland vegetation herbivory took place in experiments with "whole-lake grass carp" in the 1970s and 1980s in several lakes in eastern China (Li, 1998). Four basic concepts of management are described here:

1. *Whole-lake grass carp grazing* (WLGCG). Both carp production and aquatic plant production occur throughout a lake, with rivers to the lake cut off to prevent fish from escaping.
2. *Large-pen grass carp rotational grazing.* Grass carp are limited to pens as large as 50 to 100 ha, perhaps one-fourth of a lake. The pens are

Table 13.2 Area of wetlands and fish culture areas for the Lake Go Reed Wetland and Fish Farm System shown in Figure 13.6

	Area (ha)
Wetlands (reeds)	265
Fish ponds on edge	30
Fish culture ponds	8
Channels	7
	310

Source: Data from Mitsch (1995).

rotated from one location to another (Figure 13.7*a*) after one or two years.

3. *Small-pen high-stocking-density grass carp cultivation.* This is an intensive fish culture approach where the fish are kept in smaller 0.2 to 0.5-ha pens, and the food is harvested from the lake and brought to the pens (Figure 13.7*b*).

4. *Pond fishing in polders.* This is similar to the small-pen method except that the fish are cultivated in polders, pond–canal systems built behind dikes around the lake (Figure 13.7*c*).

A comparison of the four approaches is given in Table 13.3. Whole-lake fish production is lowest (<100 kg ha^{-1} yr^{-1}) while fish farming in adjacent

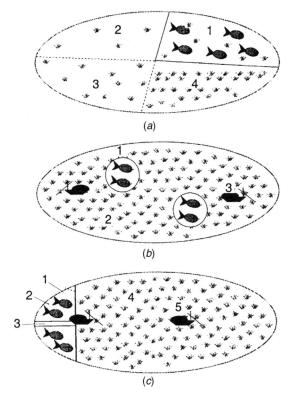

(a)

(b)

(c)

Figure 13.7 Diagrams illustrating three management schemes for grass carp-aquatic vegetation harvesting in Chinese lakes: (a) rotation of large pens with grass carp (numbers indicate sequence of location of pens); (b) small pens with high carp density: 1, carp in high-density ponds; 2, aquatic plants growing without disturbance; 3, aquatic plants are harvested and transported to pens; (c) carp in high-density polders diked around the lake edge ($<10\%$ of lake): 1, dike; 2, carp in ponds; 3, canals for transportation and drainage; 4, aquatic plants grow without disturbance; 5, aquatic plants are harvested and taken to ponds. (From Li, 1998; copyright 1998; reprinted with permission from Elsevier Science.)

Table 13.3 Comparison of four methods of grass carp production in aquatic plant-dominated lakes in China

Approach	Whole-Lake Grazing	Large-Pen Grazing	Small-Pen Grazing	Ponds Diked from Lakes
Scale	Whole lake	Pens 50–200 ha	Pens 0.2–0.5 ha	Ponds diked around lake <10% of lake area
Cultivation	Carp feed on aquatic plants	Carp feed on aquatic plants	Carp are fed harvested aquatic plants and other terrestrial plants	Carp are fed harvested aquatic plants and other terrestrial plants
Fish productivity (kg ha^{-1} yr^{-1})	<100	<1000	5000–10,000	6000–20,000
Inputs	Fingerling cost Fishing cost (high)	Fingerling cost Fishing cost Pen cost (high)	Fingerling cost Fishing cost (low) Pen cost (low) Aquatic plant harvest	Fingerling cost Fishing cost (low) Dike–pen–canal system Aquatic plant harvest
Environmental impact	Removes aquatic plants and breaks clear water	Impact restricted to pens	Causes pollution in pens	Removes aquatic plants and neither organic wastes nor nutrients returned to lake; uses up 10% of lake
Sustainability	2–3 years	Long time	Long time	Long time

Source: After Li (1998).

constructed polders was highest (6000 to 20,000 kg ha^{-1} yr^{-1}). Part of the higher productivity in the small-pen and polder systems was because the productivity of the lakes could not keep up with the demand for fish food, and other grasses, such as rye from adjacent lands, were utilized. In the two systems of high-intensity plant harvesting (Figure 13.7b,c), the amount of aquatic plants harvested from the lakes is high, and therefore the amount of nutrients taken from the lakes is also high. When the fishponds are not part of the lake itself but are adjacent to the lake in polders, the harvested nutrients are not released back to the lake, and the symptoms of eutrophication (high aquatic plant growth) are diminished. But the polders themselves take up 10 percent of the lake and a major part of its littoral zone.

Agroecological Engineering Farms

There are now thousands of sites in ecological engineering or ecoagriculture in China in more than 20 provinces (Zhang et al., 1998). Tai Chang Village, approximately 25 km west of Maanshan City on the south bank of the Yangtze River in Anhui Province, is one such agroecological engineering demonstration site. The village where the experimental site is located is approximately 700 ha in size with 390 ha under cultivation. The population is about 3408, with 894 families and 1870 workers. There were 16 village-owned enterprises producing 11 million Chinese yuan per year. The income was 1.4 million yuan per year, with a per capita income of 1030 yuan per year. Part of the village includes a demonstration site in agroecological engineering, including a pig farm, methane production and use, a fishpond, grape production, and tree plantations of oranges, *Metasequoia*, mulberry, and camphor (Figure 13.8). Silkworms consuming mulberry leaves and producing silk is another cottage industry in this agroecological engineering design. The emphasis is on recycling with pond detritus used as fertilizers for the plantations and pig waste fermented to produce methane for cooking.

Other examples of agroecological engineering are given by Wu et al. (1988) and Ma and Liu (1988) for villages in Hunan Province and Chai et al. (1988) for Jiangsu Province. For example, Wu et al. (1988) discuss the production of fertilizer, forage, and fuel at Wu Tang village (population 1505) through agroecological engineering. While 89.4 percent of the energy of the system comes from solar-based organic production, only 10.6 percent comes from industrial energy (no energy quality ratios used). Furthermore, 88 percent of the total organic production is consumed in the village, while only 12 percent is exported. This is described as a self-supporting system that does not depend greatly on outside industrial energy.

Salt Marsh Restoration

Salt marshes, dominated principally by *Spartina anglica*, have been constructed along the east coast of China to stabilize the coastline; to accelerate

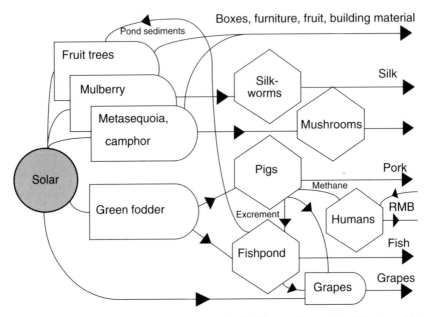

Figure 13.8 Agroecological engineering system in Tai Chang village, Anhui Province, China. A diversity of products result, and the waste product from one system provides the support for another. (From Mitsch et al., 1993; copyright 1993; reprinted with permission from Elsevier Science.)

accretion of sediments for reclamation of tidal land for agricultural or industrial development; to produce green manure, foodstuff, and fuel; and to control stream siltation and pollution (Chung, 1982, 1985, 1989). Much of the salt marsh development along the coastline of China can be traced to 21 individual plants imported from Mundon (Essex, England) by Chung in 1963 and 1964 (Chung, 1989). Other batches of plants were imported at that time from Højer marshes in western Denmark and from Poole Harbour, England. *Spartina anglica* C. E. Hubbard is an amphidiploid form of *S. townsendii* and is a hybrid between the English *S. maritima* and the North American *S. alterniflora* (Chung, 1983). It was chosen partially because of its lower-elevation habitat, its rapid growth rate, and its wider seaward distribution. Propagations from seeds, sprigs, and rhizomes were all successful in initial experiments. Large *Spartina* plantations were developed by first transplanting plants to rice paddies, a planting technique well developed in China.

The reclaimed *Spartina* marshes along China's coastline have been described as providing many benefits to humans (Figure 13.9). The planting of *Spartina* in one location led to the accretion of 80 cm of sediments in seven years (11.4 cm yr^{-1}) at one location and 66 to 68 cm over control after four years (16.7 cm yr^{-1}) at another site, leading to reclamation of the sea for crop production and coastal stabilization. The *Spartina* grass increased aeration

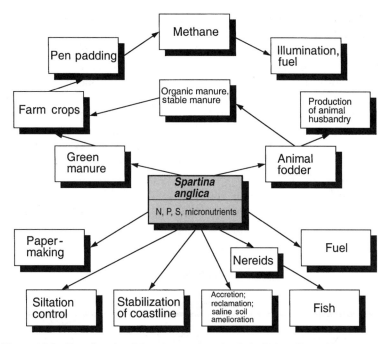

Figure 13.9 Benefits of reclaimed coastal wetlands in China. (From Chung, 1989.)

and organic content and decreased the salt content of the soil in addition to dissipating wave energy and slowing currents. The newly created salt marshes were also used as a habitat for migratory birds, waterfowl, domestic fowl, nereids (worms), and crabs, and as a pastureland for cattle and pigs. The grass is also harvested as a source of animal fodder, as an effective green manure for rice fields, as a source of fuel, and as a source of marsh gas (methane) for cooking and illumination. *Spartina alterniflora* has been investigated by the same research group in China as a source of mineral supplements (especially F, V, Cr, Mn, Fe, Co, Ni, Cu, Zn, Se, Sr, Mo, Sn, I) for beer and water. Biomineral water created from *Spartina* leaves was tested on animals and found to improve a number of health functions (Qin et al., 1998), suggesting that even more economic values could accrue from these ecologically engineered coastal marshes.

River Pollution Control

A riverine water hyacinth (*Eichhornia crassipes*) ecosystem was investigated for its role as a water pollution control system in a series of ecological engineering experiments on a section of the Fumen River in an eastern suburb of Suzhou, Jiangsu Province (Ma and Yan, 1989). From May through December 1984, approximately 2.7 ha of the river was planted in water hyacinths.

The benefits of this system includes the partial reclamation of polluted river water, particularly for nutrients, organic matter, and heavy metals, and the production of green fodder for a number of consumers (Figure 13.10). The production of hyacinths, estimated to be about 90,000 to 100,000 kg dry wt ha^{-1} yr^{-1} absorbs 1580 kg ha^{-1} nitrogen, 358 kg ha^{-1} phosphorus, and 198 kg ha^{-1} sulfur annually (Ma and Yan, 1989). In the experimental hyacinth marsh, concentrations of COD, total nitrogen, total phosphorus, ammonia-nitrogen, orthophosphates, organic nitrogen, and organic phosphorus decreased during the growing season. Microbial communities that develop on the root systems of floating water hyacinths are responsible for a reduction in organic matter in the water. Approximately 2500 mt (fresh weight) of water hyacinths was harvested from the site to be used as fodder for fish, particularly for grass carp (*Ctenopharyngodon idella*) and Wuchang fish (*Megalobrama*

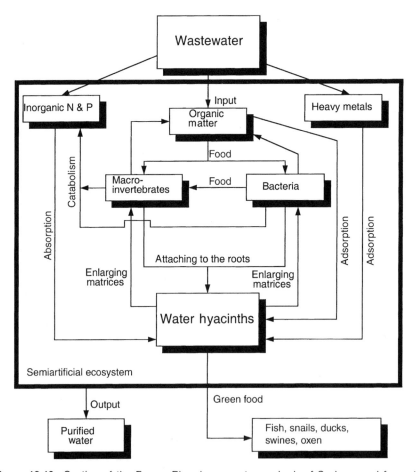

Figure 13.10 Section of the Fumen River in an eastern suburb of Suzhou used for water hyacinth river restoration experiment (From Ma and Yan, 1989.)

amblycephala) in culture ponds and for ducks, swine, and snails. Investigations of heavy metal content of fish and ducks that consumed the water hyacinths showed generally safe levels.

The Chinese Fishpond

The agriculture–aquaculture systems in China have two main components: fishponds and crop dikes (Figure 13.11; Ruddle and Zhong, 1988; Yan and Yao, 1989). The key concept is recycling. Human and livestock wastes are thrown into the ponds for nutrition, and after the growing season, pond bottom mud is applied to the dikes as fertilizers. The pond described by Yan and Yao (1989), however, receives no wastes to enhance primary production. All the inputs are direct food sources: grass, fine feeds, snails, and clams to several species of fish, each found in a unique niche in the system. For example, a unique characteristic of algae eaten by planktivorous fish is its recycling. Miura and Wang (1985) report that 35 and 50 percent of chlorophyll *a* survives after the algae is egested by silver and bighead carp. Gross production by algae after gut passage of these fish was 22 and 100 percent, respectively, of the before-passage values. Miura (1990) further estimates recycling to be around 13 and 66 percent of daily consumption by silver and bighead carp, after taking into account the amount of suspension of defecated algae. In the Chinese pond evaluated here, the principal taxa of zooplankton are protozoa, rotifera, crustacea (cladocera, copepoda), and freshwater lamellibranchia (Yan and Yao, 1989). Zooplankton is a major consumer of phytoplankton and detritus. The bulk of food ingested by zooplankton is either egested or respired. In Chinese aquaculture ponds, dissolved oxygen (DO) is probably the most critical indicator of pond characteristics. While water temperature plays a key role in influencing the growth pattern, low DO levels limit the food intake of fish. In some ponds, mechanical aeration is provided a few times a year. Due to the higher rate of photosynthesis in the upper layer and the higher rate of bacterial activity near the bottom, the pond usually has a gradient of DO: high near the surface and low near the bottom. Therefore, bottom dwellers such as common carp, black carp, and bacteria are more susceptible to DO shortage.

CASE STUDY
Simulating a Chinese Fishpond

A food web diagram of a typical Chinese fishpond is presented in Odum energy language in Figure 13.12. Model simulations described here were used to validate the published production rates in these ponds and also to see if removal of any part of the system leads to suboptimal performance (Hagiwara and Mitsch, 1994). In this pond,

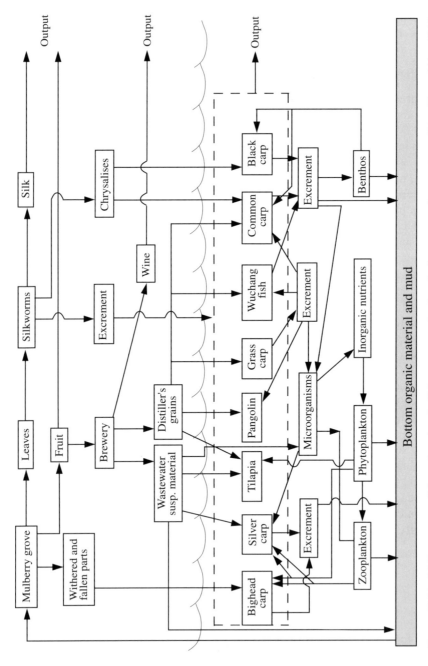

Figure 13.11 Mulberry grove–fishpond system in China as example of agriculture–aquaculture systems (After Ma and Yan, 1989, from Yan and Zhang, 1992; copyright 1992; reprinted with permission from Elsevier Science.)

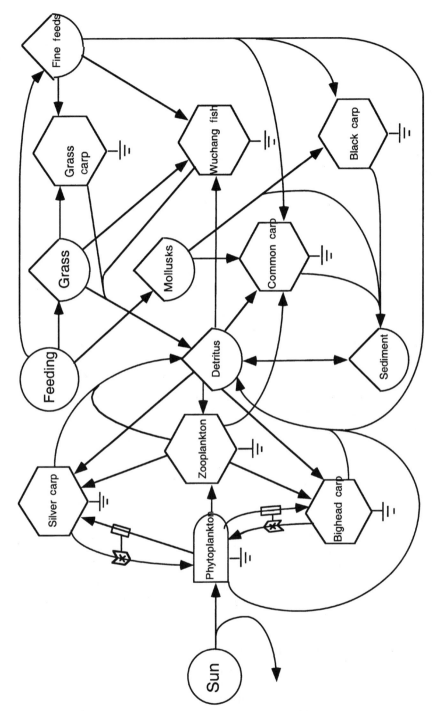

Figure 13.12 Simulation model of Chinese fishpond. (From Hagiwara and Mitsch, 1994; copyright 1994; reprinted with permission from Elsevier Science.)

327

the main fish raised are black carp (*Mylopharyngodon piceus*), common carp (*Cyprinus carpio*), Wuchang fish (*Megalobrama amblycephala*), silver carp (*Hypophthalmichtys molitrix*), grass carp (*Ctenopharyngodon idella*), and bighead carp (*Aristichthys nobilis*). In place of verification, which tests model responses to possible changes, calibration based on the energy flow data provided by Yan (1991, personal communication) and dissolved oxygen data (Yan and Yao, 1989) was performed.

In the fishpond model, the following assumptions are made (Hagiwara and Mitsch, 1994):

1. Biological state variables are expressed to be concentrations, with units of energy kJ m^{-2}.
2. Processes are temperature dependent, with specific rates of dependence.
3. Temperature is uniform throughout the pond.
4. No human intervention is made besides feeding and harvesting.
5. The volume (area and depth) of the pond is kept constant throughout the season.
6. The pond has a vertical dissolved oxygen gradient [high DO (upper column) to low DO (lower column)].
7. Food consumption by fish is influenced by an environmental factor determined by temperature, photoperiod, and dissolved oxygen concentration.
8. Food consumption of fish and zooplankton is limited by prey density, with specific half-saturation constants.
9. Fish are stocked in February through December.
10. Feeding of fish with snails, fine feeds, and grass is done daily and the amount is proportional to the biomass times the environmental factor mentioned above.
11. The amount of chemical substances such as ammonia and nitrite is negligible.
12. Fish excretion is proportional to its food intake except for bighead carp, whose assimilation decreases upon switching of food items.

Simulated growth of black carp in one season shows good agreement with data observed (Figure 13.13*a*). Filter feeders (silver carp and bighead carp) are influenced primarily by food availability. The biggest discrepancy is seen in July, when, in the model, fish suffer from lack of oxygen. The growth of grazers follows the logistic curve

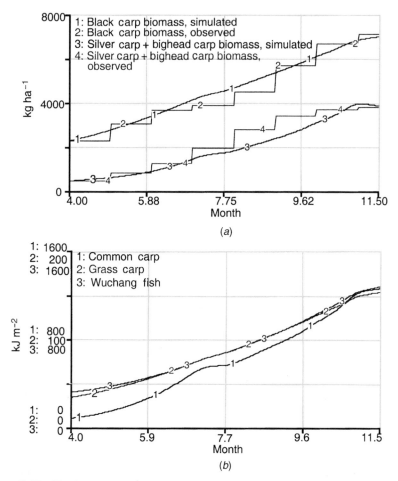

Figure 13.13 Simulated fish biomass for fishpond model shown in Figure 13.12: (a) simulated growth patterns of black carp and silver + bighead carp compared to data; (b) growth of grazing fish as predicted by the model. (Redrawn from Hagiwara and Mitsch, 1994.)

(Figure 13.13*b*). Although neither the density effect nor the effect of toxic substances are modeled, factors included are dissolved oxygen, appetite decrease due to temperature decrease, size increase, and change in photoperiod limit exponential growth, among others.

To estimate the effect of each species of fish on an entire ecosystem, the removal of each species was tested (Table 13.4). The model suggests that the growth of silver carp is retarded when no black carp is stocked. This may be partly true because silver carp exerts heavy grazing pressure on phytoplankton. The removal of silver carp leads to lower dissolved oxygen caused by higher zooplankton populations.

Table 13.4 Summary of simulated effects of removing fish species from Chinese fishpond

Species Removed	Effect
Silver carp	Lower DO, higher zooplankton biomass
	Larger fluctuation of primary production and phytoplankton biomass
	Less fish growth except bighead, which consumes zooplankton
	More phosphorus available for photosynthesis due to low DO
Bighead carp	Persistence of zooplankton, virtual absence of phytoplankton
	Much lower DO than above
	Less fish growth for all species
	Much lower level of detritus
Grass carp	Less detritus, higher DO, less phosphorus, but little overall change
Wuchang fish	Least overall change except better growth for black and common carp, due to decreased competition for fine feeds
Black carp	No silver carp growth due to decreasing detritus, which is a major source of energy (more than 70%)
	High DO, low phosphorus level (persistent limiting factor)
	Highest common carp and grass carp growth, due to less competition
	Lower bighead and Wuchang fish growth, due to resource scarcity (zooplankton and detritus)
Common carp	Higher fish growth except silver carp and Wuchang fish, due to less competition
	Higher zooplankton biomass, and lower amount of detritus, which leads to lower silver carp and Wuchang fish growth

Source: After Hagiwara and Mitsch (1994).

In almost all cases, the removal of a species causes a fishpond to behave suboptimally. In the pond, balance is maintained through multiple feedback processes generated by the complex food web structure. Removal of silver carp resulted in a fluctuating ecosystem, especially in late summer to autumn, when silver carp is supposed to exert heavy grazing pressure on phytoplankton. The addition of common carp and Wuchang fish is beneficial, as they utilize a niche that is not fully exploited otherwise. This is especially true for Wuchang fish, as very little overall change is observed. For common carp, its removal resulted in a higher growth rate for competitors.

The model appears to replicate the processes involved in the production of fish in Chinese ponds. Its most valuable contribution, however, appears to be a validation of the fluxes and operation of Chinese fishpond energetics. Furthermore, it shows that removal of any one of the diversity of fish species from the ponds leads to some suboptimal functioning of the pond ecosystem.

Wetlands and Rice Paddies

China has been inhabited and developed for thousands of years, so very little of the country's wetlands and waterways has not been utilized by humans for the production of food and fiber. Despite that it has been estimated that China has 250,000 km^2 of natural wetlands in the country out of a total of 620,000 km^2 of total wetlands (natural wetlands plus rice paddies, etc.; Lu, 1990, 1995). Thus, there are about 1500 ha of heavily managed wetlands, particularly rice paddies, for every 1000 ha of natural wetlands in China. The fact that there are managed wetlands at all is reflective of the fact that China is part of what Dugan (1993) calls *aquatic civilizations* found throughout Asia that better adapted to their surroundings of water-abundant floodplains and deltas and took advantage of nature's pulses, such as flooding. In contrast, the Europe and North America are *hydraulic civilizations* that control water flow through the use of dikes, dams, pumps, and drainage tile, partially because water was only seasonally plentiful. It is because the latter approach of controlling nature rather than working with it is so dominant around the world today that we find such high losses of wetlands worldwide.

CASE STUDY
Ecological Engineering with Wetlands in the Yangtze River Valley

The broad alluvial Jianghan-Dongting Plain in the middle Yangtze River valley of central China (Figure 13.14) is characterized by lacustrine and wetland depressions of varying size that are subject to seasonal flooding, including shallow lakes, diked agriculture, elevated river beds, and many rice paddies and fishponds. The seasonally flooded wetlands that originally typified this region have mostly been reconfigured to create artificial wetland systems—fishponds and paddy fields—that are protected by dikes, canals, and pumps (Bruins et al., 1998). Yet they are still subject to occasional inundation by rainfall in excess of drainage capacity, resulting in economic loss to crops. Intense rainfall overwhelms the drainage systems and inundates low-lying rice. A study of ecological alternatives by Bruins et al. (1998) suggested ecological engineering strategies, such as cultivation of inundation-tolerant wetland crops and passive storage of excess rainfall in wetland fields, to reduce regional vulnerability to flooding loss. Conceptual models are presented at several different scales, including a 24-km^2 state-run farm (Figure 13.15a) and a 2800-km^2 diversion area called the Honghu Flood Diversion Area (Figure 13.15b). At the scale of the farm, polder and farm boundaries coincided, and data describing land use and drainage system structure

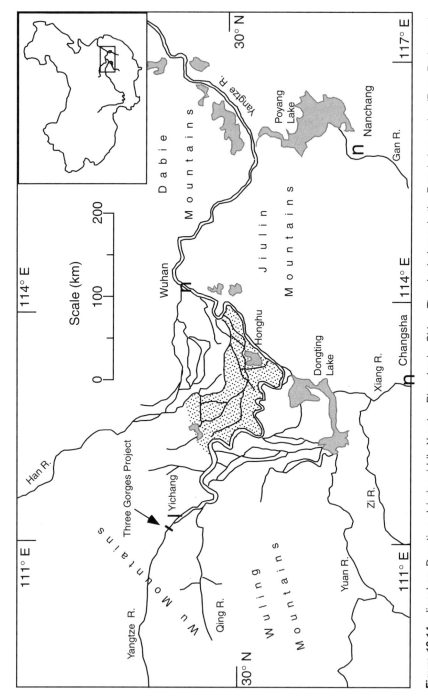

Figure 13.14 Jianghan–Dongting plain in middle Yangtze River basin, China, The shaded area is the Four Lakes region (From Bruins et al., 1998; copyright 1998; reprinted with permission from Elsevier Science.)

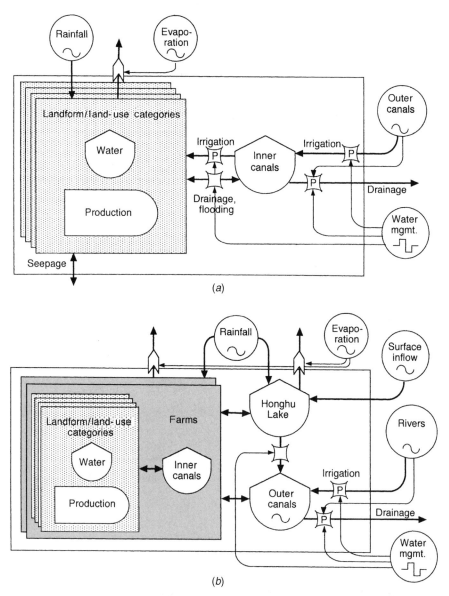

Figure 13.15 Conceptual models of Jianghan–Dongting plain hydrology and production: (a) farm scale, with imposed, farm-level water management schedule; (b) area-level-scale water management. (From Bruins, 1997; reprinted with permission.)

were readily available, and outputs reasonably represented recent-year patterns of flood damage to rice. Modeling results investigated two sets of management alternatives. In the first set of simulations, conventional engineering solutions such as increased pumping or channel deepening were suggested (Table 13.5*a*). In the second set of simulations, ecological engineering approaches (Table 13.5*b*) such as allowing seasonal flooding and converting the low areas to lotus (*Nelumbo* spp.), wild rice stem (*Zizania latifolia*), or tatami (*Juncus* spp.) cultivation. When grown in shallow wetlands, rice is susceptible to flood damage. All of these plants are obligate wetland plants that can grow in temporary or permanent standing water. Wild rice stem is a particularly delicious crop found throughout China. It grows symbiotically with a fungus in the stem that causes the stem to enlarge; it is eaten like celery after cooking and is frequently found at Chinese banquets and farm markets alike. In the ecological engineering alter-

Table 13.5   Ranking of strategies for flood resistance that rely on conventional engineering and agronomic approaches and ecological engineering approaches[a]

	Impact on Farm		Overall
Strategy	Flood Peak	Rice Damage	Effectiveness
a. Conventional Approaches			
1. Increase farm pumping capacity	Much lower	Strongly reduced	High
2. Deepen drainage canals by 20%	Slightly lower	Slightly reduced	Medium
3. Change 50% of rice to two-crop rice	Slight; variable	Usually increased	Low
b. Ecological Engineering Approaches			
4. Convert lowest areas to lotus–wild rice and raise dikes for passive storage	Higher	Strongly reduced	High
5. Convert lowest areas to lotus–wild rice	Higher	Strongly reduced	High
6. Change 50% of low area to lotus–wild rice and raise dikes	Usually higher	Variable	Medium
7. Change 50% of low area to *Juncus*–rice rotation	Little or none	Slight/variable	Low
8. Change 50% of low area to lotus or wild rice stem	Often higher	Damage increased	Low

Source: After Bruins et al. (1998).
[a] Estimates are based on simulation model results.

natives, the emphasis is on redirecting the farms to be more water tolerant rather than continually attempting to control the water with pumps, dikes, and channelization.

13.4 FUTURE OF ECOLOGICAL ENGINEERING IN CHINA

The future of ecological approaches to food production and sustainable livelihood in China is uncertain given the enormous percentage of the world's population that lives in China and the blistering pace of economic development currently occurring in that country. Many of the ecological techniques were developed over centuries of trial and error and became part of traditional Chinese culture. Modern Western agriculture and aquaculture, with extensive use of fertilizers, pesticides, energy, and monocultures, became more prevalent in China as its population soared above 1.2 billion before the twenty-first century began. Yet the sustainable and somewhat traditional approaches that are described here are exactly what China needs and what the West would be wise to emulate as well. Speaking of the situation in China and other developing countries, Wang et al. (1998a) state: "Due to economic and methodological reasons, policy makers usually deal with environmental issues in an emergency treatment way, often creating new environmental problems after solving the old ones. Fortunately, ecological engineering is just one of the instruments that is able to make trade-offs between economy and ecology, between present and future, and between local and regional development."

ECOLOGICAL
ENGINEERING TOOLS

14

MODELING IN ECOLOGICAL ENGINEERING AND ECOSYSTEM RESTORATION

A model can be considered as a synthesis of elements of knowledge and observation about a system. The quality of the model is therefore very dependent on the quality of our knowledge elements and the data available. If our knowledge of and data for a given problem are poor, we must not expect that the model of the system can fill the holes in our detailed knowledge or repair a poor set of data. On the other hand, models are able to provide new knowledge about the reactions and properties of an entire system. Because a model represents a synthesis of knowledge about the system it can provide results, particularly about the system's properties. Furthermore, when we put together the results of many different models covering different viewpoints and aspects, we get a comprehensive overall picture of the ecosystem. We can cover our observations completely only by use of a pluralistic view. Modeling is a very useful tool in our effort to achieve the best possible pluralistic view.

The emergence of very complex environmental and ecological problems has provoked the development of ecological and environmental modeling as a powerful synthesizing tool, where reactions and properties of systems are in focus. Models are, first of all, synthesizing tools, but it should not be forgotten that models can also be used to analyze and understand the properties of a system at the system level and, most important for ecological engineering, predict their behavior. In ecology, we use models not only to provide an overview of ecological problems, but also to predict the reactions of the entire system to changes, as for example, the impact on ecosystems of pollution or land-use change. This ability to predict the effects of changes may be of particular importance in ecological engineering, where we often create or restore ecosystems or utilize natural ecosystems to solve pollution problems. Models can be the primary tools that we use to predict the effects of these ecosystems before they are created or restored. The iterative modeling process is also a design tool; multiple simulations or optimization programs allow us to design ecosystems in the proper way.

Models may either be physical or mathematical. Physical models contain the main components of the real system whereby the processes and reactions of the complex system are deduced by using observations on the simpler system—the physical model. If, for instance, we want to study the interactions between a toxic substance and a system of plants, insects, and soil in nature, we may construct a simplified system containing these components, make our observations on the simpler system, and thereby facilitate our interpretation of the data. Physical models are often named *microcosms* or *mesocosms,* as they contain all major components of the larger system, but on a smaller scale. Observations from microcosms are often utilized to improve mathematical models.

Models in this chapter are understood as mathematical models based on mathematical formulations of the processes most important for the problem under consideration. The field of environmental modeling has developed very rapidly during the last two decades, due to essentially three factors: (1) the development of computer technology, which has enabled us to handle very complex mathematical systems, (2) a general understanding of pollution problems and their impacts on ecosystems, and (3) better knowledge of the characteristics and properties of ecosystems.

14.1   APPLICATION OF MODELS IN ECOTECHNOLOGY

Models are used in ecotechnology primarily as a management tool to give predictions on the effect of an ecological engineering project, to design an ecosystem, or to modify an existing ecosystem. Pollution abatement today is much more complicated than it was 20 to 30 years ago, when environmental engineering was the principal applied tool in environmental management (see Chapter 1). Today, we have a wide spectrum of abatement possibilities in addition to environmental engineering: ecological engineering, cleaner technology, environmental legislation, and politically determined tools such as green taxes (Figure 1.9).

Models are also used to perform *environmental risk assessment.* The resulting recommendations for this application may be either in the form of emission limitations or in the form of more general legislation on the type of emission in question. This is, of course, a political decision, because—even if the model were to give a rather clear answer—there are always economical aspects involved in such decisions. It is possible in some instances to construct *ecological–economic models,* which also consider the economic consequences of the various problem solutions, but these models are not yet developed sufficiently to give reliable guidelines in more than a very limited number of cases. However, these types of models will probably be developed further in the near future, and during the next decade they will be used to a greater extent in environmental management.

14.2 MODELING PROCEDURE

The difficult part of modeling is not the mathematical formulation or the translation of the mathematics into a computer language. The introduction of personal computers and easily applicable software has made it much easier to handle these steps of modeling. The more difficult part is to provide the necessary knowledge and be able to estimate which components and processes to include in the model. Ecological modeling in support of ecological engineering and ecosystem restoration therefore requires profound ecological knowledge and experience. An ecologist with knowledge of mathematics and computer science is therefore in most cases better fitted to construct ecological and environmental models than is a mathematician or engineer with only a small knowledge of ecology and environmental science. This is a much-ignored fact.

Figure 14.1 shows the procedure, but it should be considered as an iterative process, and the main requirement is to get started, and Table 14.1 lists the major components of mathematical models used in environmental sciences. The information below is taken from Jørgensen and Bendoricchio (2001) and is only summarized here.

The first modeling step is a *definition of the problem,* and the definition will need to be bound by the constituents of space, time, and subsystems. The bounding of the problem in space and time is usually easier, and consequently more explicit, than identification of the subsystems to be incorporated in the model. It is difficult to determine the optimum number of subsystems to be included in the model for an acceptable level of accuracy defined by the scope of the model. Due to lack of data it will often become necessary at a later stage to accept a lower number than intended at the start or to provide additional data for improvement of the model. It has often been argued that a more complex model should be able to account more accurately for the reactions of a real system, but this is not necessarily true. Additional factors are involved. A more complex model contains more parameters and increases the level of uncertainty, because parameters have to be estimated either by more observations in the field, by laboratory experiments, or by calibrations, which again are based on field measurements. Parameter estimations are never completely without errors, and the errors are carried through into the model and will thereby contribute to its uncertainty.

The problem of selecting the right model complexity, discussed further later in the chapter, is a problem of particular interest for modeling in ecotechnology. A first approach to the data requirement can be given at this stage, but it is most likely to be changed at a later stage, once experience with

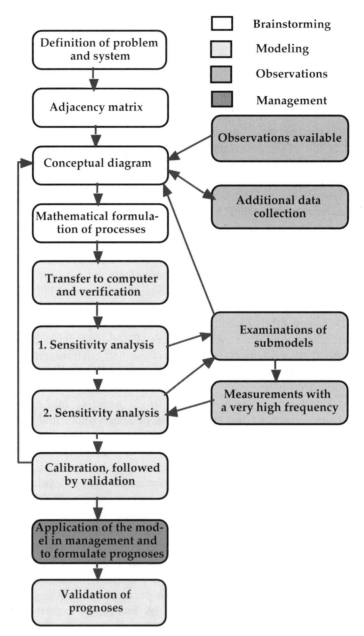

Figure 14.1 Modeling procedure that distinguishes four activities: brainstorming, when the conceptual diagram is developed; observations providing the underlying data; modeling; and translating the conceptual diagram to mathematics and a computer language, followed by a series of tests and management, when the validated model is ready to be applied. Note that there are feedback arrows from sensitivity analysis, calibration, and validation to the mathematical formulation and the conceptual diagram, indicating that modeling must be considered an iterative process. (From Jørgensen and Bendoricchio, 2001; copyright 2001; reprinted with permission from Elsevier Science.)

Table 14.1 Major components of models in environmental sciences

1. *Forcing functions, or external variables:* functions or variables of an external nature that influence the state of an ecosystem. Forcing functions under our control are often called control functions.
2. *State variables:* components of the model structure; storage of mass, energy, or information.
3. *Mathematical equations:* relationships among forcing functions and state variables.
4. *Parameters:* coefficients in the mathematical representation of processes. They may be considered constants for a specific ecosystem or part of an ecosystem or variable with time or space. A comprehensive collection of parameters in environmental science and ecology can be found in Jørgensen et al. (1997).
5. *Universal constants:* such as the gas constant and atomic weights.

Source: After Jørgensen and Bendoricchio (2001).

verification, calibration, sensitivity analysis, and validation has been gained. In principle, data for all the state variables selected should be available; in only a few cases would it be acceptable to omit measurements of selected state variables, as the success of calibration and validation is closely linked to the quality and quantity of the data.

In an *adjacency matrix* the state variables are listed vertically and horizontally, with 1 used to indicate that a direct link between the two state variables is most probable and 0 indicating that there is no link between the two components (Table 14.2). In practice it is recommended that the adjacency matrix be set up before the conceptual diagram. Thus the adjacency matrix is shown as the second step in modeling before the third step of developing a conceptual model (Figure 14.1).

A verbal model is difficult to visualize and it is therefore convenient to translate the problem definition and adjacency matrix above into a *conceptual diagram,* which contains the state variables, the forcing function, and how these components are interrelated by mathematical formulations of processes. Several other conceptual diagrams throughout this book (e.g., Figures 3.5, 4.1, 4.3, 8.14, 13.10, and 13.11) illustrate conceptual diagrams of ecosystems. Once at least a first attempt at model complexity has been selected, it is possible to conceptualize the model. It is ideal to determine which data are needed to develop a model according to a conceptual diagram (i.e., to let the conceptual model or even some first more primitive mathematical models determine the data at least within some given economic limitation), but in real life, most models have been developed *after* data collection as a compromise between model scope and data available.

The fourth step in the formulation of a model is to describe the processes as *mathematical equations.* Many processes may be described by more than one equation, and it may be of great importance for the results of the final model that the right one is selected for the case under consideration. It is recommended at this stage that all equations be checked for unit consistency.

Table 14.2 Adjacency matrix for the nitrogen model in Figure 4.3[a]

From \ To	Nitrate	Ammonium	Phytoplankton-N	Zooplankton-N	Fish-N	Detritus-N	Sediment-N
Nitrate	—	1	0	0	0	0	0
Ammonium	0	—	0	1	0	1	1
Phytoplankton-N	1	1	—	0	0	0	0
Zooplankton-N	0	0	1	—	0	0	0
Fish-N	0	0	0	1	—	0	0
Detritus-N	0	0	1	1	1	—	0
Sediment-N	0	0	1	0	0	1	—

Source: After Jørgensen and Bendoricchio (2001).

[a]State variables are listed vertically and horizontally, with 1 used to indicate a direct link between the two state variables and 0 to indicate that there is no link between the two components.

Once a system of mathematical equations is available, *verification* can be carried out. This is an important step, which unfortunately is omitted by some modelers, and consists of asking the following questions: (1) Is the model stable in the long term? and (2) Does the model react as expected? In general, playing with the model is recommended at this phase. Through such exercises the modeler gets acquainted with the model and its reactions to perturbations.

Sensitivity analysis, which follows verification, attempts to provide a measure of the sensitivity of either parameters, or forcing functions, or submodels to the state variables of greatest interest in the model. In practical modeling, sensitivity analysis is carried out by changing the parameters, the forcing functions, or the submodels. The corresponding response on the state variables selected is observed. Thus, the sensitivity, S, of a parameter, P, is defined as

$$S = \frac{\Delta x / x}{\Delta P / P} \tag{14.1}$$

where x is the state variable under consideration.

The relative change in parameter value is chosen on the basis of our knowledge of the certainty of the parameters. If, for instance, the modeler estimates the uncertainty to be about 50 percent, he or she will probably choose a change in the parameters at ± 10 percent and ± 50 percent and record the corresponding change in the state variable(s). It is often necessary to find the sensitivity at two or more levels of parameter changes, as the relation between a parameter and a state variable is rarely linear. Selection of the complexity and structure of a model should work hand in hand with the sensitivity analysis. A feedback arrow from the sensitivity analysis to the conceptual model is illustrated in Figure 14.1.

Two of the most significant steps in the modeling procedure are then calibration and validation, both of which occur while the model is being applied to the system being studied. *Calibration* is an attempt to find the best accordance between computed and observed data by variation of selected parameters. It may be carried out by trial and error or by use of software developed to find the parameters that give the best fit between observed and computed values. In some static and some simple models, which contain only a few well-defined or directly measured parameters, calibration may not be required. Even where all parameters are known within intervals, either from the literature or from estimation methods, it is usually necessary to calibrate the model, because:

1. Most parameters in environmental science and ecology are not known as exact values. Therefore even literature values for parameters have uncertainty.
2. All models in ecology and environmental sciences are simplifications of nature.

3. By far the most models in environmental sciences and ecology are *lumped models,* in which one parameter represents the average values of several species.

If it is impossible to calibrate a model properly, it is not necessarily due to an incorrect model but perhaps to a poor quality of data. The quality of the data is crucial for calibration. It is, furthermore, of great importance that observations reflect the dynamics of the system. If the objective of the model is to give a good description of one or a few state variables, it is essential that the data be able to show the dynamics of just these internal variables. The frequency of the data collection should therefore reflect the dynamics of the state variables in focus. This rule has unfortunately often been violated in modeling.

It is strongly recommended that the dynamics of all state variables be considered before the data collection program is determined in detail. Frequently, some state variables have particularly pronounced dynamics in specific periods—often in spring—and it may be a great advantage to have a dense data collection in this period in particular. Jørgensen et al. (1981) show how a dense data collection program in a certain period can be applied to provide additional certainty for the determination of some important parameters.

Calibration should always be followed by *validation.* Validation must be distinguished from verification as described above. Validation consists of an objective test as to how well the model outputs fit the data. The selection of possible objective tests will depend on the scope of the model, but the standard deviations between model predictions and observations and a comparison of observed and predicted minimum or maximum values of a particularly important state variable are frequently used. If several state variables are included in the validation, they may be given different weights. The use of validation can be summarized as follows:

1. Validation should always be required to get a picture of the reliability of a model.
2. Attempts should be made to get data for the validation that are entirely different from those used in the calibration.
3. The validation criteria should be formulated on the basis of the objectives of the model and the quality of the data available.

14.3   TYPES OF MODELS

Paired Classification

It is useful to distinguish between various types of models and to discuss briefly the selection of model types. Pairs of models are shown in a classification scheme in Table 14.3. The first division of models is based on the application: *research and management models*. The next pair is *stochastic and deterministic models*. A stochastic model contains stochastic input disturbances and random measurement errors, as shown in Figure 14.2. If they are both assumed to be zero, the stochastic model will reduce to a deterministic model, provided that the parameters are not estimated in terms of statistical distributions. A deterministic model assumes that the future response of a system is determined completely by a knowledge of the present-state and future measured inputs. Stochastic models are rarely applied in ecology today.

The third pair in Table 14.3 is *compartment and matrix models*. Compartment models are understood by some modelers to be models based on the use of compartments in the conceptual diagram, while other modelers distinguish between the two classes of models entirely by the mathematical for-

Table 14.3 Classification of models by pairs of model types

Type of Model	Characterization
Research models	These are used as a research tool.
Management models	These are used as a management tool.
Deterministic models	The values predicted are computed exactly.
Stochastic models	The values predicted depend on probability distribution.
Compartment models	The variables defining the system are quantified by means of time-dependent differential equations.
Matrix models	These use matrices in the mathematical formulation.
Reductionistic models	As many relevant details as possible are included.
Holistic models	General principles are followed.
Static models	The variables defining the system do not depend on time.
Dynamic models	The variables defining the system are a function of time (or perhaps of space).
Distributed models	The parameters are considered functions of time and space.
Lumped models	The parameters are within certain prescribed spatial locations and time, considered as constants.
Linear models	First-degree equations are used exclusively.
Nonlinear models	One or more of the equations are not first degree.
Causal models	The inputs, states, and outputs are interrelated by use of causal relations.
Black-box models	The input disturbances affect only the output responses. No causality is required.
Autonomous models	The derivatives are not explicitly dependent on the independent variable (time).
Nonautonomous models	The derivatives are explicitly dependent on the independent variable (time).

Source: After Jørgensen and Bendoricchio (2001).

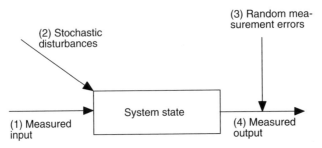

Figure 14.2 A stochastic model considers (1), (2), and (3), whereas a deterministic model assumes that (2) and (3) are zero. (From Jørgensen and Bendoricchio, 2001; copyright 2001; reprinted with permission from Elsevier Science.)

mulation, as indicated in the table. Both types of models are applied in environmental chemistry, although the use of compartment models is far more pronounced.

The classification of *reductionistic and holistic models* is based on a difference in the scientific ideas behind the model. The reductionistic modeler will attempt to incorporate as many details of a system as possible to be able to capture its behavior. The holistic modeler, on the other hand, attempts to include in the model system properties of the ecosystem working as a system by use of general principles. In this case, the properties of the system are not the sum of all the details considered, but the holistic modeler presumes that the system possesses additional properties because the subsystems are working as a unit.

Most problems in ecological engineering may be described by using a *dynamic model,* which uses *differential* or *difference equations* to describe the system response to external factors. Differential equations are used to represent continuous changes of state with time, while difference equations use discrete time steps. The steady state corresponds to the situation when all the derivatives equal zero. Oscillations around the steady state are described by use of a dynamic model (Figure 14.3), while the steady state itself can be described by use of a *static model.* As all derivatives are equal to zero in steady states, the static model is reduced to *algebraic equations.* A static model assumes, consequently, that all variables and parameters are independent of time. The advantage of the static model is its potential for simplifying subsequent computational effort through the elimination of one of the independent variables in the model relationship, but static models may give unrealistic results because oscillations caused for instance by seasonal and diurnal variations.

A *distributed model* accounts for variations of variables in time and space. A typical example would be an advection–diffusion model for transport of a dissolved substance along a stream. It might include variations in the three orthogonal directions. The analyst might, however, decide on the basis of

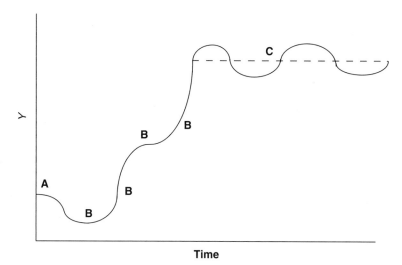

Time

Figure 14.3 *Y* is a state variable expressed as a function of time. *A* is the initial state, *B* represents transient states and *C* is the oscillation steady state. The dotted line corresponds to the steady state, which can be described by use of a static model. (From Jørgensen and Bendoricchio, 2001; copyright 2001; reprinted with permission from Elsevier Science.)

prior observations that gradients of dissolved material along one or two directions are not sufficiently large to merit inclusion in the model. The model would then be reduced by that assumption to a lumped-parameter model. Whereas a *lumped model* is frequently based on ordinary differential equations, the distributed model is usually defined by *partial differential equations.*

The *causal,* or *internally descriptive, model* characterizes the manner in which inputs are connected to states and how the states are connected to each other and to the outputs of the system, whereas the *black-box model* reflects only what changes in the input will affect the output response. In other words, the causal model describes the internal mechanisms of process behavior. The black-box model deals only with what is measurable: the input and the output. The relationship may be found by a statistical analysis. If, on the other hand, the processes are described in the model by use of equations, which cover the relationship, the model will be causal.

Autonomous models are not explicitly dependent on time (the independent variable):

$$\frac{dy}{dt} = ay^b + cy^d + e \tag{14.2}$$

Nonautonomous models contain terms, $g(t)$, that make the derivatives dependent on time: for instance,

$$\frac{dy}{dt} = ay^b + cy^d + e + g(t) \tag{14.3}$$

Functional Classification

Table 14.4 shows another classification of models. The differences among the three types of models are the choice of components used as state variables. If the model aims for a description of a number of individuals, species, or classes of species, the model will be called *biodemographic*. A model that describes the energy flows is *bioenergetic* and the state variables will typically be expressed in kilowatts or kilowatts per unit of volume or area. *Biogeochemical models* consider the flow of material, and the state variables are indicated as kilograms or kilograms per unit of volume or area. This model type is used primarily in ecotechnology.

14.4 SELECTING MODEL COMPLEXITY

A comprehensive discussion on the selection of model complexity is presented in Jørgensen and Bendoricchio (2001) and is summarized here. Among others, the following papers are devoted to this question: Jørgensen and Mejer (1977), Halfon et al. (1979), Bosserman (1980 1982), Halfon (1983, 1984), and Costanza and Sklar (1985).

The selection of model complexity is a matter of balance (Figure 14.4). On the one hand, it is necessary to include the state variables and processes essential for the problem in focus. On the other hand, it is important not to make the model more complex than the data set can bear. It is always a temptation to construct models that are too complex; it is easy to add more equations and more state variables to a computer program, but much harder to get the data needed for calibration and validation of the model. Even if we have detailed knowledge about a problem, we will never be able to develop a model that will be capable of accounting for the complete input–output behavior of a real ecosystem and be valid for all frames.

Table 14.4 Classification of models by type

Type of Model	Organization	Pattern	Measurements
Biodemographic	Conservation of genetic information	Life cycles of species	Number of species or individuals
Bioenergetic	Conservation of energy	Energy flow	Energy
Biogeochemical	Conservation of mass	Element cycles	Mass or concentrations

Source: After Jørgensen and Bendoricchio (2001).

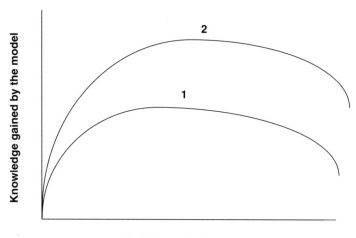

Model complexity

Figure 14.4 Knowledge gained versus model complexity, measured, for instance, by the number of state variables. The knowledge increases up to a certain level. Increased complexity beyond this level will not add to the knowledge gained about the modeled system. At a certain level the knowledge might even be decreased, due to uncertainty caused by too high a number of unknown parameters. Curve 2 corresponds to an available data set, which is more comprehensive or is of better quality than curve 1. Therefore, the knowledge gained and the optimum complexity are higher for data set (2) than for (1). (From Jørgensen and Bendoricchio, 2001; copyright 2001; reprinted with permission from Elsevier Science.)

Up to a point, a model may be made more realistic by adding ever more connections. Additions of new parameters after that point do not contribute further to improved simulation; on the contrary, more parameters imply more uncertainty because of the possible lack of information about the flows that the parameters quantify. Costanza and Sklar (1985) examined 88 different models and were able to show that the more theoretical discussion behind Figure 14.4 is actually valid in practice. Their results are summarized in Figure 14.5, where effectiveness is plotted versus articulation (i.e., expression for model complexity). *Effectiveness* is understood as a product of how much the model is able to tell and with what certainty; *articulation* is a measure of the complexity of the model with respect to the number of components, time, and space. The conclusion is, therefore, that although we can never know all that is needed to make a complete model (i.e., with inclusion of all details) we can produce good workable models that can expand our knowledge of an ecosystem, particularly of its properties as a system. This is completely consistent with the work of Ulanowicz (1979), who pointed out that the biological world is a sloppy place and that very precise predictive models will inevitably be wrong. This is a very important thing for engineers not used to ecological phenomena to learn before they attempt to fix their physical and chemical models toward ecological engineering questions. Yet models can be the most

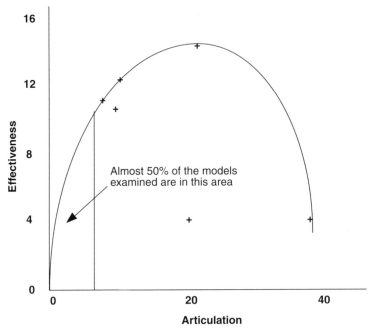

Figure 14.5 Plot of articulation index versus effectiveness = articulation times certainty for the models reviewed by Costanza and Sklar (1985). As almost 50 percent of the models were not validated, they had an effectiveness of zero. These models are not included in the figure, but are represented by the line "effectiveness = 0." Notice that almost another 50 percent of the models have a relatively low effectiveness, due to too little articulation, and that only one model had too high articulation, which implies that the uncertainty indicated by drawing the effectiveness frontier as shown in the figure is high at articulations above 25. (From Jørgensen and Bendoricchio, 2001; copyright 2001; reprinted with permission from Elsevier Science.)

useful and predictive tools in ecological engineering, both for the selection of the proper technology and for the design of created, restored, or constructed ecosystems.

14.5 CONCEPTUAL MODELS FOR ECOLOGICAL ENGINEERING

In ecological engineering, conceptual models are often used to get a first overview of how the ecosystem functions and the advantages and disadvantages that the ecotechnological solutions offer. Furthermore, conceptual diagrams can be used in the design phase of an artificial ecosystem or by modification of a natural ecosystem.

A conceptual model can be considered as a list of state variables and forcing functions of importance to an ecosystem that put the problem in focus, but it will also show how processes connect these components. It is employed as a tool to create abstractions of reality in ecosystems and to delineate the level of organization that best meets the objectives of the model. A wide spectrum of conceptualization approaches is available, and the most widely applied are presented here. Some give only the components and connections; others imply mathematical descriptions. It is hardly possible to give general recommendations as to which to apply in a given situation: It will be dependent on the problem, the ecosystem, the class of model, and to a certain extent on the habits of the modeler. Six types of conceptual diagrams are generally used in ecological engineering. These and other types of conceptual models are described by Jørgensen and Bendoricchio (2001) in detail and described briefly below:

1. *Word models* are a verbal description of model components and structure. Language is the tool of conceptualization in this case. Sentences can be used to describe a model briefly and precisely. However, word models of large complex ecosystems quickly become unwieldy and therefore are only used for very simple models. The proverb "One picture is worth a thousand words" explains why the modeler needs to use other types of conceptual diagrams to visualize the model.

2. *Picture models* use components seen in nature and place them within a framework of spatial relationships. *Box models* are simple and commonly used conceptual designs for ecosystem models. Each box represents a component in the model, and arrows between boxes indicate processes.

3. *Feedback dynamics diagrams* use a symbolic language introduced by Jay Forrester (1961), initially called *industrial dynamics* or *world dynamics* and later *Dynamo* when it was introduced as a digital simulation language in the 1980s (Figure 14.6). Rectangles represent state variables. Parameters or constants are small circles. Sinks and sources are cloudlike symbols, flows are arrows, and rate equations are the pyramids that connect state variables to the flows. The STELLA diagrams and STELLA simulation language, given as an example at the end of this chapter, are in the same category of diagrams and are modifications of Forrester's language.

4. A *computer flowchart* is sometimes used as a conceptual model. The sequence of events shown in the flowchart can be considered a conceptualization of the ordering of important ecological processes. A subcategory of computer flowcharts is that of analog computer diagrams. Analog diagrams and computers were used for simulating models before there was extensive use of digital computers. Applications of this type, useful primarily in a historical context with the availability of high-speed desktop digital machines, are described in Patten (1971–1976) and Odum (1983).

5. *Energy flow diagrams,* developed by H. T. Odum (see Chapter 4), are designed to give information on thermodynamic constraints, feedback mechanisms, and energy flows. The most commonly used symbols in this language are shown in Figure 4.5. As the symbols have an implicit mathematical mean-

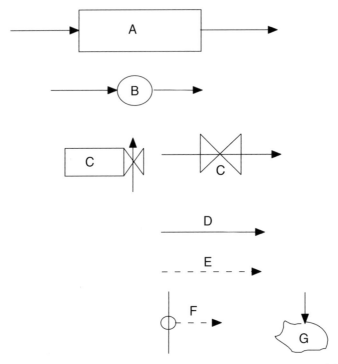

Figure 14.6   Symbolic language introduced by Jay Forrester (1961): A, state variable; B, auxiliary variable; C, rate equations; D, mass flow; E, information flow; F, parameter; G, sink. This language eventually became the basis of STELLA. (Redrawn from Jeffers, 1978.)

ing, it gives much information about the mathematics of the model. It is, furthermore, rich in conceptual information, and hierarchical levels can easily be displayed. These diagrams have found wide application for the development of ecological/economic models, where the energy is used as the translation from economy to ecology, and vice versa. Several energy flow diagrams are included in this book.

14.6   MODELING CONSTRAINTS AND RECENT DEVELOPMENTS

A modeler is very much concerned about use of the correct description of the components and processes in the models chosen. The model equations and their parameters should reflect the properties of the model components and processes as correctly as possible. The modeler must, however, also be concerned with the right description of system properties, and too little research has been done in this direction. Continuous development of models as scientific tools will need to apply constraints on models according to system properties and to take into account the limitations of our knowledge of the system.

Conservation principles are often used as modeling constraints (see Chapter 4). Biogeochemical models must follow the conservation of mass, and bio-energy models must equally obey the laws of energy and momentum conservation. These are the classical principles that are imposed on models of engineering systems, too. Boundary and initial conditions are imposed on models as mathematical constraints, based on system properties. Many biogeochemical models are given narrow bands of the chemical composition of the biomass. Eutrophication models are based on either a constant stoichiometric ratio of elements in phytoplankton or on an independent cycling of the nutrients, where, for example, the phosphorus content may vary from 0.4 to 2.5 percent, the nitrogen content from 4 to 12 percent, and the carbon content from 35 to 55 percent. Some modelers have used the second law of thermodynamics and the concept of entropy to impose thermodynamic constraints on models.

Ecological models contain many parameters and process descriptions and at least some interacting components, but the parameters and processes can hardly be given unambiguous values and equations, even by use of the model constraints already mentioned. It means that an ecological model in the initial phase of development has many degrees of freedom. It is therefore necessary to limit the degrees of freedom to come up with a workable model that is not doubtful and undeterministic. Many modelers use a comprehensive data set and a calibration to limit the number of possible models. This is, however, a cumbersome method if it is not accompanied by some realistic constraints on the model. The calibration is therefore often limited to giving the parameters realistic and literature-based intervals within which the calibration is carried out.

Far more would be gained, perhaps, if it were possible to give models more ecological properties and/or to test the model from an ecological point of view to exclude the editions of the model that are not ecologically possible. For example, how could the hierarchy of ecosystems presented in several chapters of this book be accounted for in the models? The development of models in the 1960s and 1970s was to a great extent imitating the models of physical systems. It implied that the models were given fixed (rigid) parameters. But the process coefficients in nature are indeed not fixed or rigid, adjusting according to the conditions. In principle, it is therefore wrong when we develop models with fixed parameters and use a lot of effort on assessment of the correct parameter values. *Structurally dynamic models* attempt to solve this problem by use of a goal function, which defines how the parameters should be changed to account for adaptation processes. It is important to develop such new modeling approaches to account for the real properties of ecosystems. With a few exceptions, ecological models have not been able to make very convincing predictions, which could be due to the lack of ecological properties by our present models. We know that the evolution has created very complex ecosystems with many feedback processes, regulations, and interactions. The coordinated coevolution means that rules and principles have been imposed for cooperation among the biological components. These rules and principles are the governing laws of ecosystems, which is one of the focuses of this book, and our models should of course follow these principles and laws as broadly as possible. In ecotechnology, where the effect often depends on proper development of an artificial or natural ecosystem, these approaches are of particular importance.

It also seems possible to limit the number of parameter *combinations* by use of what could be called *ecological tests* (Figure 14.7). The maximum growth rates of phytoplankton and zooplankton may, for instance, have realistic values in a eutrophication model, but the two parameters do not fit

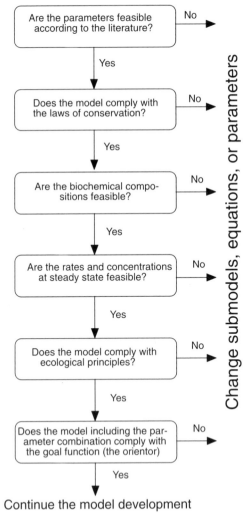

Continue the model development

Figure 14.7 Considerations on using various constraints by development of models. The range of parameter values, particularly, is limited by the procedure shown.

each other, because they will create chaos in the ecosystem, which is inconsistent with actual or general observations. Such combinations should be excluded at an early stage of model development.

Several additional developments in ecological modeling have attempted to overcome the obvious shortcomings of previous models. *Catastrophe theory* has been used to explain why ecosystems in some situations have two (or more) steady states as attractors. Furthermore, our knowledge about ecosystems will always be limited, due to the enormous complexity of nature. Ecological data bear a large inherent uncertainty, due to inaccuracy of data and lack of sufficient knowledge about the parameters and state variables. On the other hand, semiquantitative model outputs may be sufficient in many management situations. Fuzzy knowledge–based models are available in such situations. They are models that do not predict the values of the state variables exactly, but only generally, as, for instance, "high" or "low." The journal *Ecological Modeling* devoted a special issue (Volume 85, 1996) to this interesting approach, which takes into consideration the shortcomings of our ecological knowledge.

We probably need new methods to express ourselves, our observations, and the results of model simulations. We are probably too much in the habit of thinking in terms of numbers. We have therefore developed models that require numbers and generate numbers. We may, however, get more information—or more easily grasped information—by presenting the distribution of species and various levels of selected components in different parts of the system. It will tell us much more much faster than will a long table with a lot of numbers that are not exact anyhow. In systems ecology, we may reformulate the old phrase "one picture is worth a thousand words" into "a picture (pattern, model) says more than a thousand numbers." A translation of model results into geographical patterns which seems applicable in ecotechnology to find the optimum location of artificial ecosystems is already under rapid development through the use of GIS (*geographical information systems*).

14.7   MODELS AS EXPERIMENTAL TOOLS

The focus of this book is on ecological engineering and ecosystem restoration. However, modeling is an important tool when we want to reveal system properties of ecosystems. Modeling has therefore been presented in this chapter in some detail and has been mentioned in several other chapters because of its power as a scientific and engineering tool. Modeling may, to a certain extent, be considered a tool for the examination of system properties, just as statistics is a tool for performing general scientific examinations.

It is clear from the presentation of modeling in this chapter that we cannot construct a model of all details or components of an ecosystem. We have to limit ourselves to including the components and processes that are of importance for the system properties that we want to investigate. This is consistent with the application of quantum mechanical ideas in ecology (Jørgensen, 2002). It is assumed that we have been able to define a specific problem and the components of the ecosystem of importance for that problem. It implies that we are able to construct a model that to a certain extent can be used as a representation of the ecosystem in the context of the focal problem. The model has, during the construction phase, already been used as an experimental scientific tool. If the verification, calibration and validation are not running satisfactorily, we will of course ask ourselves why, and the answer may be that we need to include more feedbacks, a state variable more or less, or change a process description because it is too primitive in its present form.

When the verification, calibration, and validation have been accepted, the model is ready to be used as a tool in science or environmental management, including ecotechnology. The idea is to ask the model "scientific questions" about system properties. Patten (1991) uses models in the latter sense, asking: What is the ratio between direct and indirect effects? We can use models to test the hypothesis of ecosystem behavior, such as the various proposed thermodynamic principles of ecosystems discussed in Chapter 4. The certainty of the hypothesis test by using models is, however, not on the same level as the certainty of tests used in more reductionistic science. If a relation is found here between two or more variables on the basis of, for instance, the use of statistics on the available treatment of data, the relationship is afterward tested on a number of additional cases to increase the scientific certainty. If the results are accepted, the relationship is ready to be used to make predictions, and whether the predictions are wrong or right is examined. If the relationship still holds, we are satisfied, and a wider scientific use of the relationship is made possible.

When we are using models as scientific tools to test hypotheses, we have a "double doubt." We anticipate that the model is correct in the problem context, but the model is a hypothesis of its own. We therefore have four cases instead of two (acceptance/nonacceptance):

1. The model is correct in the problem context, and the hypothesis is correct.
2. The model is not correct, but the hypothesis is correct.
3. The model is correct, but the hypothesis is not correct.
4. The model is not correct, and the hypothesis is not correct.

To omit cases 2 and 4, only very well examined and well accepted models should be used to test hypotheses on system properties, but our experience today in modeling ecosystems is unfortunately limited. We do have some well-examined models, but we are not completely certain that they are correct in the problem context, and we would generally need a wider range of models. A wider experience in modeling may therefore be a prerequisite for further development in ecosystem research.

14.8   USING STELLA MODELING FOR ECOLOGICAL ENGINEERING

STELLA is particularly applicable to the development of small and medium-sized models based on small- and medium-sized databases. This is usually the case in ecological engineering projects. Data are not always in abundance and the time required to make predictions is relatively short. It would therefore be inappropriate to use either a model with major data requirements or one that would require a year or more of programming to develop or run. Stella is a recommended high-order language for model development in ecological engineering. STELLA modeling has been applied to ecological engineering questions in a number of situations, including evaluation of theories for pulsing (Mitsch, 1988), exergy (Jørgensen, 1988), and self-design (Metzger and Mitsch, 1997), and in numerous case studies on nutrient retention in wetlands (Mitsch and Reeder, 1991; Christensen et al., 1994; Spieles and Mitsch, 2000a; Wang and Mitsch, 2000), biogeochemical–economic models (Baker et al., 1991; Ahn and Mitsch, 2002b), Chinese ecological engineering systems (Hagiwara and Mitsch, 1994; Bruins et al., 1998), mine drainage wetlands (Flanagan et al., 1994; Mitsch and Wise, 1998), and functional analysis of mitigation wetlands (Niswander and Mitsch, 1995), among others. STELLA is based on the following key components:

1. A conceptual diagram is developed where the state variables are boxes, the flows are arrows, and the information controlling the flows is represented by thin arrows. Differential equations are set up automatically by the software, based on conservation principles and on the conceptual model that is "drawn" on the computer screen. It implies that it is not possible to shift from nitrogen in one state variable, for example, to biomass in another state variable. The mass conservation must be maintained either by consistent use of nitrogen or account for the making up of biomass by considering inflows of P, C, and perhaps other elements that may contribute to the biomass.
2. Flows must be defined by use of equations, tables, or graphs describing the relationships between flows and other variables.
3. Forcing functions are flows of inputs and outputs. They can be formulated as equations, tables, or graphs.

The application of STELLA is illustrated next by an example of nitrogen removal by a wetland.

CASE STUDY
Ecological Modeling of Nitrogen Retention by Wetlands

Nonpoint sources have been in focus since the late 1970s. Nitrogen and phosphorus balances have shown that agriculture and other nonpoint sources contribute significantly to overall pollution, particularly to eutrophication. Environmental technology is not sufficient but must be supplemented with other methods to cope with the problems of nonpoint sources. As described in Chapter 10, wetlands often are very effective, at least where nitrogen plays a role in eutrophication. Nitrogen balances for agricultural regions have revealed that nitrogen from nonpoint sources plays a major role in causing eutrophication problems in freshwater and marine ecosystems, and solutions cannot be found without solving the problems associated with nonpoint pollution. The entire spectrum of available ecological engineering methods touched on above has been implemented so far to solve the problem. In this context, there is a need for a wetland model that is able to make predictions as to the nitrogen removal capacity of a wetland on the basis of certain information about an existing or planned wetland.

The model presented here is based on previous model approaches by Jørgensen et al. (1988) and Dørge (1991). The model differs from the previous models in being simpler, which was needed to make the model more general. Furthermore, the model is dynamic as to both hydrology and biology, whereas Dørge's model is a steady-state model for the biological components. A dynamic model is more difficult to calibrate, but calibration of a dynamic model will often reveal bias relations more clearly. This feature of dynamic models has been used to make a site-specific calibration. The results of model applications in two case studies are presented.

A conceptual diagram of the model is illustrated with STELLA figures in Figure 14.8, STELLA equations are presented in Table 14.5, and parameters are presented in Table 14.6. The climatic forcing functions are precipitation, evaporation, temperature, and solar radiance. The last is given as a cosine function and the three former functions, as tables in the programming. The same functions are applied to both case studies. The site-specific forcing functions are the nitrate and ammonium concentrations in the in-flowing water and the rate of inflow.

The model construction considers 1 m² of wetland and looks into the conversion of nitrogen in this area. The result of the model will therefore be how much nitrogen can be removed, accumulated, and/or released per unit area. Two hydrological state variables are applied, one representing the surface layer, where nitrification can take place, and the active zone, where a pronounced denitrification and accumulation take place. The depth of this layer is not very important because in the great majority of cases, the limiting factor is the hydraulic conductivity. The amount of organic matter and the available space for denitrifying microorganisms in this zone are not limiting under any circumstances.

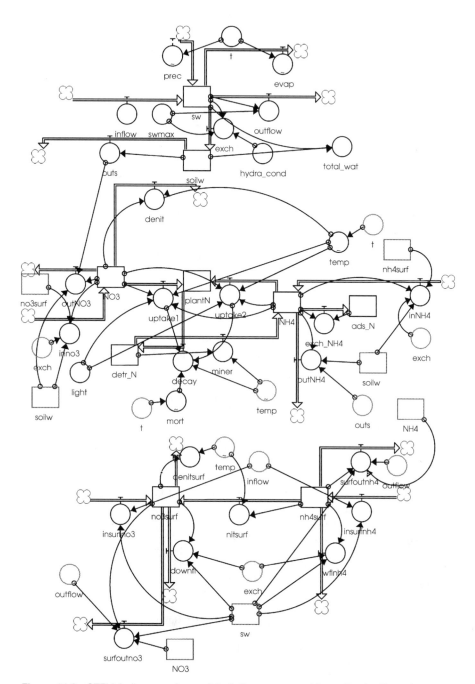

Figure 14.8 STELLA diagram of a model of nitrogen removal by wetlands. (From Jørgensen and Bendoricchio, 2001; copyright 2001; reprinted with permission from Elsevier Science.)

Table 14.5 Model equations in STELLA for the nitrogen model illustrated in Figure 14.8

ads_N = ads_N + dt * (exch_NH4)
INIT(ads_N) = 200/9
detr_N = detr_N + dt * (decay − miner)
INIT(detr_N) = 1200
NH4 = NH4 + dt * (−uptake2 + miner − exch_NH4 − outNH4 + inNH4)
INIT(NH4) = 1.0
nh4surf = nh4surf + dt * (−nitsurf + insurfnh4 − wflnh4 − surfoutnh4)
INIT(nh4surf) = 0.1
NO3 = NO3 + dt * (−uptake1 − outNO3 − denit + inno3)
INIT(NO3) = 10
no3surf = no3surf + dt * (insurfno3 + nitsurf − downfl − denitsurf − surfoutno3)
INIT(no3surf) = 5
plantN = plantN + dt * (uptake1 + uptake2 − decay)
INIT(plantN) =20
soilw = soilw + dt * (exch − outs)
INIT(soilw) = 2.0
sw = sw + dt * (inflow − outflow + prec − evap − exch)
INIT(sw) = 0.015
decay = (1.04^(temp-20))*mort*(uptake1+uptake2)
denit = (1.12^(temp-20))*8*NO3/(12+NO3)
denitsurf = (1.12^(temp-20)) * 8 * no3surf/(12+no3surf)
downfl = exch * no3surf/sw
exch = IF sw > swmax THEN hydra_cond ELSE sw * hydra_cond/swmax
exch_NH4 = IF ads_N < 200 * NH4/(8+NH4) THEN NH4/(8+NH4) ELSE 0
hydra_cond = 0.09
inflow = 0.035
inNH4 = (exch * nh4surf+0.01 * (nh4surf-NH4))/soilw
inno3 = (exch * no3surf+0.01 * (no3surf-NO3))/soilw
insurfnh4 = inflow * 0.2/sw
insurfno3 = inflow * 5/sw
light = 1.91 − 1.68 * COS(6.1 * (TIME-355)/365)
miner = 0.0001 * detr_N * 1.07^(temp-20)
nitsurf = 8 * (1.12^(temp-20)) * nh4surf/(8+nh4surf)
outflow = IF sw > swmax THEN 1.0 * (sw-swmax) ELSE 0
outNH4 = outs * NH4/soilw
outNO3 = outs * NO3/soilw
outs = IF soilw > 2.45 THEN 0.1 ELSE 0
surfoutnh4 = (nh4surf * outflow+0.01 * (nh4surf-NH4))/sw
surfoutno3 = (outflow * no3surf+0.01 * (no3surf-NO3))/sw
swmax = 0.05
t = TIME
total_wat = soilw+sw
uptake1 = IF NO3 > 0.05 THEN light * 0.15*(1.05^(temp-20)) * NO3/
  (NO3+NH4) ELSE 0
uptake2 = IF NH4 > 0.05 THEN light * 0.15 * (1.05^(temp-20)) * NH4/
  (NO3+NH4) ELSE 0
wflnh4 = exch * nh4surf/sw
evap = graph(t)
mort = graph(t)
prec = graph(t)
temp = graph(t)

The nitrogen state variables are nitrate and ammonium in the surface layer and nitrate, ammonium, detritus-N, plant-N, and adsorbed-N in the active layer. Cycling of nitrogen takes place in the active layer; ammonium and nitrate are taken up by plants. Plant-N form detritus-N by decay and after mineralization is ammonium-formed. Nitrification and denitrification are described by Michaelis–Menten equations, and the uptake of nitrate and ammonium by the plants is formulated by first-order kinetics and is proportional to the light. There are no differences between the uptake rates for ammonium and nitrate. The uptake is therefore proportional to the concentration of inorganic nitrogen = ammonium + nitrate. The mineralization also follows first-order kinetics.

The decay depends on the uptake and a mortality function, which can be formulated in a table according to the seasonal variations generally observed in a given area. All biological rates are dependent on temperature, with a more pronounced dependence for nitrification and denitrification. The following site-specific parameters are used: hydrological conductivity, nitrification capacity, denitrification capacity, detritus-N pool (the initial value of this state variable), and initial and maximum values of plant-N. The parameters are calibrated: uptake rates for nitrate and ammonium and the mineralization rate. These parameters are adjusted to give the observed trends in detritus-N and the aforementioned maximum value of plant-N. The model has been used in several case studies, of which two are shown. Site-specific parameters, which are the basis for the model application, are shown in Table 14.6.

Uptake rates for nitrate and ammonium and the mineralization rate are found by calibration. These two parameters are presented in Table 14.7. Calibration of the two case studies was easy and gave a reasonable validation, as shown in Table 14.8.

The most interesting results of the model applications are the nitrate concentration in the outflowing water (Figure 14.9) and the

Table 14.6   Wetland properties and wetland model parameters (based on 1 m²)

Parameter	Rabis Wet Meadow	Glumsø Reed Swamp
Hydraulic conductivity (m day⁻¹)	0.009	0.009
Production-N, (yr⁻¹)	7.0	40.0
Detritus-N (g)	800	1200
Maximum nitrification (g-N day⁻¹)	11	7
Maximum denitrification (g-N day⁻¹)	22	72

Table 14.7　Parameters calibrated for wetland nitrogen model

Parameter	Rabis Wet Meadow	Glumsø Reed Swamp
Uptake rate (day^{-1})	0.025	0.125
Mineralization rate (day^{-1})	0.00005	0.00025

comparison of nitrogen balances obtained by the validation and corresponding observations (shown in parentheses in Table 14.8). The agreement between the validation and observations is fully acceptable, particularly in the light of the uncertainty, which must always be accepted in environmental planning.

It was the aim of the described model to obtain general applicability. The idea may be expressed as follows: Give pertinent information about the wetland and the model will give you the capability of the wetland to remove nitrogen. The ecological engineer will thus be able to assess how much wetland area is needed to achieve certain goals for the removal of nitrogen from nonpoint sources on a regional basis. The model has been used in several case studies with acceptable results, which is promising for general use of the model for assessment of the size of a constructed or modified wetland. The model can also be used in the siting-sizing procedure. Yet it is recommended that more experience be obtained from even more case studies before use on a regional basis.

The procedure for a wider application seems clear from the experience already obtained in the case studies. A tentative procedure is summarized in the flowchart in Figure 14.10. The methods to be used if a wetland does not exist but is planned for construction are similar. The climatic forcing functions used are regionally based, but the wetland properties for a nonexisting wetland can, of course, not be found. The hydrological conductivity can probably be estimated

Table 14.8　Nitrogen balance (based on 1 m²) determined in wetland model

Nitrogen Flow $(g\text{-}N\ yr^{-1})$	Rabis Wet Meadow[a]	Glumsø Reed Swamp[a]
Loading (L)	55	64
Removed by denitrification (1)	24 (20)	89 (92)
Released (2)	0 (0)	37 (40)
Accumulated (3)	3 (5)	7 (5)
Percent removal = [(1) + (3) − (2)]/L	49 (45)	92 (89)

[a]The numbers indicate simulation results; the numbers in parentheses are data.

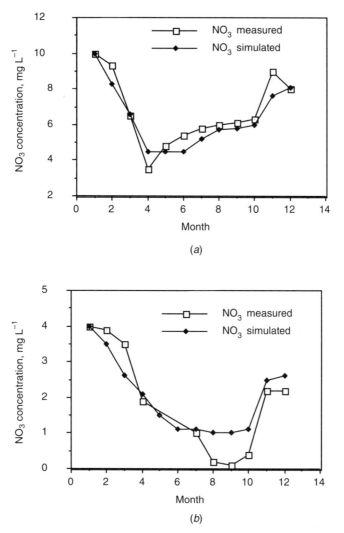

Figure 14.9 Comparison of measured and simulated values of outflow nitrate for Danish wetlands: (a) wet meadow; (b) Glumsø reed swamp. (From Jørgensen and Bendoricchio, 2001; copyright 2001; reprinted with permission from Elsevier Science.)

from the soil characteristics and by comparison with wetlands of similar vegetation and soil types. The initial and maximum values of plant-N and the trends in detritus-N are estimated from a wetland in the region that has similar vegetation. The depth of the surface layer is estimated from wetlands with similar vegetation and from the slope of the landscape where the wetland is planned.

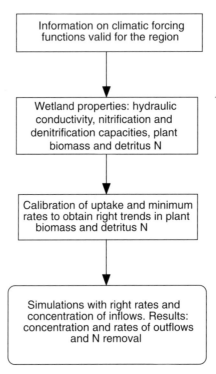

Figure 14.10 Procedure applicable for development of a wetland model for a specific site, from the general model presented in the text. (From Jørgensen and Bendoricchio, 2001; copyright 2001; reprinted with permission from Elsevier Science.)

REFERENCES

Aanen, P., W. Alberts, G. J. Bekker, H. D. van Bohemen, P. J. M. Melman, J. van der Sluijs, G. Veenbaas, H. J. Verkaar, and C. F. van de Watering. 1991. *Nature Engineering and Civil Engineering Works.* Center for Agricultural Publishing and Documentation (Pudoc), Wageningen, The Netherlands.

Aber, J. D. and W. R. Jordan III. 1985. Restoration ecology: an environmental middle ground. *BioScience* 35:399.

Adey, W. H. and K. Loveland. 1991. *Dynamic Aquaria: Building Living Ecosystems.* Academic Press, New York. 643 pp.

Ahl, T. and T. Weiderholm. 1977. *Svenska Vattenkvalitetskriterier: Eurofierande Ammen.* Report SNV PM 918, Swedish National Environmental Protection Board, Solna, Sweden.

Ahn, C. and W. J. Mitsch. 2001. Chemical analysis of soil and leachate from experimental wetland mesocosms lined with coal combustion products. *Journal of Environmental Quality* 30:1457–1463.

Ahn, C. and W. J. Mitsch. 2002a. Scaling considerations of mesocosm wetlands in simulating large created freshwater marshes. *Ecological Engineering* 18:327–342.

Ahn, C. and W. J. Mitsch. 2002b. Evaluating the use of recycled coal combustion products in constructed wetlands: An ecologic–economic modeling approach. *Ecological Modelling* 150:117–140.

Ahn, C., W. J. Mitsch, and W. E. Wolfe. 2001. Effects of recycled FGD liner material on water quality and macrophytes of constructed wetlands: a mesocosm experiment. *Water Research* 35:633–642.

Allen, P. M. 1988. Ecology, thermodynamics, and self-organization: towards a new understanding of complexity. In: R. E. Ulanowicz and T. Platt, eds., *Ecosystem Theory for Biological Oceanography. Canadian Bulletin of Fisheries and Aquatic Sciences,* 123:3–26.

Allen, T. F. H. and T. B. Starr. 1982. *Hierarchy: Perspectives for Ecological Complexity.* University of Chicago Press, Chicago. 310 pp.

Alper, J. 1998. Ecosystem "engineers" shape habitats for other species. *Science* 280: 1195–1196.

Arheimer, B. and H. B. Wittgren. 1994. Modelling the effects of wetlands on regional nitrogen transport. *Ambio* 23:378–386.

Aronstein, B. N., Y. M Calvillo, and M. Alexander. 1991. Effects of surfactants at low concentration on the desorption and biodegradation of sorbed aromatic compounds in soil. *Environmental Sciences & Technology* 25:1728–1731.

Australian Nature Conservation Agency. 1996. *Wetlands Are Important.* National Wetlands Program, ANCA, Canberra, Australia. 2 pp.

Baker, K., S. Fennessy, and W. J. Mitsch. 1991. Designing wetlands for controlling coal mine drainage: an ecologic–economic modelling approach. *Ecological Economics* 3:1–24.

Bayley, P. B. 1995. Understanding large river–floodplain ecosystems. *BioScience* 45: 153–158.

Bedinger, M. S. 1981. Hydrology of bottomland hardwood forests of the Mississippi embayment. Pages 161–176. In: J. R. Clark and J. Benforado, eds., *Wetlands of Bottomland Hardwood Forests*. Elsevier, Amsterdam.

Beeftink, W. G. 1977. Salt marshes. Pages 93–121. In: R. S. K. Barnes, ed., *The Coastline*. Wiley, New York.

Bendoricchio, G. 1988. An application of the theory of catastrophe to the eutrophication of the Venice Lagoon. Pages 156–166. In: A. Marani, ed., *Advances in Environmental Modelling*. Elsevier, Amsterdam. 690 pp.

Benndorf, J., 1990. Conditions for effective biomanipulation: conclusions derived from whole-lake experiments in Europe. *Hydrobiologia* 200/201:187–203.

Benthem, W., L. P. van Lavieren, and W. J. M. Verheugt. 1999. Mangrove rehabilitation in the coastal Mekong delta, Vietnam. Pages 29–36. In: W. Streever, ed., *An International Perspective on Wetland Rehabilitation*. Kluwer Academic, Dordrecht, The Netherlands.

Bergen, S. D., S. M. Bolton, and J. L. Fridley. 2001. Design principles for ecological engineering. *Ecological Engineering* 18:201–210.

Best, G. R., P. M. Wallace, J. J. Dunn, and H. T. Odum. 1988. *Enhancing Ecological Succession Following Phosphate Mining*. Publication 03-008-064. Florida Institute of Phosphate Research Bartow, FL.

Boltzmann, L. 1905. *The Second Law of Thermodynamics*. Populare Schriften, Essay 3 (address to Imperial Academy of Science in 1886). Reprinted in English in: *Theoretical Physics and Philosophical Problems: Selected Writings of L. Boltzmann*. D. Reidel, Dordrecht, The Netherlands. 364 pp.

Boon, P. J., B. R. Davies, and G. E. Petts, eds. 2000. *Global Perspectives on River Conservation: Science, Policy, and Practice*. Wiley, Chichester, West Sussex, England.

Bosserman, R. W. 1980. Complexity measures for assessment of environmental impact in ecosystem networks. *Proceedings of the Pittsburgh Conference on Modelling and Simulation,* Pittsburgh, PA, April 20–23, 1980.

Bosserman, R. W. 1982. Structural comparison for four lake ecosystem models. Pages 559–568 In: L. Troncale, ed., *A General Survey of Systems Methodology— Proceedings of the 26th Annual Meeting of the Society for General Systems Research,* Washington, DC.

Boule, M. E. 1988. Wetland creation and enhancement in the Pacific Northwest. Pages 130–136. In: J. Zelazny and J. S. Feierabend, eds., *Proceedings of the Conference on Wetlands–Wetlands: Increasing Our Wetland Resources*. Corporate Conservation Council, National Wildlife Federation, Washington, DC.

Boumans, R. M. J., J. W. Day, G. P. Kemp, and K. Kilgen. 1997. The effect of intertidal sediment fences on wetland surface elevation, wave energy and vegetation establishment in two Louisiana coastal marshes. *Ecological Engineering* 9:37–50.

Boustany, R. G., C. R. Crozier, J. M. Rybczyk, and R. R. Twilley. 1997. Denitrification in a south Louisiana wetland forest receiving treated sewage effluent. *Wetlands Ecology and Management* 4:273–283.

Bradshaw, A. D. 1983. The reconstruction of ecosystems. *Journal of Applied Ecology* 20:1–17.

Bradshaw, A. D. 1987. Restoration: the acid test for ecology. Pages 23–29. In: W. R. Jordan. III, M. E. Gilpin, and J. D. Aber, eds., *Restoration Ecology: A Synthetic Approach to Ecological Research.* Cambridge University Press, Cambridge.

Bradshaw, A. D. 1992. The biology of land restoration. Pages 25–44. In: S. K. Jain and L. W. Botsford, eds., *Applied Population Biology.* Kluwer Academic, Dordrecht, The Netherlands.

Bradshaw, A. D. 1996. Underlying principles of restoration. *Canadian Journal of Fisheries and Aquatic Sciences* 53(Suppl. 1):3–9.

Bradshaw, A. D. 1997. Restoration of mined lands: using natural processes. *Ecological Engineering* 8:255–269.

Bradshaw, A. D. and R. F. Hüttl. 2001. Future minesite restoration involves a broader approach. *Ecological Engineering* 17:87–90.

Braskerud, B. C. 2002a. Factors affecting nitrogen retention in small constructed wetlands treating agricultural non-point source pollution. *Ecological Engineering* 18:351–370.

Braskerud, B. C. 2002b. Factors affecting phosphorus retention in small constructed wetlands treating agricultural non-point source pollution. *Ecological Engineering* 19:41–61.

Brierley, C. L. 1995. Bioremediation of metal-contaminated surface and ground waters. *Geomicrobiology Journal* 8:201–223.

Brinson, M. M., B. L. Swift, R. C. Plantico, and J. S. Barclay. 1981. *Riparian Ecosystems: Their Ecology and Status.* FWS/OBS-81/17. U.S. Fish and Wildlife Service, Washington, DC. 151 pp.

Brix, H. 1987. Treatment of wastewater in the rhizosphere of wetland plants: the root-zone method. *Water Science and Technology* 19:107–118.

Brix, H. 1994. Use of constructed wetlands in water pollution control: historical development, present status, and future perspectives. *Water Science and Technology* 30:209–223.

Brix, H. 1998. Denmark. Pages 123–152. In: J. Vymazal, H. Brox, P. F. Cooper, M. D. Green, and R. Haberl, eds., *Constructed Wetlands for Wastewater Treatment in Europe.* Backhuys Publishers, Leiden, The Netherlands.

Brix, H. and H.-H. Schierup. 1989a. The use of aquatic macrophytes in water-pollution control. *Ambio* 18:100–107.

Brix, H. and H.-H. Schierup. 1989b. Sewage treatment in constructed reed beds: Danish experiences. *Water Science and Technology* 21:1655–1668.

Brodie, G. A., D. A. Hammer, and D. A. Tomljanovich. 1988. Constructed wetlands for acid drainage control in the Tennessee valley. Pages 325–331. In: *Mine Drainage and Surface Mine Reclamation.* Circular 9183. Bureau of Mines, Pittsburgh, PA.

Broome, S. W. 1990. Creation and restoration of tidal wetlands of the southeastern United States. Pages 37–72. In: J. A. Kusler and M. E. Kentula, eds., *Wetland Creation and Restoration.* Island Press, Washington, DC.

Broome, S. W., E. D. Seneca, and W. W. Woodhouse, Jr. 1988. Tidal salt marsh restoration. *Aquatic Botany* 32:1–22.

Brown, D. S. and E. W. Flagg. 1981. Empirical prediction of organic pollutant sorption in natural sediments. *Journal of Environmental Quality* 10:382–386.

Brown, M. T. 1987. *Conceptual Design for a Constructed Wetlands System for the Renovation of Treated Effluent.* Report from the Center for Wetlands, University of Florida, Gainesville, FL. 18 pp.

Brown, M. T., R. E. Tighe, T. R. McClanahan, and R. W. Wolfe. 1992. Landscape reclamation at a central Florida phosphate mine. *Ecological Engineering* 1:323–354.

Bruins, R. J. F. 1997. Modeling of floodplain response and ecological engineering in an agricultural wetland region of central China. Ph.D. dissertation. Ohio State University, Columbus, OH.

Bruins, R. J. F., S. Cai, S. Chen, and W. J. Mitsch. 1998. Ecological engineering strategies to reduce flooding damage to wetland crops in central China. *Ecological Engineering* 11:231–259.

Buckley, G. P., ed. 1989. *Biological Habitat Reconstruction.* Belhaven Press, London. 363 pp.

Burton, T. M. and H. H. Prince, 1995a. A landscape approach to wetlands restoration research along Saginaw Bay, Michigan: baseline data collection and project description. *Proceedings of the 38th Conference of the International Association of Great Lakes Research.* International Association of Great Lakes Research, Ann Arbor, MI.

Burton, T. M. and H. H. Prince. 1995b. Restoration of Saginaw Bay coastal wetlands in Michigan. In: M. C. Landin, ed., *Proceedings of the National Interagency Workshop on Wetlands,* New Orleans, LA.

Busnardo, M. J., R. M. Gersberg, R. Langis, T. L. Sinicrope, and J. B. Zedler. 1992. Nitrogen and phosphorus removal by wetland mesocosms subjected to different hydroperiods. *Ecological Engineering* 1:287–307.

Cairns, J., Jr. 1980. *The Recovery Process in Damaged Ecosystems.* Ann Arbor Press, Ann Arbor, MI.

Cairns, J., Jr. 1988a. Restoration ecology: The new frontier. Pages 1–11. In: J. Cairns, Jr., ed., *Rehabilitating Damaged Ecosystems,* Vol. I. CRC Press, Boca Raton, FL.

Cairns, J., Jr., ed. 1988b. *Rehabilitating Damaged Ecosystems,* Vol. I. CRC Press, Boca Raton, FL.

Calow, P. and G. E. Petts, eds. 1994. *The Rivers Handbook,* Vol. 2. Blackwell Scientific, London. 523 pp.

Carpenter, S. R. 1998. The need for large-scale experiments to assess and predict the response of ecosystems to perturbation. Pages 287–312. In: M. L. Pace and P. M. Groffman, eds., *Successes, Limitations and Frontiers of Ecosystem Science.* Springer-Verlag, New York.

CH2M-Hill and Payne Engineering. 1997. *Constructed Wetlands for Livestock Wastewater Management: Literature Review, Database, and Research Synthesis.* Mississippi Nutrient Enrichment Committee, U.S. Environmental Protection Agency, Stennis Space Center, MS.

Chai T., W. Shi, T. Lu, and M. Ye. 1988. Benefit analyses of agroecological engineering in Dongxu Village of Jiangsu Province. In: M. Shijun, A. Jiang, R. Xu, and D. Li, eds., *Proceedings of the International Symposium on Agro-Ecological Engineering,* August 1988. Ecological Society of China, Beijing.

Chapelle, F. H., P. M. Bradley, D. R. Lovley, and D. A. Vloblesky. 1996. Measuring rates of biodegradation in contaminated aquifer using field and laboratory methods. *Ground Water* 34:691–698.

Christensen, N., W. J. Mitsch, and S. E. Jørgensen. 1994. A first generation ecosystem model of the Des Plaines River experimental wetlands. *Ecological Engineering* 3: 495–521.

Christensen, T. H. 1981. *The Application of Sludge as Soil Conditioner,* Vol. 3. Polyteknisk Forlag, Copenhagen, pp. 19–47.

Christensen, T. H. 1984. Cadmium soil sorption at low concentrations: (1) effect of time, cadmium load, pH and calcium and (2) reversibility, effect of changes in solute composition, and effect of soil ageing. *Water Air and Soil Pollution* 21:105–125.

Chubin, R. G. and J. J. Street. 1981. Adsorption of cadmium on soil constituents in the presence of complexing agents. *Journal of Environmental Quality* 10:225–228.

Chung, C. H. 1982. Low marshes, China. Pages 131–145. In: R. R. Lewis, ed., *Creation and Restoration of Coastal Plant Communities.* CRC Press, Boca Raton, FL.

Chung, C. H. 1983. Geographical distribution of *Spartinas anglica* Hubbard in China. *Bulletin of Marine Science* 33:753–758.

Chung, C. H. 1985. The effects of introduced *Spartina* grass on coastal morphology in China. *Zeitschrift fuer Geomorphologie N.F. Supplementband* 57:169–174.

Chung, C. H. 1989. Ecological engineering of coastlines with salt marsh plantations. Pages 255–289. In: W. J. Mitsch and S. E. Jørgensen, eds., *Ecological Engineering: An Introduction to Ecotechnology.* Wiley, New York.

Chung, C. H. 1993. Thirty years of ecological engineering with *Spartina* plantations in China. *Ecological Engineering* 2:261–289.

Chung, Y. P., B. J. McCoy, and K. M. Scow. 1993. Criteria to assess when biodegradation is kinetically limited by intraparticle diffusion and sorption. *Biotechnology and Bioengineering* 41:625–632.

Clewell, A. F. 1999. Restoration of riverine forest at Hall Branch on phosphate-mined land, Florida. *Restoration Ecology* 7:1–14.

Cloud, P. E., Jr. 1971. Resources, population, and quality of life. Pages 124–152. In: S. F. Singer, ed., *Is There an Optimum Level of Population?* McGraw-Hill, New York. 478 pp.

Cole, C. A., and D. Shafer. 2002. Section 404 wetland mitigation and permit success criteria in Pennsylvania, USA, 1986–1999. *Environmental Management* 30:508–515.

Colinvaux, P. 1993. *Ecology 2.* Wiley, New York.

Comin, F. A., J. A. Romero, V. Astorga, and C. Garcia. 1997. Nitrogen removal and cycling in restored wetlands used as filters of nutrients for agricultural runoff. *Water Science and Technology* 35:255–261.

Conway, T. E. and J. M. Murtha. 1989. The Iselin marsh pond meadow. Pages 139–144. D. A. Hammer, ed., *Constructed Wetlands for Wastewater Treatment.* Lewis Publishers, Chelsea, MI.

Cooke, J. G. 1992. Phosphorus removal processes in a wetland after a decade of receiving a sewage effluent. *Journal of Environmental Quality* 21:733–739.

Cooper, P. F. and B. C. Findlater, eds. 1990. *Constructed Wetlands in Water Pollution Control.* Pergamon Press, Oxford. 605 pp.

Cooper, P. F. and J. A. Hobson. 1989. Sewage treatment by reed bed systems: the present situation in the United Kingdom. Pages 153–172. In: D. A. Hammer, ed., *Constructed Wetlands for Wastewater Treatment.* Lewis Publishers, Chelsea, MI.

Costanza, R. and F. H. Sklar. 1985. Articulation, accuracy and effectiveness of mathematical models: a review of freshwater wetland applications. *Ecological Modelling* 27:45–69.

Costanza, R., R. d'Arge, R. de Groot, S. Farber, M. Grasso, B. Hannon, K. Limburg, S. Naeem, R. V. O'Neill, J. Paruelo, R. G. Raskin, P. Sutton, and M. van den Belt. 1997. The value of the world's ecosystem services and natural capital. *Nature* 387: 253–260.

Cox, J. L. 1970. Accumulation of DDT residues in *Triphoturus mexicanus* from the Gulf of California. *Nature* 227:192–193.

Cronk, J. K. 1996. Constructed wetlands to treat wastewater from dairy and swine operations: a review. *Agriculture Ecosystems and Environment* 58:97–114.

Crusberg, T. C., G. Gudmonsson, S. C., Moore, P. J. Weathers, and R. R. Biederman. 1994. Resistance to arsenic compounds in microorganisms. *FEMS Microbiology Reviews* 15:366–367.

Cunningham, J. A., C. J. Werth, M. Reinhard, and P. V. Roberts. 1997. Effects of grain-scale mass transfer on the transport of volatile organics through sediments. 1. Model development. *Water Resources Research* 33:2713–2726.

Dahl, T. E. 1990. *Wetland Losses in the United States, 1780s to 1980s.* U.S. Fish and Wildlife Service, Washington, DC. 21 pp.

Danish Ministry of Environment and Energy. 1999. *The Skjern River Restoration Project.* DMEE and National Forest and Nature Agency, Copenhagen. 32 pp.

Das, S. and Jana, B. B. 1999. Dose dependent uptake and *Eichhornia*-induced elimination of cadmium in various organs of the freshwater mussel, *Lamellidens marginalis. Ecological Engineering* 12:207–230.

Day, J. W., Jr., C. A. S. Hall, W. M. Kemp, and A. Yanez-Arancibia. 1989. *Estuarine Ecology.* Wiley, New York, 558 pp.

de Bernardi, R. and G. Giussani. 1995. Biomanipulation: Bases for a top-down control. Pages 1–14. In: R. De Bernardi and G. Giussani, eds., *Guidelines of Lake Management,* Vol. 7, *Biomanipulation in Lakes and Reservoirs.* International Lake Environment Committee, Kusatsu, Japan and United Nations Environmental Programme, Nairobi, Kenya. 211 pp.

DeJong, J. 1976. The purification of wastewater with the aid of rush or reed ponds. Pages 133–139. In: J. Tourbier and R. W. Pierson, eds., *Biological Control of Water Pollutions.* University of Pennsylvania, Philadelphia.

De Leon, R. O. D. and A. T. White. 1999. Mangrove rehabilitation in the Philippines. Pages 37–42. In: W. Streever, ed., *An International Perspective on Wetland Rehabilitation.* Kluwer Academic, Dordrecht, The Netherlands.

Deshmukh, A. P. and W. J. Mitsch. 2000. Hydric soil development in the Olentangy River Experimental Wetlands after five years of inundation. Pages 113–120. In: W. J. Mitsch and L. Zhang, eds., *Olentangy River Wetland Research Park Annual Report,* 1999. Ohio State University, Columbus, OH.

Dierberg, F. E. and P. L. Brezonik. 1985. Nitrogen and phosphorus removal by cypress swamp sediments. *Water Air and Soil Pollution* 24:207–213.

Dørge, J. 1991. Model for nitrogen cycling in freshwater wetlands. Master's thesis. University of Copenhagen, Denmark.

Dubnyak, S. and V. Timchenko. 2000. Ecological role of hydrodynamic processes in the Dnieper reservoirs. *Ecological Engineering* 16:181–188.

Dugan, P. 1993. *Wetlands in Danger.* Michael Beasley, Reed International Books, London. 192 pp.

Edmondson, W. T. and J. T. Lehman. 1981. The effects of changes in the nutrient income on the condition of Lake Washington. *Limnology and Oceanography* 26: 1–29.

EPA Denmark. 1979. *Lead Contamination in Denmark.* 145 pp.

Erlich, H. L. and C. L. Brierley. 1990. *Microbial Mineral Recovery.* McGraw-Hill, New York. 240 pp.

Erwin, K. L., G. R. Best, W. J. Dunn, and P. M. Wallace. 1984. Marsh and forested wetland reclamation of a central Florida phosphate mine. *Wetlands* 4:87–104.

Etnier, C. and B. Guterstam, eds. 1991. *Ecological Engineering of Wastewater Treatment—International Conference,* Stensund Folk College, Trosa, Sweden. Bokskogen, Gothenburg, Sweden.

Etnier, C. and B. Guterstam, eds. 1997. *Ecological Engineering for Wastewater Treatment,* 2nd ed. CRC Press/Lewis Publishers, Boca Raton, FL.

Ewel, K. C. and H. T. Odum, eds. 1984. *Cypress Swamps.* University Presses of Florida, Gainesville, FL. 472 pp.

Faber, P. A., E. Keller, A. Sands, and B. M. Masser. 1989. *The Ecology of Riparian Habitats of the Southern California Coastal Region: A Community Profile.* Biological Report 85(7.27). U.S. Fish and Wildlife Service, Washington, DC. 152 pp.

Fanta, J. 1994. Forest ecosystem development on degraded and reclaimed sites. *Ecological Engineering* 3:1–3.

Fanta, J. 1997. Rehabilitating degraded forests in central Europe into self-sustaining forest ecosystems. *Ecological Engineering* 8:289–297.

Federal Interagency Stream Restoration Working Group. 2001. *Stream Corridor Restoration: Principles, Processes, and Practices.* NISR Working Group, Part 653 of *National Engineering Handbook.* USDA–Natural Resources Conservation Service, Washington, DC.

Fennessy, M. S. 1988. Reclamation of coal mine drainage using a created wetland: exploring ecological treatment systems. M.S. thesis. Ohio State University, Columbus, OH.

Fennessy, M. S. and W. J. Mitsch. 1989. Treating coal mine drainage with an artificial wetland. *Research Journal of the Water Pollution Control Federation* 61:1691–1701.

Fennessy, M. S., J. K. Cronk, and W. J. Mitsch. 1994. Macrophyte productivity and community development in created freshwater wetlands under experimental hydrologic conditions. *Ecological Engineering* 3:469–484.

Fiechter, A. 1992. Biosurfactant: moving towards industrial application. *Trends in Biotechnology* 10:208–217.

Fink, D. F. 2001. Efficacy of a newly created wetland at reducing nutrient loads from agricultural runoff. Master's thesis. Environmental Science Graduate Program, Ohio State University, Columbus, OH.

Flanagan, N. E., W. J. Mitsch, and K. Beach. 1994. Predicting metal retention in a constructed mine drainage wetland. *Ecological Engineering* 3:135–159.

Fogel, S., R. Lancione, and A. Sewall. 1981. *A Literature and Laboratory Investigation of the Influence of Water Solubility on the Biodegradation of Organic Chemicals.* Report 560/5-82-015. U.S. Environmental Protection Agency, Washington, DC, pp. 1–21.

Forrester, J. W. 1961. *Industrial Dynamics.* MIT Press, Cambridge, MA.

Galatowitsch, S. M. and A. G. van der Valk. 1994. *Restoring Prairie Wetlands: An Ecological Approach.* Iowa State University Press, Ames, IA. 246 pp.

Garbisch, E. W. 1977. *Recent and Planned Marsh Establishment Work Throughout the Contiguous United States: A Survey and Basic Guidelines.* CR D-77-3. U.S. Army Corps of Engineers Waterways Experiment Station, Vicksburg, MS.

Garbisch, E. W., P. B. Woller, and R. J. McCallum. 1975. *Salt Marsh Establishment and Development.* Technical Memorandum 52. U.S. Army Coastal Engineering Research Center, Fort Belvoir, VA.

Gerheart, R. A. 1992. Use of constructed wetlands to treat domestic wastewater, city of Arcata, California. *Water Science and Technology* 26:1625–1637.

Gerheart, R. A., F. Klopp, and G. Allen. 1989. Constructed free surface wetlands to treat and receive wastewater: pilot project to full scale. Pages 121–137. In: D. A. Hammer, ed., *Constructed Wetlands for Wastewater Treatment.* Lewis Publishers, Chelsea MI.

Gifford, A. M. 2002. The effect of macrophyte planting on amphibian and fish community use of two created wetland ecosystems in central Ohio. Master's thesis. Environmental Science Graduate Program, Ohio State University, Columbus, OH.

Godfrey, P. J., E. R. Kaynor, S. Pelczarski, and J. Benforado, eds. 1985. *Ecological Considerations in Wetlands Treatment of Municipal Wastewaters.* Van Nostrand Reinhold, New York. 474 pp.

Gore, J. A., ed. 1985. *The Restoration of Rivers and Streams.* Butterworth, Boston.

Graedel, T. E. and B. R. Allenby. 1995. *Industrial Ecology.* Prentice Hall, Englewood Cliffs, NJ.

Greer, L. E. and D. R. Shelton. 1992. Effect of inoculant strain and organic matter content on kinetics of 2,4-dichloropehnoxyacetic acid degradation in soil. *Applied Environmental Microbiology* 58:1459–1465.

Gumbricht, T. 1992. Tertiary wastewater treatment using the root-zone method in temperate climates. *Ecological Engineering* 1:199–212.

Guterstam, B. and J. Todd. 1990. Ecological engineering for wastewater treatment and its application in New England and Sweden. *Ambio* 19:173–175.

Hagiwara, H. and W. J. Mitsch. 1994. Ecosystem modelling of an integrated aquaculture in South China. *Ecological Engineering* 72:41–73.

Halfon, E., 1983. Is there a best model structure? II. Comparing the model structures of different fate models. *Ecological Modelling* 20:153–163.

Halfon, E. 1984. Error analysis and simulation of *Mirex* behavior in Lake Ontario. *Ecological Modelling* 22:213–253.

Halfon, E., H. Unbehauen, and C. Schmid. 1979. Model order estimation and system identification theory to the modelling of 32P kinetics within the trophogenic zone of a small lake. *Ecological Modelling* 6:1–22.

Hall, C. A. S. 1995a. Introduction: What is maximum power? Pages xiii–xvi. In: C. A. S. Hall, ed., *Maximum Power: The Ideas and Applications of H. T. Odum.* University Press of Colorado, Niwot, CO.

Hall, C. A. S., ed. 1995b. *Maximum Power: The Ideas and Applications of H. T. Odum.* University Press of Colorado, Niwot, CO. 393 pp.

Hammer, D. A., ed. 1989. *Constructed Wetlands for Wastewater Treatment.* Lewis Publishers, Chelsea, MI. 831 pp.

Hammer, D. A. 1997. *Creating Freshwater Wetlands,* 2nd ed. CRC Press/Lewis Publishers, Boca Raton, FL. 406 pp.

Hansen, J. A. and J. C. Tjell. 1981. *The Application of Sludge as Soil Conditioner,* Vol. 2. Polyteknisk Forlag, Copenhagen, pp. 137–181.

Hansen, H. O., ed. 1996. *River Restoration: Danish Experience and Examples.* Ministry of Environment and Energy, National Environmental Research Institute, Silkekborg, Denmark. 99 pp.

Hansen, H. O. and B. L. Madsen. 1998. *River Restoration '96: Session Lecture Proceedings.* Ministry of Environment and Energy, National Environmental Research Institute, Silkekborg, Denmark. 293 pp.

Hansen, H. O., B. Kronvang, and B. L. Madsen. 1996. Classification system for watercourse rehabilitation. Pages 73–79. In: H. O. Hansen, ed., *River Restoration: Danish Experience and Examples.* Ministry of Environment and Energy, National Environmental Research Institute, Silkekborg, Denmark.

Hansen, H. O., P. J. Boon, B. L. Madsen, and T. M. Iversen. 1998. River restoration: The physical dimension. Special issue. *Aquatic Conservation: Marine and Freshwater Ecosystems* 8:1–264.

Hart, D. D. and N. L. Poff, eds. 2002. Dam removal and river restoration: a special section. *BioScience* 52:653–747.

Hart, D. D., T. E. Johnson, K. L. Bushaw-Newton, R. J. Horiwitz, A. T. Bednarek, D. F. Charles, D. A. Kreeger, and D. J. Velinsky. 2002. Dam removal: challenges and opportunities for ecological research and river restoration. *BioScience* 52:669–681.

Henry, C. P. and C. Amoros. 1995. Restoration ecology of riverine wetlands. I. A scientific basis. *Environmental Management* 19:891–902.

Henry, C. P., C. Amoros, and N. Roset. 2002. Restoration ecology of riverine wetlands: a 5-year post-operation survey on the Rhône River, France. *Ecological Engineering* 18:543–554.

Hey, D. L. and N. S. Philippi. 1995. Flood reduction through wetland restoration: the upper Mississippi River basin as a case study. *Restoration Ecology* 3:4–17.

Hey, D. L., M. A. Cardamone, J. H. Sather, and W. J. Mitsch. 1989. Restoration of riverine wetlands: the Des Plaines River wetland demonstration project. Pages 159–183. In: W. J. Mitsch and S. E. Jørgensen, eds., *Ecological Engineering: An Introduction to Ecotechnology.* Wiley, New York.

Hickman, S. C. and V. J. Mosca. 1991. *Improving Habitat Quality for Migratory Waterfowl and Nesting Birds: Assessing the Effectiveness of the Des Plaines River*

Wetlands Demonstration Project. Technical paper 1. Wetlands Research, Chicago. 13 pp.

Higgins, C. R. 2002. Ecosystem engineering by muskrats (*Ondatra zibethicus*) in created freshwater marshes. Master's thesis. Environmental Science Graduate Program, Ohio State University, Columbus, OH.

Hoagland, C. R., L. E. Gentry, M. B. David, and D. A. Kovacic. 2001. Plant nutrient uptake and biomass accumulation in a constructed wetland. *Journal of Freshwater Ecology* 16:527–540.

Holan, Z. R., B. Volesky, and I. Prasetyo. 1993. Biosorption of cadmium by biomass of marine algae. *Biotechnology and Bioengineering* 41:819–825.

Holbein, B. E. 1990. Immobilization of metal-binding compounds. Pages 327–349. In: B. Voleksy, ed., *Biosorption of Heavy Metals.* CRC Press, Boca Raton, FL.

Holling, C. S. 1986. The resilience of terrestrial ecosystems: local surprise and global change. Pages 292–317. In: W. C. Clark and R. E. Munn, eds., *Sustainable Development of the Biosphere.* Cambridge University Press, Cambridge.

Hosper, S. H. and E. Jagtman. 1990. Biomanipulation additional to nutrient control for restoration of shallow lakes in the Netherlands. *Hydrobiologia* 200/201:523–524.

Hosper, S. H. and M.-L. Meijer. 1993. Biomanipulation, will it work for your lake? A simple test for the assessment of chances for clear water following drastic fish-stock reduction in shallow, eutrophic lakes. *Ecological Engineering* 2:63–72.

Hubbell, S. P. 1997. A unified theory of biogeography and relative species abundance and its application to tropical rain forests and coral reefs. *Proceedings of the 7th International Coral Reef Symposium,* Vol. I, pp. 33–42.

Hunter, I. R., M. Hobley, and P. Smale. 1998. Afforestation of degraded land: pyrrhic victory over economic, social and ecological reality? *Ecological Engineering* 10:97–106.

Hüttl, R. F. and A. B. Bradshaw, eds. 2001. Ecology of post-mining landscapes. Special issue. *Ecological Engineering* 17:87–330.

Hüttl, R. F. and B. U. Schneider. 1998. Forest ecosystem degradation and rehabilitation. *Ecological Engineering* 10:19–31.

Hynes, H. B. N. 1970. *The Ecology of Running Water.* University of Toronto Press, Toronto, Ontario, Canada.

IWA Specialists Group on Use of Macrophytes in Water Pollution Control. 2000. *Constructed Wetlands for Pollution Control.* Scientific and Technical Report 8. International Water Association, London. 156 pp.

Jacks, G., A. Joelsson, and S. Fleischer. 1994. Nitrogen retention in forested wetlands. *Ambio* 23:358–362.

Jackson, J. 1989. Man-made wetlands for wastewater treatment: two case studies. Pages 57–580. In: D. A. Hammer, ed., *Constructed Wetlands for Wastewater Treatment.* Lewis Publishers, Chelsea, MI.

Jana, B. B. and S. Das. 1997. Potential of freshwater mussel for cadmium clearance in a model system. *Ecological Engineering* 8:179–194

Japp, W. C. 2000. Coral reef restoration. *Ecological Engineering* 15:345–364.

Jeffers, N. R. J. 1978. *An Introduction to Systems Analysis with Ecological Applications.* Edward Arnold, London.

Jensen, K. and J. C. Tjell. 1981. *The Application of Sludge as Soil Conditioner,* Vol. 3. Polyteknisk Forlag, Copenhagen, pp. 121–147.

Jiang, M., X. Zhang, and R. Wang. 1992. The ecological significance of Chinese ancient philosophy. Paper presented at International Studies Program, *China's Environment: Meeting Local and Global Challenges,* May 1992. Portland State University, Portland, OR.

Johansson, A. 1992. *Clean Technology.* Lewis Publishers, Boca Raton, FL.

Johengen, T. H. and P. A. LaRock. 1993. Quantifying nutrient removal processes within a constructed wetland designed to treat urban stormwater runoff. *Ecological Engineering* 2:347–366.

Johnson, R. R. and J. F. McCormick, tech. coords. 1979. *Strategies for the Protection and Management of Floodplain Wetlands and Other Riparian Ecosystems–Proceedings of the Symposium,* Callaway Gardens, GA, December 11–13, 1978. General Technical Report WO-12. U.S. Forest Service, Washington, DC. 410 pp.

Johnson, B. L., W. B. Richardson, and T. J. Naimo. 1995. Past, present, and future concepts in large river ecology. *BioScience* 45:134–141.

Johnston, C. A. 1991. Sediment and nutrient retention by freshwater wetlands: effects on surface water quality. *Critical Reviews in Environmental Control* 21:491–565.

Jones, C. G., J. H. Lawton, and M. Shachak. 1994. Organisms as ecosystem engineers. *Oikos* 69:373–386.

Jones, C. G., J. H. Lawton, and M. Shachak. 1997. Positive and negative effects of organisms as physical ecosystem engineers. *Ecology* 78:1946–1957.

Jordan, W. R., III, M. E. Gilpin, and J. D. Aber, eds. 1987. *Restoration Ecology: A Synthetic Approach to Ecological Research.* Cambridge University Press, Cambridge. 342 pp.

Jørgensen, S. E. 1975. Do heavy metals prevent the agricultural use of municipal sludge? *Water Research* 9:163–170.

Jørgensen, S. E. 1976. A eutrophication model for a lake. *Ecological Modelling* 2:147–165.

Jørgensen, S. E. 1981. *Application of Ecological Modelling in Environmental Management.* Elsevier, Amsterdam.

Jørgensen, S. E. 1982. Exergy and buffering capacity in ecological systems. Pages 61–72. In: W. J. Mitsch, R. K. Ragade, R. W. Bosserman, and J. A. Dillon, Jr., eds., *Energetics and Systems,* Ann Arbor Science, Ann Arbor, MI.

Jørgensen, S. E. 1986. Structural dynamic model. *Ecological Modelling* 31:1–9.

Jørgensen, S. E. 1988. Use of models as experimental tools to show that structural changes are accompanied by increased exergy. *Ecological Modelling* 41:117–126.

Jørgensen, S. E. 1990. *Modelling in Ecotoxicology.* Elsevier, Amsterdam. 350 pp.

Jørgensen, S. E. 1992. Development of models able to account for changes in species composition. *Ecological Modelling* 62:195–208.

Jørgensen, S. E. 1993. Removal of heavy metal from compost and soil by ecotechnological methods. *Ecological Engineering* 2:89–100

Jørgensen, S. E. 1994. Review and comparison of goal functions in system ecology. *Vie Milieu* 44:11–20.

Jørgensen, S. E. 2000. *Principles of Pollution Abatement.* Elsevier, Amsterdam. 526 pp.

Jørgensen, S. E. 2002. *Integration of Ecosystem Theories: A Pattern,* 3rd ed. Kluwer Academic, Dordrecht, The Netherlands. 428 pp.

Jørgensen, S. E. and G. Bendoricchio. 2001. *Fundamentals of Ecological Modelling,* 3rd ed. Elsevier, Amsterdam. 530 pp.

Jørgensen, S. E. and R. de Bernardi. 1997. The use of structural dynamic models to explain the success and failure of biomanipulation. *Hydrobiologia* 379:147–158.

Jørgensen, S. E. and J. F. Mejer. 1977. Ecological buffer capacity. *Ecological Modelling* 3:39–61.

Jørgensen, S. E. and J. F. Mejer. 1979. A holistic approach to ecological modeling. *Ecological Modelling* 7:169–189.

Jørgensen, S. E., O. S. Jacobsen, and I. Hoi. 1973. A prognosis for a lake. *Vatten* 29: 382–404.

Jørgensen, S. E., L. A. Jørgensen, L. Kamp Nielsen, and H. F. Mejer. 1981. Parameter estimation in eutrophication modelling. *Ecological Modelling* 13:111–129.

Jørgensen, S. E., C. C. Hoffman, and W. J. Mitsch. 1988. Modelling nutrient retention by reedswamp and wetland meadow in Denmark. Pages 133–151. In: W. J. Mitsch, M. Straskraba, and S. E. Jørgensen, eds., *Wetland Modelling.* Elsevier, Amsterdam.

Jørgensen, S. E., B. Halling-Sørensen, and H. Mahler. 1997. *Handbook of Estimation Methods in Ecotoxicology and Environmental Chemistry.* Lewis Publishers, Boca Raton, FL. 230 pp.

Josselyn, M., J. Zedler, and T. Griswold. 1990. Wetland mitigation along the Pacific coast of the United States. Pages 3–36. In: J. A. Kusler and M. E. Kentula, eds., *Wetland Creation and Restoration.* Island Press, Washington, DC.

Junk, W. J., P. B. Bayley, and R. E. Sparks. 1989. The flood pulse concept in river–floodplain systems. Pages 11–127. In: P. P. Dodge, ed., *Proceedings of the International Large River Symposium,* Special publication. *Journal of Canadian Fisheries and Aquatic Sciences* 106:11–127.

Kadlec, R. H. 1999. Constructed wetlands for treating landfill leachate. Pages 17–31. In: G. Mulamoottil, E. A. McBean, and F. Rovers, eds., *Constructed Wetlands for the Treatment of Landfill Leachates.* Lewis Publishers, Boca Raton, FL.

Kadlec, R. H. and D. L. Hey. 1994. Constructed wetlands for river water quality improvement. *Water Science and Technology* 29:159–168.

Kadlec, R. H. and R. L. Knight. 1996. *Treatment Wetlands.* CRC Press/Lewis Publishers, Boca Raton, FL. 893 pp.

Kadlec, R. H. and D. L. Tilton. 1979. The use of freshwater wetlands as a tertiary treatment alternative. *CRC Critical Reviews in Environmental Control* 9:185–212.

Kalin, M. 2001. Biogeochemical and ecological considerations in designing wetland treatment systems in post-mining landscapes. *Waste Management* 21:191–196.

Kangas, P. C. 1981. Succession as an alternative for reclaiming phosphate spoil mounds. Pages 11–43. In: H. T. Odum, ed., *Studies on Phosphate Mining, Reclamation, and Energy.* Center for Wetlands, University of Florida, Gainesville, FL.

Kapoor and T. Viraraghavan. 1995. Fungal biosorption: an alternative treatment option for heavy metal bearing wastewaters–A review. *Bioresource Technology* 53:195–206.

Kemp, W. M., J. E. Petersen, and R. H. Gardner. 2001. Scale-dependence and the problem of extrapolation: implications for experimental and natural coastal ecosystems. Pages 3–57. In: R. H. Gardner, W. M. Kemp, V. S. Kennedy, and J. Petersen, eds., *Scaling Relations in Experimental Ecology*. Columbia University Press, New York.

Kentula, M. E., J. C. Sifneos, J. W. Good, M. Rylko, and K. Kuntz. 1992. Trends and patterns in Section 404 permitting requiring compensatory mitigation on Oregon and Washington, USA. *Environmental Management* 16:109–119.

Kilian, W. 1998. Forest site degradation: temporary deviation from the natural site potential. *Ecological Engineering* 10:5–18.

Kilian, W. and J. Fanta, eds. 1998. Degradation and restoration of forests. Special issue. *Ecological Engineering* 10:1–106.

Klostermann, J. E. M. and A. Tukker. 1998. *Product Innovation and Eco-efficiency*. Kluwer Academic, Boston. 224 pp.

Knight, R. L. 1990. Wetland systems. Pages 211–260. In: *Natural Systems for Wastewater Treatment*. Manual of Practice FD-16. Water Pollution Control Federation, Alexandria, VA.

Knight, R. L., T. W. McKun, and H. R. Kohl. 1987. Performance of a natural wetland treatment system for wastewater management. *Journal of the Water Pollution Control Federation* 59:746–754.

Knox, R. C., D. A. Sabatini, and L. W. Canter. 1993. Subsurface transport and fate processes. Pages 55–112. In: R. C. Knox and L. W. Canter, eds., *Ecotoxicological Processes*. Lewis Publishers, Boca Raton, FL. 328 pp.

Kolka, R. K., E. A. Nelson, and C. C. Trettin. 2000. Conceptual assessment framework for forested wetland restoration: the Pen Branch experience. *Ecological Engineering* 15:S17–S21.

Kompare, B. 1995. The use of artificial intelligence to estimate ecotoxicological and ecological parameters. Thesis. Royal Danish School of Pharmacy, Copenhagen.

Kovacic, D. A., M. B. David, L. E. Gentry, K. M. Starks, and R. A. Cooke. 2000. Effectiveness of constructed wetlands in reducing nitrogen and phosphorus export from agricultural tile drainage. *Journal of Environmental Quality* 29:1262–1274.

Kusler, J. A., and M. E. Kentula, eds. 1990. *Wetland Creation and Restoration: The Status of the Science*. Island Press, Washington, DC. 594 pp.

Kuyucak, N. and B. Voleksy. 1989. The mechanism of cobalt bio-sorption. *Biotechnology and Bioengineering* 33:815–822.

Lamb, D. 1994. Reforestation of degraded tropical forest lands in the Asia–Pacific region. *Journal of Tropical Forestry Science* 7:1–7.

Lambert, J. M. 1964. The *Spartina* story. *Nature* 204:1136–1138.

Lane, R. R., J. W. Day, Jr., and B. Thibodeaux. 1999. Water quality analysis of a freshwater diversion at Caernarvon, Louisiana. *Estuaries* 22:327–336

Larson, A. C., L. E. Gentry, M. B. David, R. A. Cooke, and D. A. Kovacic. 2000. The role of seepage in constructed wetlands receiving tile drainage. *Ecological Engineering* 15:91–104.

Lavoie, C., and L. Rochefort. 1996. The natural revegetation of a harvested peatland in southern Quebec: a spatial and dentroecological analysis. *Ecoscience* 3:101–111.

Laws, E. A. 1993. *Aquatic Pollution*, 2nd ed. Wiley, New York.

Leeuwen, van C. J. and J. L. M. Hermens, eds. 1995. *Risk Assessment of Chemicals: An Introduction*. Kluwer Academic, Dordrecht, The Netherlands. 374 pp.

Lefeuvre, J. C., W. J. Mitsch, and V. Bouchard, eds. 2002. Ecological engineering applied to river and wetland restoration. Special issue. *Ecological Engineering* 18: 529–658.

Lemons, J., L. Westra, and R. Goodland. 1998. *Ecological Sustainability and Integrity: Concepts and Approaches.* Kluwer Academic, Boston.

Leonardson, L., L. Bengtsson, T. Davidsson, T. Persson, and U. Emanuelsson. 1994. Nitrogen retention in artificially flooded meadows. *Ambio* 23:332–341.

Leopold, A. as edited by L. B. Leopold. 1972. *Round River.* Oxford University Press, Oxford. 286 pp.

Leopold, L. B. 1994. *A View of the River.* Harvard University Press, Cambridge, MA.

Leopold, L. B., M. G. Wolman, and J. E. Miller. 1964. *Fluvial Processes in Geomorphology.* W. H. Freeman, San Francisco. 522 pp.

Letterman, R. D. and W. J. Mitsch. 1978. Impact of mine drainage on mountain streams in Pennsylvania. *Environmental Pollution* 17:53–73.

Lewis, R. R. 1990a. Wetland restoration/creation/enhancement terminology: suggestions for standardization. Pages 1–7. In: J. A. Kusler and M. E. Kentula, eds., *Wetland Creation and Restoration.* Island Press, Washington, DC.

Lewis, R. R. 1990b. Creation and restoration of coastal plain wetlands in Florida. Pages 73–101. In: J. A. Kusler and M. E. Kentula, eds., *Wetland Creation and Restoration.* Island Press, Washington, DC.

Lewis, R. R. 1990c. Creation and restoration of coastal plain wetlands in Puerto Rico and the U.S. Virgin Islands. Pages 103–123. In: J. A. Kusler and M. E. Kentula, eds., *Wetland Creation and Restoration.* Island Press, Washington, DC.

Lewis, R. R. 2000. Ecologically based goal setting in mangrove forest and tidal marsh restoration. *Ecological Engineering* 15:191–198.

Lewis, R. R., J. A. Kusler, and K. L. Erwin. 1995. Lessons learned from five decades of wetland restoration and creation in North America. In: C. Montes, G. Oliver, F. Molina, and J. Cobos, eds., *Bases Ecológicas para la Restauración de Humedales en la Cuenca Mediterránea.* Consejería de Medio Ambiente, Junta de Andalucía, Andalucía, Spain.

Li, W. 1998. Utilization of aquatic macrophytes in grass carp farming in Chinese shallow lakes. *Ecological Engineering* 11:61–72.

Lin, O. and I. A. Mendelssohn. 1998. The combined effect of phytoremediation and biostimulation in enhancing habitat restoration and oil degradation of petroleum contaminated wetlands. *Ecological Engineering* 10:263–274.

Litchfield, D. K. and D. D. Schatz. 1989. Constructed wetlands for wastewater treatment at Amoco Oil Company's Mandan, North Dakota, refinery. Pages 101–119. In: D. A. Hammer, ed., *Constructed Wetlands for Wastewater Treatment,* Lewis Publishers, Chelsea, MI.

Liu, S., X. Li, and L. Niu. 1998. The degradation of soil fertility in pure larch plantations in the northeastern part of China. *Ecological Engineering* 10:75–86.

Lotka, A. J. 1922. Contribution to the energetics of evolution. *Proceedings of the National Academy of Sciences USA* 8:147–150.

Louisiana Coastal Wetlands Conservation and Restoration Task Force. 1998. *Coast 2050: Toward a Sustainable Coastal Louisiana–An Executive Summary.* Louisiana Department of Natural Resources, Baton Rouge, LA. 12 pp.

Lowrance, R., B. R. Stinner, and G. J. House, eds. 1984. *Agricultural Ecosystems: Unifying Concepts.* Wiley, New York. 233 pp.

Lu, J. 1990. *Wetlands in China.* East China Normal University Press, Shanghai. 177 pp. (in Chinese).

Lu, J. 1995. Ecological significance and classification of Chinese wetlands. *Vegetatio* 118:49–56.

Luckett, C., W. H. Adey, J. Morrissey, and D. M. Spoon. 1996. Coral reef mesocosms and microcosms: successes, problems, and the future of laboratory models. *Ecological Engineering* 6:57–72.

Ma, S. 1985. Ecological engineering: application of ecosystem principles. *Environmental Conservation* 12:331–335.

Ma, S. 1988. Development of agro-ecological engineering in China. Pages 1–13. In: S. Ma, A. Jiang, R. Xu, and D. Li, eds., *Proceedings of the International Symposium on Agro-Ecological Engineering,* August 1988. Ecological Society of China, Beijing.

Ma, S. and H. Liu. 1988. Analysis of the functions of paddy-field ecosystem engineering in southern mountain and hilly areas of Hunan Province, China. In: S. Ma, A. Jiang, R. Xu, and D. Li, eds., *Proceedings of the International Symposium on Agro-Ecological Engineering,* August 1988. Ecological Society of China, Beijing.

Ma, S. and R. Wang. 1984. Social-economic natural complex ecosystem. *Acta Ecologica Sinica* 4(1):1–9 (in Chinese).

Ma, S. and R. Wang. 1989. Social–economic natural complex ecosystem and sustainable development. In: R. Wang, ed., *Human Ecology in China.* China Science and Technology Press, Beijing (in Chinese).

Ma, S. and J. Yan. 1989. Ecological engineering for treatment and utilization of wastewater. Pages 185–217. In: W. J. Mitsch and S. E. Jørgensen, eds., *Ecological Engineering: An Introduction to Ecotechnology.* Wiley, New York.

Ma, S., A. Jiang, R. Xu, and D. Li, eds. 1988. *Proceedings of the International Symposium on Agro-Ecological Engineering,* August 1988. Ecological Society of China, Beijing.

Macaskie, L. E. 1991. The application of biotechnology to the treatment of wastes produced from the nuclear fuel cycle. *Critical Reviews in Biotechnology* 11:41–112.

Maier, R. M. 2000. Bioavailability and its importance to bioremediation. Pages 58–78. In: J. J. Valdes, ed., *Bioremediation.* Kluwer Academic, Dordrecht, The Netherlands. 169 pp.

Malakoff, D. 1998. Restored wetlands flunk real-world test. *Science* 280:371–372.

Mann, K. H. 1975. Patterns of energy flow. Pages 248–263. In: B. A. Whitton, *River Ecology.* Blackwell Scientific, Oxford.

Manyin, T., F. M. Williams, and L. R. Stark. 1997. Effects of iron concentration and flow rate on treatment of coal mine drainage in wetland mesocosms: an experimental approach to sizing of constructed wetlands. *Ecological Engineering* 9:171–185.

Maragos, J. E. 1992. Restoring coral reefs with emphasis on Pacific reefs. Pages 141–221. In: G. W. Thayer, ed., *Restoring the Nation's Marine Environment.* Maryland Sea Grant College, College Park, MD.

Marino, B. D. V. and H. T. Odum, eds. 1999. Biosphere 2: Research past and present. Special issue. *Ecological Engineering* 13:1–356.

Massey, B. 2000. *Wetlands Engineering Manual.* Ducks Unlimited, Southern Regional Office, Hackson, MS. 16 pp.

May, R. M. 1977. *Stability and Complexity in Model Ecosystems,* 3rd ed. Princeton University Press, Princeton, NJ. 530 pp.

May, R. M. 1981. *Theoretical Ecology: Principles and Applications,* 2nd ed. Blackwell Scientific, Oxford. 489 pp.

McCarty, P. L. 1997. Breathing with chlorinated solvents. *Science* 276:1521–1522.

Meade, R. and R. Parks. 1985. Sediment in rivers of the United States. Pages 49–60. In: *National Water Summary.* Water Supply Paper 2275. U.S. Geological Survey, Washington, DC.

Mejer, H. F. and S. E. Jørgensen. 1979. Energy and ecological buffer capacity. Pages 829–846. In: S. E. Jørgensen, ed., *State-of-the-Art of Ecological Modelling– Proceedings of a Conference on Ecological Modelling,* August 28–September 2, 1978. International Society for Ecological Modelling, Copenhagen.

Metzger, K. and W. J. Mitsch. 1997. Modelling self–design of the aquatic community in a newly created freshwater wetland. *Ecological Modelling* 100:61–86.

Meyer, J. L. 1985. A detention basin/artificial wetland treatment system to renovate stormwater runoff from urban, highway, and industrial areas. *Wetlands* 5:135–145.

Middleton, B. A. 1999. *Wetland Restoration: Flood Pulsing and Disturbance Dynamics.* Wiley, New York. 388 pp.

Miljøstyrelsen. 2000. Environmental risk assessment by purification of a contaminated plot by the use of adapted microorganisms. *Videnskabelige fra Miljøstyrelsen* 3:27–30.

Miljøstyrelsen. 2002. Removal of PAHS from contaminated soil. *Videnskabelige fra Miljøstyrelsen* 4(1):17–20.

Miller, R. M. 1995. Surfactant–enhanced bioavailability of slightly soluble organic compounds. Pages 33–54. In: H. Skipper and R. Turco, eds., *Bioremediation: Science and Application.* Special publication. Soil Science Society of America, Madison, WI.

Miller, M. E. and M. Alexander. 1991. Kinetics of bacterial degradation of benzylamine in a montmorillonie suspension. *Environmental Science and Technology* 25:240–245.

Minshall, G. W., R. C. Peterson, K. W. Cummins, T. L. Bott, J. R. Sedall, C. E. Cushing, and R. L. Vannote. 1983. Inter-biome comparison of stream ecosystem dynamics. *Ecological Monographs* 53:1–25.

Minshall, G. W., K. W. Cummins, R. C. Peterson, C. E. Cushing, D. A. Bruins, J. R. Sedall, and R. L. Vannote. 1985. Development in stream ecosystem theory. *Canadian Journal of Fisheries and Aquatic Sciences* 37:130–137.

Mitchell, D. S., A. J. Chick, and G. W. Rasin. 1995. The use of wetlands for water pollution control in Australia: an ecological perspective. *Water Science and Technology* 32:365–373.

Mitsch, W. J. 1977. Water hyacinth (*Eichhornia crassipes*) nutrient uptake and metabolism in a north-central Florida marsh. *Archivfuer Hydrobiologia* 81:188–210.

Mitsch, W. J. 1988. Productivity–hydrology–nutrient models of forested wetlands. Pages 115–132. In: W. J. Mitsch, M. Straskraba, and S. E. Jørgensen, eds., *Wetland Modelling.* Elsevier, Amsterdam.

Mitsch, W. J. 1991. Ecological engineering: Approaches to sustainability and biodiversity in the U.S. and China. Pages 428–448. In: R. Costanza, ed., *Ecological Economics: The Science and Management of Sustainability.* Columbia University Press, New York.

Mitsch, W. J. 1992. Landscape design and the role of created, restored, and natural riparian wetlands in controlling nonpoint source pollution. *Ecological Engineering* 1:27–47.

Mitsch, W. J. 1993. Ecological engineering: A cooperative role with the planetary life-support systems. *Environmental Science and Technology* 27:438–445.

Mitsch, W. J. 1995. Ecological engineering: From Gainesville to Beijing—A comparison of approaches in the United States and China. Pages 109–122. In: C. A. S. Hall, ed., *Maximum Power.* University Press of Colorado, Niwot, CO.

Mitsch, W. J. 1996. Ecological engineering: A new paradigm for engineers and ecologists. Pages 111–128. In: P. C. Schulze, ed., *Engineering within Ecological Constraints.* National Academy Press, Washington, DC.

Mitsch, W. J. 1998. Ecological engineering: The seven-year itch. *Ecological Engineering* 10:119–138.

Mitsch, W. J. 1999. Preface: Biosphere 2. Special issue. *Ecological Engineering* 13: 1–2.

Mitsch, W. J. and V. Bouchard, eds. 1998. Great Lakes coastal wetlands: their potential for restoration. Special issue. *Wetlands Ecology and Management* 6:1–82.

Mitsch, W. J. and D. L. Fink. 2001. *Wetlands for Controlling Nonpoint Source Pollution from Agriculture: Indian Lake Wetland Demonstration Project, Logan County, OH.* Final report submitted to Indian Lake Watershed Project, Bellfontaine, OH. School of Natural Resources, Ohio State University, Columbus, OH.

Mitsch, W. J. and J. G. Gosselink. 2000. *Wetlands,* 3rd ed. Wiley, New York. 920 pp.

Mitsch, W. J. and S. E. Jørgensen. 1989. *Ecological Engineering: An Introduction to Ecotechnology.* Wiley, New York.

Mitsch, W. J. and K. S. Kaltenborn. 1980. Effects of copper sulfate application on diel dissolved oxygen and metabolism in the Fox Chain of Lakes. *Transactions of the Illinois State Academy of Science* 73:55–64.

Mitsch, W. J. and U. Mander, eds. 1997. Ecological engineering in central and eastern Europe: remediation of ecosystems damaged by environmental contamination. Special issue. *Ecological Engineering* 8:247–346.

Mitsch, W. J. and B. C. Reeder. 1991. Modelling nutrient retention of a freshwater coastal wetland: estimating the roles of primary productivity, sedimentation, resuspension and hydrology. *Ecological Modelling* 54:151–187.

Mitsch, W. J. and N. Wang. 2000. Large-scale coastal wetland restoration on the Laurentian Great Lakes: determining the potential for water quality improvement. *Ecological Engineering* 15:267–282.

Mitsch, W. J. and R. F. Wilson. 1996. Improving the success of wetland creation and restoration with know-how, time, and self-design. *Ecological Applications* 6:77–83.

Mitsch, W. J. and K. M. Wise. 1998. Water quality, fate of metals, and predictive model validation of a constructed wetland treating acid mine drainage. *Water Research* 32:1888–1900.

Mitsch, W. J. and L. Zhang. 2002. Floodplain enhancement as an approach for restoring rivers: examples at the Olentangy River Wetland Research Park. Poster presentation. American Ecological Engineering Society Annual Meeting, Burlington, VT.

Mitsch, W. J., M. A. McPartlin, and R. D. Letterman. 1978. Energetic evaluation of a stream ecosystem affected by coal mine drainage. *Verhandlungen des Internationale Vereinigung fuer Limnologie* 21:1388–1395.

Mitsch, W. J., R. K. Ragade, R. W. Bosserman, and J. A. Dillon, Jr., eds. 1982. *Energetics and Systems.* Ann Arbor Science, Ann Arbor, MI. 132 pp.

Mitsch, W. J., J. Yan, and J. Cronk, eds. 1993. Ecological engineering in China. Special issue. *Ecological Engineering* 2:177–307.

Mitsch, W. J., J. K. Cronk, X. Wu, R. W. Nairn, and D. L. Hey. 1995. Phosphorus retention in constructed freshwater riparian marshes. *Ecological Applications* 5: 830–845.

Mitsch, W. J., X. Wu, R. W. Nairn, P. E. Weihe, N. Wang, R. Deal, and C. E. Boucher. 1998. Creating and restoring wetlands: a whole-ecosystem experiment in self-design. *BioScience* 48:1019–1030.

Mitsch, W. J., J. W. Day, Jr., J. W. Gilliam, P. M. Groffman, D. L. Hey, G. W. Randall, and N. Wang. 1999. *Reducing Nutrient Loads, Especially Nitrate-Nitrogen, to Surface Water, Groundwater, and the Gulf of Mexico.* Final report to the National Oceanic and Atmospheric Association Coastal Program, Silver Spring, MD.

Mitsch, W. J., A. Horne, and R. W. Nairn, eds. 2000. Nitrogen and phosphorus retention in wetlands. Special issue. *Ecological Engineering* 14:1–206.

Mitsch, W. J., J. W. Day, Jr., J. W. Gilliam, P. M. Groffman, D. L. Hey, G. W. Randall, and N. Wang. 2001. Reducing nitrogen loading to the Gulf of Mexico from the Mississippi River basin: strategies to counter a persistent ecological problem. *BioScience* 51:373–388.

Mitsch, W. J., N. Wang, V. Bouchard, L. Zhang, X. Wu, R. Deal, A. Gifford, C. Higgins, and A. Zuwernik. Work in progress. A whole-ecosystem wetland experiment illustrates effects of macrophyte planting and community diversity on ecosystem function.

Miura, T. 1990. The effects of planktivorous fishes on the plankton community in a eutrophic lake. *Hydrobiologia,* 200/201:567–579.

Miura, T. and J. Wang. 1985. Chlorophyll *a* found in feces of planktivorous cyprinids and its photosynthetic activity. *Verhandlungen Internationale Vereinigung fuer Limnologie* 22:2636–2642.

Miyawaki, A. and K. Fujiwara. 1988. Vegetation mapping in Japan. Pages 427–441. In: A. W. Kuchler and I. S. Zonnenveld, eds., *Vegetation Mapping.* Kluwer Academic, Dordrecht, The Netherlands.

Miyawaki, A. and F. B. Golley. 1993. Forest reconstruction as ecological engineering. *Ecological Engineering* 2:333–345.

Moore, D. R. J., P. A. Keddy, C. L. Gaudet, and I. C. Wisheu. 1989. Conservation of wetlands: do infertile wetlands deserve a higher priority? *Biological Conservation* 47:203–217.

Moravcik, P. 1994. Development of new forest stands after a large scale forest decline in the Kršnéhory Mountains. *Ecological Engineering* 3:57–69.

Morowitz, H. J. 1968. *Energy Flow in Biology: Biological Organization as a Problem in Thermal Physics.* Academic Press, New York. 179 pp.

Moser, A. 1994. Trends in biotechnology: From high-tech to eco-tech. *Acta Biotechnologica* 14:315–335.

Moser, A. 1996. Ecotechnology in industrial practice: implementation using sustainability indices and case studies. *Ecological Engineering* 7:117–138.

Moshiri, G. A., ed. 1993. *Constructed Wetlands for Water Quality Improvement.* Lewis Publishers, Boca Raton, FL.

Moustafa, M. Z. 1999. Nutrient retention dynamics of the Everglades nutrient removal project. *Wetlands* 19:689–704.

Moustafa, M. Z., M. J. Chimney, T. D. Fontaine, G. Shih, and S. Davis. 1996. The response of a freshwater wetland to long-term "low level" nutrient loads: marsh efficiency. *Ecological Engineering* 7:15–33.

Mulamoottil, G., E. A. McBean, and F. Rovers, eds. 1999. *Constructed Wetlands for the Treatment of Landfill Leachates.* Lewis Publishers, Boca Raton, FL. 281 pp.

Nairn, R. W. and W. J. Mitsch. 2000. Phosphorus removal in created wetland ponds receiving river overflow. *Ecological Engineering* 14:107–126.

National Research Council. 1992. *Restoration of Aquatic Ecosystems.* National Academy Press, Washington, DC.

National Research Council. 1993. *In Situ Bioremediation: When Does It Work?* National Academy Press, Washington, DC.

National Research Council. 1995. *Wetlands: Characteristics and Boundaries.* National Academy Press, Washington, DC. 306 pp.

National Research Council. 2001. *Compensating for Wetland Losses under the Clean Water Act.* National Academy Press, Washington, DC. 158 pp.

National Wetlands Working Group. 1988. *Wetlands of Canada.* Ecological Land Classification Series 24. Environment Canada, Ottawa, Ontario, Canada and Polyscience Publications, Montreal, Quebec, Canada. 452 pp.

Nelson, E. A., R. K. Kolka, C. C. Trettin, and J. Wisniewski. 2000. Restoration of the severely impacted riparian wetland system. Special issue. *Ecological Engineering* 15:S1–S187.

Newman, S. and K. Pietro. 2001. Phosphorus storage and release in response to flooding: implications for Everglades stormwater treatment areas. *Ecological Engineering* 18:33–39.

Newman, J. M., J. C. Clausen, and J. A. Neafsey. 2000. Seasonal performance of a wetland constructed to process dairy milkhouse wastewater in Connecticut. *Ecological Engineering* 14:181–198.

Nguyen, L. M. 2000. Phosphate incorporation and transformation in surface sediments of a sewage-impacted wetland as influenced by sediment sites, sediment pH and added phosphate concentration. *Ecological Engineering* 14:139–155.

Nguyen, L. M., J. G. Cooke, and G. B. McBride. 1997. Phosphorus retention and release characteristics of sewage-impacted wetland sediments. *Water Air and Soil Pollution* 100:163–179.

Niswander, S. F. and W. J. Mitsch. 1995. Functional analysis of a two-year-old created in-stream wetland: hydrology, phosphorus retention, and vegetation survival and growth. *Wetlands* 15:212–225.

Odum, E. P. 1969. The strategy of ecosystem development. *Science* 164:262–270.

Odum, E. P. 1971. *Fundamentals of Ecology,* 3rd ed. W.B. Saunders, Philadelphia. 544 pp.

Odum, E. P. 1981. Foreword. Pages xi–xiii. In: J. R. Clark and J. Benforado, eds., *Wetlands of Bottomland Hardwood Forests.* Elsevier, Amsterdam.

Odum, E. P. 2000. Tidal marshes as outwelling/pulsing systems. Pages 3–8. In: M. P. Weinstein and D. A. Kreeger, eds., *Concepts and Controversies in Tidal Marsh Ecology.* Kluwer Academic, Dordrecht, The Netherlands.

Odum, H. T. 1962. Man in the ecosystem. Pages 57–75. In: *Proceedings of the Lockwood Conference on the Suburban Forest and Ecology.* Bulletin 652 Connecticut Agricultural Station, Storrs, CT.

Odum, H. T. 1971. *Environment, Power, and Society.* Wiley, New York.

Odum, H. T. 1973. *Energy Basis for Man and Nature.* McGraw-Hill, New York.

Odum, H. T. 1982. Pulsing, power, and hierarchy. Pages 33–59. In: W. J. Mitsch, R. K. Ragade, R. W. Bosserman, and J. A. Dillon, Jr., eds., *Energetics and Systems.* Ann Arbor Science, Ann Arbor, MI.

Odum, H. T. 1983. *Systems Ecology.* Wiley, New York. Reprinted in 1994 by University Press of Colorado, Niwot, CO.

Odum, H. T., ed. 1985. *Self Organization of Ecosystems in Marine Ponds Receiving Treated Sewage.* UNC-SG-85-04. North Carolina Sea Grant Office, North Carolina State University, Raleigh, NC. 250 pp.

Odum, H. T. 1989a. Ecological engineering and self-organization. Pages 79–101. In: W. J. Mitsch and S. E. Jørgensen, eds., *Ecological Engineering: An Introduction to Ecotechnology.* Wiley, New York.

Odum, H. T. 1989b. Experimental study of self-organization in estuarine ponds. Pages 291–340. In: W. J. Mitsch and S. E. Jørgensen, eds., *Ecological Engineering: An Introduction to Ecotechnology.* Wiley, New York.

Odum. H. T. 1996. *Environmental Accounting: Energy and Environmental Decision Making.* Wiley, New York. 370 pp.

Odum, H. T. and R. C. Pinkerton. 1955. Time's speed regulator, the optimum efficiency for maximum output in physical and biological systems. *American Scientist* 43: 331–343.

Odum, H. T., W. L. Siler, R. J. Beyers, and N. Armstrong. 1963. Experiments with engineering of marine ecosystems. *Publications of the Institute of Marine Science University of Texas* 9:374–403.

Odum, H. T., B. J. Copeland, and E. A. McMahan, eds. 1974. *Coastal Ecological Systems of the United States,* 4 vols. Conservation Foundation, Washington, DC.

Odum, H. T., K. C. Ewel, W. J. Mitsch, and J. W. Ordway. 1977. Recycling treated sewage through cypress wetlands. Pages 35–67. In: F. M. D'Itri, ed., *Wastewater Renovation and Reuse.* Marcel Dekker, New York.

Odum, H. T., G. R. Best, M. A. Miller, B. T. Rushton, R. Wolfe, C. Bersok, and J. Feiertag. 1990. *Accelerating Natural Processes for Wetland Restoration after Phos-*

phate Mining. FIPR Publication 03-041-086. Florida Institute of Phosphate Research, Bartow, FL. 408 pp.

Odum, W. E. 1987. Predicting ecosystem development following creation and restoration of wetlands. Pages 67–70. In: J. Zelazny and J. S. Feierabend, eds., *Proceedings of the Conference on Wetlands–Wetlands: Increasing Our Wetland Resources.* Corporate Conservation Council, National Wildlife Federation, Washington, DC.

Odum, W. E., E. P. Odum, and H. T. Odum. 1995. Nature's pulsing paradigm. *Estuaries* 18:547–555.

Ohlendorf, H. M., D. J. Hoffman, M. K. Saiki, and T. W. Aldrich. 1986. Embryonic mortality and abnormalities of aquatic birds: apparent impacts of selenium from irrigation drainwater. *Science of the Total Environment* 52:49–63.

Ohlendorf, H. M., R. L. Hothem, C. M. Bunck, and K. C. Marois. 1990. Bioaccumulation of selenium in birds at Kesterson Reservoir, California. *Archives of Environmental Contamination and Toxicology* 19:495–507.

Olson, R. K., ed. 1992. The role of created and natural wetlands in controlling nonpoint source pollution. *Ecological Engineering* 1:1–170.

O'Neill, R. V. 1976. Ecosystem persistence and heterotrophic regulation. *Ecology* 57:1244–1253.

O'Neill, R. V., D. L. DeAngelis, J. B. Waide, and T. F. H. Allen. 1986. *A Hierachical Concept of Ecosystems.* Princeton University Press, Princeton, NJ. 253 pp.

Orlob, G. 1981. *State of the Art of Water Quality Modelling.* International Institute for Applied Systems Analysis, Laxenburg, Austria.

Özesmi, U. and W. J. Mitsch. 1997. A spatial model for the marsh-breeding red-winged blackbird (*Agelaius phoeniceus* L.) in coastal Lake Erie wetlands. *Ecological Modelling* 101:139–152.

Pahl-Wostl, C. 1995. *The Dynamic Nature of Ecosystems: Chaos and Order Entwined.* Wiley, New York. 267 pp.

Park, R. A. et al., 1978. The aquatic ecosystem model MS. CLEANER. *Proceedings of the International Conference on Ecological Modelling,* August 28–September 2, Copenhagen, Denmark, p. 579.

Parrotta, J. A. and O. H. Knowles. 1999. Restoration of tropical moist forests on bauxite mined lands in the Brazilian Amazon. *Restoration Ecology* 7:103–116.

Parrotta, J. A. and O. H. Knowles. 2001. Restoring tropical forests on lands mined for bauxite: examples from the Brazilian Amazon. *Ecological Engineering* 17:219–239.

Parrotta, J. A., O. H. Knowles, and J. M. Wunderle, Jr. 1997. Development of floristic diversity in 10–year–old restoration forests on a bauxite mined site in Amazonia. *Forest Ecology and Management* 99:21–42.

Patten, B. C. 1971–1976. *Systems Analysis and Simulation in Ecology,* Vols. 1–4. Academic Press, New York.

Patten, B. C. 1991. Network ecology: indirect determination of the life-environment relationship in ecosystems. Pages 288–351. In: M. Higashi and T. P. Burns, eds., *Theoretical Studies of Ecosystems: The Network Perspective.* Cambridge University Press, Cambridge. 364 pp.

Petersen, J. E., C. C. Chen, and W. M. Kemp. 1997. Scaling aquatic primary productivity: experiments under nutrient and light-limited conditions. *Ecology* 78:2326–2328.

Phipps, R. G. and W. G. Crumpton. 1994. Factors affecting nitrogen loss in experimental wetlands with different hydrologic loads. *Ecological Engineering* 3:399–408.

Poff, N. L. and D. D. Hart. 2002. How dams vary and why it matters for the emerging science of dam removal. *BioScience* 52:659–668.

Price, J., L. Rochefort, and F. Quinty. 1998. Energy and moisture considerations on cutover peatlands: surface microtopography, mulch cover and *Sphagnum* regeneration. *Ecological Engineering* 10:293–312.

Prigogine, I. 1980. *From Being to Becoming: Time and Complexity in the Physical Sciences.* W. H. Freeman, San Francisco. 260 pp.

Prigogine, I. 1982. Order out of chaos. Pages 13–32. In: W. J. Mitsch, R. K. Ragade, R. W. Bosserman, and J. A. Dillon, Jr., eds., *Energetics and Systems.* Ann Arbor Science, Ann Arbor, MI.

Pringle, C. M. 1997. Exploring how disturbance is transmitted upstream: going against the flow. *Journal of the North American Benthological Society* 16:425–438.

Pritchard, D. 1967. Observations of circulation in coastal plain estuaries. Pages 37–44. In: G. Lauff, ed., *Estuaries.* Publication 83. American Association for the Advancement of Science, Washington, DC.

Qi, Y. and H. Tian. 1988. Some views on ecosystem design. In: M. Shijun, A. Jiang, R. Xu, and L. Dianmo, eds., *Proceedings of the International Symposium on Agro-Ecological Engineering,* August 1988. Ecological Society of China, Beijing.

Qin, P., M. Xie, Y. Jiang, and C.-H. Chung. 1997. Estimation of the ecological–economic benefits of two *Spartina alterniflora* plantations in North Jiangsu, China. *Ecological Engineering* 8:5–17.

Qin, P., M. Xie, and Y. Jiang. 1998. *Spartina* green food ecological engineering. *Ecological Engineering* 11:147–156.

Quinty, F., and L. Rochefort. 1997. Plant reintroduction on a harvested peat bog. Pages 133–145. In: C. C. Trettin, M. F. Jurgensen, D. F. Grigal, M. R. Gale, and J. K. Jeglum, eds., *Northern Forested Wetlands: Ecology and Management.* CRC Press/Lewis Publishers, Boca Raton, FL.

Rabalais, N. N., W. J. Wiseman, R. E. Turner, B. K. Sengupta, and Q. Dortch. 1996. Nutrient changes in the Mississippi River and system responses on the adjacent continental shelf. *Estuaries* 19:386–407.

Rabalais, N. N., R. E. Turner, W. J. Wiseman, and Q. Dortch. 1998. Consequences of the 1993 Mississippi River flood in the Gulf of Mexico. *Regulated Rivers* 14:161–177.

Raisin, G. W. and D. S. Mitchell. 1995. The use of wetlands for the control of non-point source pollution. *Water Science and Technology* 32:177–186.

Raisin, G. W., D. S. Mitchell, and R. L. Croome. 1997. The effectiveness of a small constructed wetland in ameliorating diffuse nutrient loadings from an Australian rural catchment. *Ecological Engineering* 9:19–35.

Ranwell, D. S. 1967. World resources of *Spartina townsendii* and economic use of *Spartina* marshland. *Coastal Zone Management Journal* 1:65–74.

Reddy, K. R. and W. H. Smith, eds. 1987. *Aquatic Plants for Water Treatment and Resource Recovery.* Magnolia Publishing, Orlando, FL.

Redfield, A. C. 1958. The biological control of chemical factors in the environment. *American Scientist* 46:206–226.

Reed, S. C., R. W. Crites, and E. J. Middlebrooks. 1995. *Natural Systems for Waste Management and Treatment,* 2nd ed. McGraw-Hill, New York. 433 pp.

Reinartz, J. A. and E. L. Warne. 1993. Development of vegetation in small created wetlands in southeast Wisconsin. *Wetlands* 13:153–164.

Reinelt, L. E. and R. R. Horner. 1995. Pollutant removal from stormwater runoff by palustrine wetlands based on comprehensive budgets. *Ecological Engineering* 4: 77–97.

Rheinhardt, R. D., M. M. Brinson, and P. M. Farley. 1997. Applying wetland reference data to functional assessment, mitigation, and restoration. *Wetlands* 17:195–215.

Richardson, C. J. and C. B. Craft. 1993. Effective phosphorus retention in wetlands: fact or fiction? Pages 271–282. In: G. A. Moshiri, ed., *Constructed Wetlands for Water Quality Improvement.* CRC Press, Boca Raton, FL.

Richardson, C. J., S. Qian, C. B. Craft, and R. G. Qualls. 1997. Predictive models for phosphorus retention in wetlands. *Wetlands Ecology and Management* 4:159–175.

Robinson, K. G., W. S. Farmer, and J. T. Novak. 1990. Availability of sorbed toluene in solids for biodegradation by acclimatized bacteria. *Water Research* 24:345–350.

Rochefort, L. and S. Campeau. 1997. Rehabilitation work on post-harvested bogs in south eastern Canada. Pages 287–284. In: L. Parkyn, R. E. Stoneman, and H. A. P. Ingram, eds., *Conserving Peatlands.* CAB International, Wallingford, Berkshire, England.

Rosenber, E. 1986. Microbial surfactants. *Critical Reviews in Biotechnology* 3:109–132.

Roszak, D. B. and R. Colwell. 1987. Survival strategies of bacteria in natural environment. *Microbiological Reviews* 51:365–379.

Ruddle, K. and G. Zhong. 1988. *Integrated Agriculture–Aquaculture in South China: The Dike-Pond System of the Zhujiang Delta.* Cambridge University Press, Cambridge. 173 pp.

Russell, R. C. 1999. Constructed wetlands and mosquitoes: health hazards and management options–an Australian perspective. *Ecological Engineering* 12:107–124.

Sanville, W. and W. J. Mitsch, eds. 1994. Creating freshwater marshes in a riparian landscape: research at the Des Plaines River Wetland Demonstration Project. Special issue. *Ecological Engineering* 3:315–521.

Sartoris, J. J., J. S. Thullen, L. B. Barber, and D. E. Salas. 2000. Investigation of nitrogen transformations in a southern California constructed wastewater treatment wetland. *Ecological Engineering* 14:49–65.

Schaafsma, J. A., A. H. Baldwin, and C. A. Streb. 2000. An evaluation of a constructed wetland to treat wastewater from a dairy farm in Maryland, USA. *Ecological Engineering* 14:199–206.

Scheffer, M., S. Carpenter, J. A. Foley, C. Folke, and B. Walker. 2001. Catastrophic shifts in ecosystems. *Nature* 413:591–596.

Schiechtl, H. 1980. *Bioengineering for Land Reclamation and Conservation.* University of Alberta Press, Edmonton, Alberta, Canada. 404 pp.

Schiewer, S. and B. Volesky. 1995. Modeling of the proton ion exchange in biosorption. *Environmental Science & Technology* 29:3049–3058.

Schiewer, S., and B. Volesky. 1997. Ionic strength and electrostatic effects in biosorption of divalent metal ions and protons. *Environmental Science & Technology* 30: 2478–2485.

Schindler, D. W. 1998. Replication versus realism: the need for ecosystem-scale experiments. *Ecosystems* 1:323–334.

Schipper, L. A., B. R. Clarkson, M. Vojvodic-Vukovic, and R. Webster. 2002. Restoring cut-over restiad peat bogs: a factorial experiment of nutrients, seed, and cultivation. *Ecological Engineering* 19:29–40.

Schlesinger, W. H. 1997. *Biogeochemistry: An Analysis of Global Change,* 2nd ed. Academic Press, San Diego, CA. 680 pp.

Schueler, T. R. 1992. *Design of Stormwater Wetland Systems: Guidelines for Creating Diverse and Effective Stormwater Wetlands in the Mid-Atlantic Region.* Metropolitan Washington Council of Governments, Washington, DC. 133 pp.

Seidel, K. 1964. Abbau von Bacterium Coli durch höhere Wasserpflanzen. *Naturwissenschaften* 51:395.

Seidel, K. 1966. Reinigung von Gewässern durch höhere Pflanzen. *Naturwissenschaften* 53:289–297.

Seidel, K. and H. Happl. 1981. Pflanzenkläranlage "Krefelder system." *Sicherheit in Chemie und Umbelt* 1:127–129.

Seip, H. M., L. Pawlowski, and T. J. Sullivan. 1994. Environmental degradation due to heavy metals and acidifying deposition: a Polish–Scandinavian workshop. Special issue. *Ecological Engineering* 3:205–312.

Sharma, S. S. and J. P. Gaur. 1995. Potential of *Lemna polyrrhiza* for removal of heavy metals. *Ecological Engineering* 4:37–44.

Sheehy, D. J. and S. F. Vik. 1992. Developing prefabricated reefs: an ecological and engineering approach. Pages 543–221. In: G. W. Thayer, ed., *Restoring the Nation's Marine Environment.* Maryland Sea Grant College, College Park, MD.

Shisler, J. K. 1990. Creation and restoration of coastal wetlands of the northeastern United States. Pages 143–170. In J. A. Kusler and M. E. Kentula, eds., *Wetland Creation and Restoration.* Island Press, Washington, DC.

Shutes, R. B., J. B. Ellis, D. M. Revitt, and T. T. Zhang. 1993. The use of *Typha latifolia* for heavy metal pollution contol in urban wetlands. Pages 407–414. In: G. A. Mosheri, ed., *Constructed Wetlands for Water Quality Improvement.* Lewis Publishers, Boca Raton, FL.

Sinicrope, T. L., P. G. Hine, R. S. Warren, and W. A. Niering. 1990. Restoring of an impounded salt marsh in New England. *Estuaries* 13:25–30.

Sinicrope, T. L., R. Langis, R. M. Gersberg, M. J. Busnardo, and J. B. Zedler. 1992. Metal removal by wetland mesocosms subjected to different hydroperiods. *Ecological Engineering* 1:309–322.

Soldatov, V. S., L. Pawlowski, E. Kloc, I. Symanska, and V. V. Maushevich. 1997. Remediation of depleted soils by addition of ion exchange resins. *Ecological Engineering* 8:337–346.

Spieles, D. J. and W. J. Mitsch. 2000a. The effects of season and hydrologic and chemical loading on nitrate retention in constructed wetlands: a comparison of low and high nutrient riverine systems. *Ecological Engineering* 14:77–91.

Spieles, D. J. and W. J. Mitsch. 2000b. Macroinvertebrate community structure in high- and low-nutrient constructed wetlands. *Wetlands* 20:716–729.

Stark, L. R. and F. M. Williams. 1995. Assessing the performance indices and design parameters of treatment wetlands for H+, Fe, and Mn retention. *Ecological Engineering* 5:433–444.

Steffan, R. J., K. L. Sperry, M. T. Walsh, S. Vainberg, and C. W. Condee. 1999. Field scale evaluation of in situ bioaugmentation for remediation of chlorinated solvents. *Environmental Science & Technology* 33:2771–2791.

Stein, E. D. and R. F. Ambrose. 1998. A rapid impact assessment method for use in a regulatory context. *Wetlands* 18:379–392.

Steinberg, S. M., J. J. Pignatello, and B. L. Sawhney. 1987. Persistence of 1,2-dibromoethane in soils. *Environmental Science & Technology* 21:1201–1209.

Steiner, G. R. and R. J. Freeman, Jr. 1989. Configuration and substrate design considerations for constructed wetlands for wastewater treatment. Pages 363–378. In: D. A. Hammer, ed., *Constructed Wetlands for Wastewater Treatment.* Lewis Publishers, Chelsea, MI.

Steiner, G. R., J. T. Watson, D. Hammer, and D. F. Harker, Jr. 1987. Municipal wastewater treatment with artificial wetlands: a TVA/Kentucky demonstration. In: K. R. Reddy and W. H. Smith, eds., *Aquatic Plants for Wastewater Treatment and Resource Recovery.* Magnolia Publishing, Orlando, FL. 923 pp.

Straskraba, M. 1980. The effects of physical variables on freshwater production: analyses based on models. Pages 13–31. In: E. D. Le Cren and R. H. McConnell, eds., *The Functioning of Freshwater Ecosystems.* International Biological Programme 22. Cambridge University Press, Cambridge.

Straskraba, M. 1984. New ways of eutrophication abatement. Pages 37–45. In: M. Straskraba, Z. Brandl, and P. Procalova, eds., *Hydrobiology and Water Quality of Reservoirs.* Academy of Science, České Budějovice, Czechoslovakia.

Straskraba, M. 1985. *Simulation Models as Tools in Ecotechnology Systems: Analysis and Simulation,* Vol. II. Akademie Verlag, Berlin.

Straskraba, M. 1993. Ecotechnology as a new means for environmental management. *Ecological Engineering* 2:311–331.

Straskraba, M. and A. H. Gnauck. 1985. *Freshwater Ecosystems: Modelling and Simulation.* Elsevier, Amsterdam. 305 pp.

Streever, W., ed. 1999. *An International Perspective on Wetland Rehabilitation.* Kluwer Academic, Dordrecht, The Netherlands. 338 pp.

Svengsouk, L. M. and W. J. Mitsch. 2001. Dynamics of mixtures of *Typha latifolia* and *Schoenoplectus tabernaemontani* in nutrient-enrichment wetland experiments. *American Midland Naturalist* 145:309–324.

Tang, S.-Y. 1993. Experimental study of a constructed wetland for treatment of acidic wastewater from an iron mine in China. *Ecological Engineering* 2:253–259.

Tanner, C. C. 1996. Plants for constructed wetland treatment systems: a comparison of the growth and nutrient uptake of eight emergent species. *Ecological Engineering* 7:59–83.

Tanner, C. C., J. S. Clayton, and M. P. Upsdell. 1995. Effect of loading rate and planting on treatment of dairy farm wastewaters in constructed wetlands. II. Removal of nitrogen and phosphorus. *Water Research* 29:27–34.

Tanner, C. C., G. Raisin, G. Ho, and W. J. Mitsch. 1999. Constructed and natural wetlands for pollution control. Special issue. *Ecological Engineering* 12:1–170.

Tarutis, W. J., L. R. Stark, and R. M. Williams. 1999. Sizing and performance estimation of coal mine drainage wetlands. *Ecological Engineering* 12:353–372.

Teal, J. M. and S. B. Peterson. 1991. The next generation of septage treatment. *Research Journal of the Water Pollution Control Federation* 63:84–89.

Teal, J. M. and M. P. Weinstein. 2002. Ecological engineering, design, and construction considerations for marsh restorations in Delaware Bay, USA. *Ecological Engineering* 18:607–618.

Thayer, G. W., ed. 1992. *Restoring the Nation's Marine Environment.* Maryland Sea Grant Project, College Park, MD. 716 pp.

Thofelt, L. and A. Englund, eds. 1996. *Ecotechniques for a Sustainable Society.* Mid Sweden University Press, Östersund, Sweden.

Thullen, J. S., J. J. Sartoris, and W. E. Walton. 2002. Effects of vegetation management in constructed wetland treatment cells on water quality and mosquito production. *Ecological Engineering* 18:441–450.

Todd, J. H. and B. Josephson. 1996. The design of living technologies for waste treatment. *Ecological Engineering* 6:109–136.

Toth, L. A., D. A. Arrington, M. A. Brady, and D. A. Muszick. 1995. Conceptual evaluation of factors potentially affecting restoration of habitat structure within the channelized Kissimmee River ecosystem. *Restoration Ecology* 3:160–180.

Turner, R. E. and M. E. Boyer. 1997. Mississippi River diversions, coastal wetland restoration/creation, and an economy of scale. *Ecological Engineering* 8:117–128.

Uhlmann, D. 1983. Entwicklungstendenzen der Okotechnologie. *Wissenschaftliche Zeitschrift der Technischen Universitaet Dresden* 32:109–116.

Ulanowicz, R. E. 1979. Prediction chaos and ecological perspective. Pages 107–117. In: E. A. Halfon, ed., *Theoretical Systems Ecology.* Academic Press, New York.

Ulanowicz, R. E., 1986. *Growth and Development: Ecosystem Phenomenology.* Springer-Verlag, New York. 203 pp.

Ulanowicz, R. E. 1997. *Ecology: The Ascendent Perspective.* Columbia University Press, New York. 201 pp.

U.S. Environmental Protection Agency. 1991. Summary report: High-priority research in bioremediation. Presented at the Bioremediation Research Needs Workshop, April 15–16, 1991, Washington, DC.

U.S. Environmental Protection Agency. 1993. *Constructed Wetlands for Wastewater Treatment and Wildlife Habitat: 17 Case Studies.* EPA832-R-93-005. U.S. EPA, Washington, DC. 174 pp.

Van der Valk, A. G. 1998. Succession theory and restoration of wetland vegetation. Pages 657–667. In: A. J. McComb and J. A. Davis, eds., *Wetlands for the Future.* Gleneagles Publishing, Adelaide, Australia.

Van der Valk, A. G. and C. B. Davis. 1978. Primary production of prairie glacial marshes. Pages 21–37. In: R. E. Good, D. F. Whigham, and R. L. Simpson, eds., *Freshwater Wetlands: Ecological Processes and Management Potential.* Academic Press, New York.

Vannote, R. L., G. W. Minshall, K. W. Cummins, J. R. Sedell, and C. E. Cushing. 1980. The river continuum concept. *Canadian Journal of Fisheries and Aquatic Sciences* 37:130–137.

Van Slyke, L. P. 1988. *Yangtze: Nature, History and the River.* Additon-Wesley, Reading, MA. 211 pp.

Vitousek, P. M., H. A. Mooney, J. Lubchenck, and J. M. Mellilo. 1997. Human domination of the Earth's ecosystems. *Science* 277:494–499.

Volkering, F., A. M. Breure, J. G. van Andel, and W. H. Rulkens. 1995. Influence of nonionic surfactants on bioavailability and biodegradation of polycyclic aromatic hydrocarbons. *Applied and Environmental Microbiology* 61:1699–1705.

Vollenweider, R. A. 1969. Möglichkeiten und Grenzen elementarer Modelle der Stoffbilanz von Seen. *Archiv fuer Hydrobiologie* 66:1–136.

Vollenweider, R. A. 1975. Input–output models with special reference to the phosphorus loading concept in limnology. *Schweizerische Zeitschrift fuer Hydrologie* 37:53–83.

Vymazal, J. 1995. Constructed wetlands for wastewater treatment in the Czech Republic: State of the art. *Water Science and Technology* 32:357–364.

Vymazal, J. 1998. Czech Republic. Pages 95–121. In: J. Vymazal, H. Brix, P. F. Cooper, M. B. Green, and R. Haberl, eds., *Constructed Wetlands for Wastewater Treatment in Europe.* Backhuys Publishers, Leiden, The Netherlands.

Vymazal, J. 2002. The use of sub-surface constructed wetlands for wastewater treatment in the Czech Republic: 10 years' experience. *Ecological Engineering* 18:633–646.

Vymazal, J., H. Brix, P. F. Cooper, M. B. Green, and R. Haberl, eds. 1998. *Constructed Wetlands for Wastewater Treatment in Europe.* Backhuys Publishers, Leiden, The Netherlands.

Wali, M. K., ed. 1992. *Environmental Rehabilitation: Preamble to Sustainable Development,* 2 vols. SPB Academic Publishing, The Hague, The Netherlands.

Wali, M. K., 1999. Ecological succession and the rehabilitation of disturbed terrestrial ecosystems. *Plant and Soil* 213:195–220.

Wang, N. and W. J. Mitsch. 1998. Estimating phosphorus retention of existing and restored wetlands in a tributary watershed of the Laurentian Great Lakes in Michigan, USA. *Wetlands Ecology and Management* 6:69–82.

Wang, N. and W. J. Mitsch. 2000. A detailed ecosystem model of phosphorus dynamics in created riparian wetlands. *Ecological Modelling* 126:101–130.

Wang, N., W. J. Mitsch, S. Johnson, and W. T. Acton. 1997. Early hydrology of a newly constructed riparian mitigation wetland at the Olentangy River Wetland Research Park. Pages 247–254. In: W. J. Mitsch, ed., *Olentangy River Wetland Research Park 1996 Annual Report.* School of Natural Resources, Ohio State University, Columbus, OH.

Wang, R. and J. Yan. 1998. Integrating hardware, software, and mindware for sustainable ecosystem development: principles and methods of ecological engineering in China. *Ecological Engineering* 11:277–289.

Wang, R., J. Yan, and W. J. Mitsch. 1998a. Ecological engineering: a promising approach towards sustainable development in developing countries. *Ecological Engineering* 11:1–15.

Wang, R., J. Yan, and W. J. Mitsch, eds. 1998b. Ecological engineering in developing countries. Special issue. *Ecological Engineering* 11:1–313.

Ward, A. D. and W. J. Elliot, eds. 1995. *Environmental Hydrology.* CRC Press/Lewis Publishers, Boca Raton, FL.

Ward, J. V. and J. A. Stanford. 1983. The immediate-disturbance hypothesis: an explanation for biotic diversity patterns in lotic ecosystems. Pages 347–356. In: T. D. Fontaine and S. M. Bartell, eds., *Dynamics of Lotic Systems.* Ann Arbor Science, Ann Arbor, MI.

Ward, J. V. and J. A. Stanford. 1995. The serial discontinuity concept: extending the model to floodplain rivers. *Regulated Rivers* 10:1598.

Webster, J. R. 1979. Hierarchical organization of ecosystems. Pages 119–131. In: E. Halfon, ed., *Theoretical Systems Ecology.* Academic Press, New York.

Weiderholm, T. 1980. Use of benthos in lake monitoring. *Journal of the Water Pollution Control Federation* 52:537.

Weinstein, M. P. and D. A. Kreeger, eds. 2000. *Concepts and Controversies in Tidal Marsh Ecology.* Kluwer Academic, Amsterdam, The Netherlands.

Weinstein, M. P., J. H. Balletto, J. M. Teal, and D. F. Ludwig. 1997. Success criteria and adaptive management for a large–scale wetland restoration project. *Wetlands Ecology and Management* 4:111–127.

Weinstein, M. P., J. M. Teal, J. H. Balletto, and K. A. Strait. 2001. Restoration principles emerging from one of the world's largest tidal marsh restoration projects. *Wetlands Ecology and Management* 9:387–407.

Weller, M. W. 1994. *Freshwater Marshes,* 3rd ed. University of Minnesota Press, Minneapolis, MN. 192 pp.

Wetzel, P. R., A. G. van der Valk, and L. A. Toth. 2001. Restoration of wetland vegetation on the Kissimmee River flood plain: potential role of seed banks. *Wetlands* 21:189–198.

White, C., S. C. Wilkinson, and G. M. Gadd. 1995. The role of microorganisms in biosorption of toxic metals and readionuclides. *International Biodeterioration and Biodegradation* 35:17–40.

White, J. S., S. E. Bayley, and P. J. Curtis. 2000. Sediment storage of phosphorus in a northern prairie wetland receiving municipal and agro–industrial wastewater. *Ecological Engineering* 14:127–138.

Whitton, B. A., ed. 1975. *River Ecology.* Blackwell Scientific, Oxford.

Wilber, P., G. Thayer, M. Croom, and G. Mayer, eds. 2000. Goal setting and success criteria for coastal habitat restoration. Special issue. *Ecological Engineering* 15:165–395.

Wieder, R. K. 1989. A survey of constructed wetlands for acid coal mine drainage treatment in eastern United States. *Wetlands* 9:299–315.

Wieder, R. K., M. N. Linton, and K. P. Heston. 1990. Laboratory mesocosm studies of Fe, Al, Mn, Ca, and Mg dynamics in wetlands exposed to synthetic acid coal mine drainage. *Water Soil and Air Pollution* 51:181–196.

Wilhelm, M., S. R. Lawry, and D. D. Hardy. 1989. Creation and management of wetlands using municipal wastewater in northern Arizona: a status report. Pages 179–185. In: D. A. Hammer, ed., *Constructed Wetlands for Wastewater Treatment.* Lewis Publishers, Chelsea, MI.

Wilson, R. F. and W. J. Mitsch. 1996. Functional assessment of five wetlands constructed to mitigate wetland loss in Ohio, USA. *Wetlands* 16:436–451.

Wilson, J. T. and M. D. Jawson. 1995. Science needs for implementation of biore-mediation. Pages 293–303. In: H. D Skipper and R. F. Turco, eds., *Bioremediation Science and Applications.* Soil Science Society of America, Madison, WI.

Wind-Mulder, H. L., L. Rochefort, and D. H. Vitt. 1996. Water and peat chemistry comparisons of natural and post-harvested peatlands across Canada and their rele-vance to peatland restoration. *Ecological Engineering* 7:161–181.

Wisheu, I. C. and P. A. Keddy. 1992. Competition and centrifugal organization of plant communities: theory and tests. *Journal of Vegetation Science* 3:147–156.

Wood, T. S. and M. L. Shelley. 1999. A dynamic model of bioavailability of metals in constructed wetland sediments. *Ecological Engineering* 12:231–252.

Woodhouse, W. W., Jr. 1979. *Building Salt Marshes along the Coasts of the Conti-nental United States.* Special Report 4. U.S. Army, Coastal Engineering Research Center, Fort Belvoir, VA.

World Commission on Environment and Development. 1987. *Our Common Future.* Oxford University Press, Oxford, p. 11.

Wu Jin Fu, Yan Fu, Yang Shi Jie, Liang Hai Quan, and Yu Wu Jiao. 1988. A primary study on coordinated ecological engineering of fertilizer and forage and fuel in Wu Tang Village of Changsha County. In: Ma Shijun, Jiang Ailiang, Xu Rumei, and Li Dianmo, eds., *Proceedings of the International Symposium on Agro-Ecological Engineering,* August 1988. Ecological Society of China, Beijing.

Yan, J. 1992. In memoriam: dedicated to the memory of Professor Ma. *Ecological Engineering* 1:ix–x.

Yan, J. and H. Yao. 1989. Integrated fish culture management in China. Pages 375–408. In: W. J. Mitsch, and S. E. Jørgensen, eds., *Ecological Engineering: An In-troduction to Ecotechnology.* Wiley, New York.

Yan, J. and Y. Zhang. 1992. Ecological techniques and their application with some case studies in China. *Ecological Engineering* 1:261–285.

Yan, J., Y. Zhang, and X. Wu. 1993. Advances of ecological engineering in China. *Ecological Engineering* 2:193–215.

Yao, H. 1993. Phytoplankton production in integrated fish culture high-output ponds and its status in energy flow. *Ecological Engineering* 2:217–229.

Yuan, C., Q. Zhao, and J. Zhen. 1993. Comparing crop-hog-fish agroecosystems with conventional fish culturing in China. *Ecological Engineering* 2:231–242.

Zalewski, M. 2000a. Ecohydrology: The scientific background to use ecosystem prop-erties as management tools toward sustainability of water resources. *Ecological Engineering* 16:1–8.

Zalewski, M., ed. 2000b. Ecohydrology. Special issue. *Ecological Engineering* 16:1–188.

Zalewski, M. and T. Wagner. 2000. *Ecohydrology.* Technical Documents in Hydrology 34. UNESCO, Paris.

Zalewski, M., G. A. Janauer, and G. Jolankai. 1997. *Ecohydrology: A New Paradigm for the Sustainable Use of Aquatic Resources.* Technical Documents in Hydrology 7. UNESCO-IHP, Paris.

Zedler, J. B. 1988. Salt marsh restoration: lessons from California. Pages 123–138. In: J. Cairns, ed., *Rehabilitating Damaged Ecosystems,* vol. I. CRC Press, Boca Raton, FL.

Zedler, J. B. 1996a. Coastal mitigation in southern California: the need for a regional restoration strategy. *Ecological Applications* 6:84–93.

Zedler, J. B. 1996b. *Tidal Wetland Restoration: A Scientific Perspective and Southern California Focus.* Report T-038. California Sea Grant College System, University of California, La Jolla, CA. 129 pp.

Zedler, J. B., ed. 2001. *Handbook for Restoring Tidal Wetlands.* CRC Press, Boca Raton, FL. 439 pp.

Zeigler, B. P. 1976. *Theory of Modelling and Simulation.* Wiley, New York. 435 pp.

Zhang, J., S. E. Jørgensen, M. Beklioglu, and O. Ince. 2002. Hysteresis in catastrophic shift: Lake Mogan prognoses. *Ecological Modelling* 164:227–238.

Zhang, L. and W. J. Mitsch. 2001. Hydrologic budgets for the ORW mitigation wetland, 2000. Pages 115–125. In: W. J. Mitsch and L. Zhang, eds., *Olentangy River Wetland Research Park 2000 Annual Report.* School of Natural Resources, Ohio State University, Columbus, OH.

Zhang, R., W. Ji, and B. Lu. 1998. Emergence and development of agro–ecological engineering in China. *Ecological Engineering* 11:17–26.

Zhang, Y. and R. M. Miller. 1992. Enhanced octadecane dispersion and biodegradation by *Pseudomonas rhamnolipid* surfactant. *Applied and Environmental Microbiology* 58:3276–3282.

Zwolinski, J. 1994. Rates of organic matter decomposition in forests polluted with heavy metals. *Ecological Engineering* 3:17–26.

ORGANISM INDEX

Acadia, 301
Acanthaster, 228
Agelaius phoeniceus, 173
Aix sponsa, 173
Alligator weed, 186
Alnus, 300
Alternanthera philoxeroides, 186
Anas
   discors, 173
   platyrhynchos, 173
Ardea herodias, 173
Aristichthys nobilis, 328

Bacillis sphaericus, 258
Bats, 258, 301
Baumea articulata, 256
Beaver, 181, 257
Beech, European, 297
Berula erecta, 159
Betula, 300
   verrucosa, 300
Birch, 300
Blue jay, 173
Blue-winged teal, 173
Bombycilla cedorum, 173
Bowfin, 143
Branata canadensis, 173, 257
Bubulcus ibis, 143
Bullfrog, 189

Bulrush, 256
   soft-stem, 187
Buteo jamaicensis, 173
Butorides virescens, 173

Caddis fly, 143
Callitriche platycarpa, 159
Cardinal, northern, 173
Cardinalis cardinalis, 173
Cardoparus mexicanus, 173
Carex, 185, 191
Carp, 224, 257, 318–320
   bighead, 326
   black, 325–330
   common, 325–330
   grass, 316, 318, 324, 326–330
   silver, 326–327
Casmerodius albus, 173
Castor canadensis, 181, 257
Cathartes aura, 173
Cattail, 174, 185, 187, 256
Cedar waxwing, 173
Ceratophyllum demersum, 159
Chaetura plagica, 173
Charadrius vociferous, 173
Chen, 257
Chimney swift, 173
Circus cyaneus, 173
Cladium jamaicense, 174

397

Cladocera, 325
Colaptes auratus, 173
Coot, American, 173
Copepoda, 325
Coral, 201–202
Corvus
   brachyrnchos, 173
   corax, 173
Crabs, 201
Crocodile, American, 216–217
Crocodylus acutus, 217
Crow, 173
Crown-of-thorns starfish, 228
Ctenopharyngodon idella, 316, 324,
   328
Cyanocitta cristata, 173
Cymodocea, 200
Cyperus involucratus, 256
Cyprinus carpio, 224, 257, 328

Distichlis spicata, 198
Duckweed, 234, 256

Eagles, 66
Eelgrass, 200
Egret
   cattle, 143
   great, 173
Eichhornia crassipes, 53, 119, 121,
   143, 186, 234, 256, 323–325
Epidonax, 173
Esox lucius, 49
Eucalyptus, 301

Fagus sylvatica, 300
Falco sparverius, 173
Falcons, 66
Finch, house, 173
Flicker, common, 173
Flycatcher, Epidonax, 173
Fulica Americana, 173

Gambusia affinis, 258
Gar, Florida, 143
Glyceria maxima, 256
Goldfinch, American, 173
Goose
   Canada, 173, 257

   snow, 257
Grackle, common, 173
Groenlanda densa, 159

Halodule, 200
Hawk
   marsh, 173
   red-tailed, 173
Heron
   great blue, 173
   green, 173
Hirundo, 173
   rustica, 173
Hypophthalmichtys molitrix, 328

Indigo bunting, 173
Iridoprocne, 173
   bicolor, 173

Junco hyemalis, 173
Junco, northern, 173
Juncu, 334
   effuses, 256
   roemerianus, 198

Kestrel, American, 173
Killdeer, 173
Kingfisher, belted, 173

Lamelbibranchia, 325
Lanius, 173
Larch, 297
Larix
   deciduas, 300
   gmelinii, 297
   leptolepis, 297
   olgensis, 297
Larus delawarensis, 173
Least bittern, 258
Lemna, 159, 234, 256
   trisulca, 120
Lepomis, 188
Lobster, 201
Lotus, 334
Lythrum salicaria, 187, 224

Mallard, 173
Manatee, 201

Mayfly, 143
Megaceryle alcyon, 173
Megalobrama amblycephala, 316, 324, 328
Melospiza
  Georgiana, 173
  melodia, 173
Metasequoia, 321–322
Mosquito fish, 258
Mourning dove, 173
Muskrat, 181, 189, 245, 257, 260
Mylopharyngodon piceus, 328

Nelumbo, 256, 334
Nuphar, 185, 256
Nutria, 260
Nymphaea, 185, 256

Ondatra zibethicus, 181, 189, 245, 257

Pangolin, 326
Parus bicolor, 173
Passerina cyanea, 173
Philohela minor, 173
Phragmites, 54, 198, 205–206, 210, 212–215, 254, 316
  australis, 7, 186, 203–204, 206, 209, 232, 234, 256
Picea
  abies, 298, 300
  pungens, 300
Pike, 49, 117
Pistia, 256
  stratiotes, 143
Purple loosestrife, 187

Quiscalus quiscalus, 173

Rallus limicola, 173
Rana catesbeiana, 189
Raven, northern, 173
Red-winged blackbird, 173
Reed grass, 186, 209, 234, 256, 316
Rhizobium, 285
Ring-billed gull, 173
Robin, 173

Sagittaria, 191

Salicornia, 198, 212
Salix, 187
  caroliniana, 174
Salvinia, 285
Sawgrass, 174
Schoenoplectus, 185, 256
  lacustris, 233
  tabernaemontani, 187
  validus, 256
Scirpus, 185, 256
  acutus, 191
  fluviatilis, 187, 256
  validus, 187, 191 (=*Schoenoplectus tabernaemontani*)
Sea turtles, 201
Sedges, 185, 191
Shrike, 173
Shrimp, 201
Sorbus aucuparia, 300
Sparganium eurycarpum, 187
Sparrow
  song, 173
  swamp, 173
Spartina, 7, 54, 210, 212–213, 215, 322–323
  alterniflora, 7, 196, 202–204, 209, 211–212, 282, 322–323
  anglica, 198, 202, 321–323
  patens, 198, 205–206, 282
  pectinata, 187
  townsendii, 198, 202, 322–323
Spatterdock, 185
Spinus tristis, 173
Spirodela, 285
Sponges, 201
Spruce, Norway, 297–298, 300
Starling, European, 173
Sturnus vulgaris, 173
Swallows (Hirundinidae), 66, 173, 258
  barn, 173
  tree, 173
Swift (Apodidae), 258
  chimney, 173
Syringodium, 200

Tatami, 334
Taxodium, 174, 200
Thalassia testudinum, 200

Tilapia, 326
Tufted titmouse, 173
Turdis migratorius, 173
Turkey vulture, 173
Turtlegrass, 200
Typha, 53, 82, 153, 170, 184–188,
    190–191, 234, 256, 292-294, 296,
    314
    angustifolia, 203–204
    domingensis, 174

Virginia rail, 173

Water hyacinth, 119–121, 143, 186,
    234, 256, 323–325

Water lettuce, 143, 256
Water lily, 185
Wild rice, 334
Willow, 187
Woodcock, 173
Wood duck, 173
Wuchang fish, 316, 318, 324, 326–330

Yellow-headed blackbird, 258

Zenaida macroura, 173
Zizania latifolia, 256, 334
Zooxanthellae, 201
Zostera, 200

SUBJECT INDEX

Acid mine drainage, 292–296
modeling, 294–296, 359
wetland treatment, 237
Adaptive management, 215
Adjacency matrix, 343–344
Agricultural runoff, 174, 238–242
Agroecological engineering, 25–26, 54
China, 310, 321–322
Agroecosystems, 24
Alberta, Canada, 235, 250
Algae
biosorption, 283–284
uptake of metals, 283–284
Amazon River, 301–302
Arizona, 5, 8, 51, 235, 258
Artificial ecology, 19
Aspect ratio, treatment wetlands, 246
Atchafalaya River, 219
Australia, 238, 250
wetland loss, 165
Austria, 114

Back swamp, 148
Baltic Sea, 3, 5, 166, 175
Before–after study, restoration success,
159
Billabong, 148–149, 151, 166
Bioaccumulation, 77–79
plants, 285–286

Bioaccumulators, 285
Bioavailability, 263–270
factors that affect, 265–270
Biobarriers, 281
Biodegradation
contaminants, 270–274
example calculation, 273–274
primary, 271
rate, 271–273
ultimate, 271
Bioengineering, 24
Biological Concentration Factor, 268–269
Biological magnification, 65–66
Biomagnification, 77
Biomanipulation, 19, 24, 49–50, 117–119
Biomineral water, 323
Bioremediation, 263–286
bioavailability, 263–270
Biosphere, 2, 5–6, 8–9, 41, 51–52
Biospherics, 24
Biotechnology, comparison with
ecological engineering, 35–36
Biotraps, microbial, 284–285
Black triangle, 298–300
BOD, wetland removal, 251–252
Bog, 164
Boltzman, 90–91

Bottomland forest(s), 97
    restoration, 151, 155–159
Brazil, 97, 301–302
Bregnier–Cordon plain, France, 159–162
Buffer capacity, 84

Cadmium, 253, 286
    bioremediation, 275–279
California, 52, 168, 202, 235, 253
Canada, 242
    Great Lakes Water Quality
        Agreement, 224
    wetland loss, 165
Case studies
    Agricultural runoff wetland, Ohio,
        239–242
    Biomanipulation, Bautzen Reservoir,
        Germany, 119
    Bottomland hardwood forest
        restoration, Ohio, 153–159
    Chinese fishpond, 325–330
    Dam removal, Manatawny Creek,
        Pennsylvania, 137–140
    Delaware Bay salt marsh restoration,
        203–215
    Florida Everglades restoration, 174
    Forested wetland restoration, Florida,
        172, 174
    Great Lakes delta restoration,
        Michigan, 222, 224–227
    Lake restoration comparison,
        Denmark, 123–124
    Louisiana delta restoration, 218–223
    Mangrove restoration, Florida, 216–217
    Mine drainage wetland, Ohio, 294–296
    Mitigation wetland, Central Ohio,
        168–172
    Nitrogen retention by wetlands,
        Denmark, 359–366
    Oxbow creation, Ohio, 149, 151–155
    Phosphate mine restoration, Florida,
        303–305
    Recycling metals in agricultural
        cropland, 275–279
    Restoring Black Triangle, Central
        Europe, 298–300
    Restoring flood pulses, Rhône River,
        France, 159–162
    River channel restoration,
        Kissimmee River, Florida, 141–145
    River channel restoration, Skjern
        River. Denmark, 145–147
    Solving the Gulf of Mexico hypoxia,
        USA, 175
    Treatment wetland, Michigan, 235–237
    Wetland planting experiment, Ohio,
        187–189
    Wetlands in Yangtze River valley,
        China, 331–335
Catastrophe theory, 119, 357
Catchment, 126
Chesapeake Bay, 51, 175, 202
China, 25–27, 38, 51, 53, 54
    coal reserves, 292
    ecological engineering, 309–335
    wetland loss, 165
Chromium, 253, 285
Civilizations
    aquatic, 331
    hydraulic, 331
Clean technology, 10–13
Clean Water Act, 9
Climate change, 11, 16–17
Coal mine
    drainage control, 53
    restoration, 290, 292–296
Coastal ecosystems, 196–202
    Odum classification, 196
Coastal restoration, 5, 7, 195–229
    techniques, 202–229
Connecticut, 243
Conservation Reserve Program (CRP),
    167
Constructed wetland, see also
        Treatment wetland, 5–6, 166, 232
    root-zone method, 234, 254
    subsurface flow, 232, 254
    surface-flow, 232, 254
Control structures, wetlands, 181–183
Copper, 253, 284–286
Coral reefs, 201–202
    bleaching, 202

restoration, 227–229
restoration techniques, 227–228
Creative destruction, 86
Crevasses, 218
Czech Republic, 235, 298–300

Dairy wastewater, 242–243
Dams, removal, 133–140
Danbue delta, 163
DDT, 66, 74, 77
Dead zone, coastal, see also Hypoxia, 195
Delaware, 7, 212, 203, 205
Delaware Bay, 5, 7, 54, 166, 203–215
Delta restoration, 217–227
Denitrification, 43–44, 183, 255, 363
Denmark, 5–6, 99, 107, 123–124, 145–147, 234, 235
Des Plaines River Wetland Demonstration Project, Illinois, 178, 239, 247–248, 258
Design graphs, treatment wetlands, 247–248
Design principles
  Chinese ecological engineering, 312
  ecological engineering, 94–102
Designer ecosystems, 30, 32
Detention time, 245–246
Development, see also Succession, 84–87, 100
Dissapative structures, 89–90, 95
Dissolved oxygen, Chinese fishponds, 328–330
Diversions, see also Mississippi River diversions, 222
Diversity
  and self-design, 99
  species, 80–84
  and stability, 81–84

Eatout, 257
Ecoefficiency, 18, 39
Ecohydrology, 24–25, 94, 101
Ecological design principles, 94–102
Ecological economics, 37–38
Ecological engineering
  basic concepts, 27–35
  China, 25, 309–335
  classification, 40–55
classification by function, 42–50
classification by scale, 50–54
comparison with biotechnology, 35–36
comparison of West and China, 311–316
current practicing, 19–20
definition, 23
design principles, 94–102
ecology basis, 36–38
education, 20–21
future, 21–22
goals, 23, 312–314
history, 25–27
professional societies, 27
reasons for, 14–20
spectrum, 40–41
timing, 19–20
when to use, 51, 55
why needed, 13–19
Ecological hierarchy, 61–62
Ecological restoration, see also Ecosystem restoration; Restoration, 24
Ecology
  alliance with engineering, 18
  comparison with ecological engineering, 36–38
  contribution to ecological engineering, 20
  definition, 56
  development of science, 21–22
  reductionistic, 56–57
  systems, see Systems Ecology
  terrestrial restoration, 289–291
  theories, 32
  in universities, 21
Economics
  Biosphere 2, 5–6, 9
  comparison of China and West, 312, 315
  lake restoration, 122–124
Ecosystem
  complexity, 79–84
  conservation, 34–35
  coupling, 99–100
  diversity, 80–81
  ecology, 56
  engineers, 101

Ecosystem (*continued*)
  feedback, 59
  hierarchy, 101
  rehabilitation
  restoration, 19, 23–24
  stability, 81–84
  stoichiometry, 69–70
  structure, 314–315
  theories, a pattern, 93
Ecotechniques, 24, 38
Ecotechnology, 19, 24, 26
Ecotones, 87, 99
Education and ecological engineering,
    20–21
Egypt, 217
Einstein's Law, 62–63
El Niño, 202, 227
Emergy, 90–91
Empirical models, treatment wetlands,
    249, 251–252
Energese, See also Odum energy
    diagrams, 71–72
Energy
  availability in oxidation, 69, 71
  content in biological material, 67
  conversion factors, 70
  in ecosystems, 65–69
  flow models, 68, 192, 353–354
  language, see also Odum energy
      diagram, 71–72
  laws, 62, 65–71, 95
  succession, 86
  use in Chinese ecological
      engineering, 321
Engineering ecology, 24
Engineering
  alliance with ecology, 18
  comparison of conventional and
      ecological, 30–31
  and self-design, 30
  in universities, 21
England, 107, 322
Environmental engineering
  comparison with ecological
      engineering, 35
Environmental management, 9–13
Environmental risk assessment, 340
Environmental technology, 9–13, 44
Epilimnion, 106

Equations
  algbraic, 348
  differential, 348, 359
  partial differential, 349
Estonia, 27
Estuarine ponds, experimental, 52
Estuarine pulsing, 97
Estuary, definition, 197
Europe, wetland loss, 165
Eutrophication, 16–17, 43, 82, 105–
    108, 158
  process, 107–108
  modeling, 108–109
Evapotranspiration, 244
Everglades, Florida, 3–4, 51, 163, 174
  model, 52
  nutrient removal project, 239
  restoration, see also Case study,
      3–4
Evolution, 87–89
Exergy, 359
  maximization theory, 90–92
Exotic plants, 186–187

Feedback, 59
Fen, 164
Finland, 176
Fishponds, China, 45–46, 325–330
Fish
  carnivorous, 118
  planktivorous, 117–119
  production systems in China, 316–
      321
Five elements, China, 310–311
Flood Pulse Concept (FPC), 132–133
Floodplains
  aggradation, 128
  degradation, 128
  terraces, 128
Florida, 3–4, 51, 53–54, 120–121, 125,
    141–145, 168, 172, 174, 202,
    216, 232, 235, 238, 239, 250,
    303–305
Forcing functions, 95
Forest
  degradation, 296–297
  restoration, 296–303
  urban environments, 306–308

France, 27, 51, 54, 159–162
Freundlich adsorption isotherms, 267–268

Gene pool, 88–89
Geographical information systems, 357
Germany, 27, 176, 233, 254, 298–300
Gravel bed wetlands, 233, 254
Great Lakes, 222
Green taxes, 13
Groundwater, in wetlands, 181–182
Growth forms, 85–86
Gulf of Mexico, 3–5, 166, 175–176, 219

Herbicides, 212–214
Hierarchy
  of ecosystems, 101
  feedback mechanisms, 59
  theory, 61–62
Holism, 56, 61, 102
Homeostasis, 96
Houghton Lake, Michigan, 51, 53, 235–237
Hydraulic Loading Rate (HLR), 245, 252, 259
Hydric soil, see also Wetland(s), soils, 182–183
Hydroecology, 19
Hydrology
  budget, 149, 152
  created and restored wetlands, 181–182
  treatment wetlands, 243–246, 259
Hydroperiod, wetlands, 153, 244, 170–171
Hyperaccumulator, 285
Hypolimnion, 105–106
  siphoning, 113–114
Hypoxia, Gulf of Mexico, 175

Illinois, 51, 53–54, 117, 178, 238, 247–248, 250, 294
Indian Lake wetland, Ohio, 239–242
Indirect effects, 100
Indonesia, mangrove loss, 216
Industrial ecology, 11, 38–39
Information, in ecosystems, 102

Iowa, 294
Iron, 253

Japan, 216, 306–308

$k–C^*$ model, 249, 251–252
Kentucky, 235
Kesterton Wildlife Refuge, California, 253
Kissimmee River, Florida, 3–4, 125, 141–145, 174, 239
Krefeld system, 233
K-selection, 81, 85

Lake Erie, 222
Lake Okeechobee, 4
Lake and reservoir restoration, 105–124
  aeration, 115
  algacides, 116–117
  biological control, 119–120
  biomanipulation, 117–119
  covering of sediments, 113
  diversion of wastewater, 111
  eutrophication, 105–108
  fertilizer control, 115–116
  flocculation of phosphorus, 114–115
  hydrologic regulation, 115
  hypolimnetic water siphoning, 113–114
  macrophyte removal, 113
  method selection, 120, 123
  neutralization, 116
  removal of sediments, 111–113
  restoration techniques, 110–124
  shoreline vegetation, 116
  submerged vegetation recovery, 120
Lake Victoria, 120
Lake Washington, 111–112
Landscape
  placement of wetlands, 178–180
Langmuir adsorption isotherms, 267–268, 283
Lead, 78, 253, 286
LeChâtelier's principle, 92–93
Levees, 147–148
  coastal removal, 204–205
  river removal, 155–159
Liebig's law of the minimum, 76

Life-cycle
  analyses, 13
  assessments, 39
Limiting factors, 74–77, 106
  multiple, 76–77
Linewaver–Burke plot, 75
Litter raking, 297
Lotka's principle, 90
Louisiana, 3, 51, 54, 175–176, 217–
  223, 235, 243, 260
  coastal wetland loss, 218
Louisiana Coastal Area (LCA), 219–
  220

Mangal, see also Mangroves, 198
Manganese, 253
Mangroves, 198–200
  decline, 216
  restoration, 215–217
Marsh, see also Wetland(s), 164
Maryland, 243
Massachusetts, 3, 52
Mass conservation, 63–65, 95
Maximum power principle, 90
Max-Planck-Institute process, 233
Meander scrolls, 148
Mercury, 253
Mesocosm(s), 52, 58, 340
Mesohaline estuary, 197
Metals, 10, 77, 79–80
  bioremediation, 264, 282–286
  effects on water solubility, 266
  plant bioaccumulation, 285–286
  uptake by plants, 264, 274–279
Methane, from wetlands, 183
Meuse River, 125
Michaelis–Menten equation, 74–76,
  266, 268, 270, 363
Michigan, 51, 222, 224–226, 232,
  235–237
Microbial
  adaptations, 270
  biobarriers, 281
  biomass and biodegradation, 270–
  271
  biotraps, 284
  decomposition of xenobiotic
    compounds, 271
Microcosm, see also Mesocosm, 340

Microorganisms, see also Microbial
  adaptations, 270
  ecotechnological approaches, 280
Mine land restoration, 287–306
Mire, 164
Mississippi River, 128
  basin, 3, 168, 175–176
  delta restoration, 166, 217–223
  diversions, 219–223
Mitigation, see also Wetland(s),
  mitigation
  ratio, 166
  wetland, 45–47
  of wetland loss, 166–167
Modeling, 61
  adjacency matrix, 343
  calibration, 345
  conceptual diagram, 343
  constraints, 354–357
  defining the problem, 341
  ecological engineering, 339–366
  ecological tests, 356
  mathematical equations, 343–345
  nitrogen retention by wetlands, 359–
  366
  procedure, 341–346
  recent developments, 354–357
  sensitivity analysis, 345
  validation, 346
  verification, 345
Model(s), see also Modeling; STELLA
  model
  acid mine drainage control, 294–
  296, 359
  application in ecotechnology, 340–
  341
  articulation, 351–352
  autonomous, 347, 349
  biodemographic, 350
  bioenergetic, 350
  biogeochemical, 63, 350
  black-box, 347, 349
  cadmium recycling, 275–279
  causal, 347, 349
  Chinese fishpond, 325–330
  compartment, 347
  complexity, 350–352
  conceptual, 352–354, 359
  deterministic, 347

distributed, 347–348
dynamic, 347–348
ecological–economic, 340
effectiveness, 351–352
experimental tools, 357–358
holistic, 347–348
lumped, 346–347, 349
management, 347
matrix, 347
nonautonomous, 347, 349
nutrient, 226–227
reductionistic, 347–348
research, 347
static, 347–348
structurally dynamic, 355
types, 346–350
Mosquito control
    wetlands, 258
Mulberry production, 321–322
Mutations, 88

Natural Resources Conservation
    Service (NRCS), 168
Nature engineering, 19, 24
Net efficiency, 65
Net primary productivity, see also
    Productivity, 192
Netherlands, 176, 234
New Brunswick, Canada, 176
New Jersey, 5, 7, 54, 203, 205, 210,
    212
New Orleans, 219
New Zealand, 52, 235
    wetland loss, 165–166
Nickel, 253, 285
Nile
    delta, 217
    river, 125
Nitrification, 363
Nitrogen
    cycle, 64
    fixation, 17, 289–290
    limitation, 106
    pollution, 16–17
    removal by wetlands, 153, 155,
        175–176, 238, 241–243,
        247–249, 251–252, 359–366
    terrestrial restoration, 289–291

No net loss, see also Wetland(s),
    mitigation, 167
Nonpoint source pollution, 10, 43, 53,
    95, 237, 244, 249, 359
Nonrenewable resource conservation,
    33–34
North Carolina, 52, 202
North Dakota, 235
North Sea, 79–80, 145
Norway, 53, 250
Nutrient spiraling, 132–133

Odum energy diagram
    agroecological engineering in
        China, 322
    Chinese fishpond, 327
    Chinese floodplain hydrology, 333
    ecological engineering, 31
    ecosystem energy flow, 68
    experimental wetlands, 192
    mine drainage wetland, 295
    symbols, 72
    water hyacinth marsh, 121
Odum energy language, 71–72, 353
Odum maximum power principle, 90
Odum, H.T., and ecological
    engineering, 25
Ohio, 6, 52, 53, 58, 149, 151, 158–
    172, 184–189, 168–169, 178, 222,
    227, 235, 238–242, 247–248, 250,
    294–296
Ohio River, 125
Olentangy River Wetland Research
    Park, Ohio, 58, 149, 151–159,
    178, 184–189, 239, 247–248
Oligohaline estuary, 197
Oligotrophic–mesotrophic–eutrophic
    series, 108
Order
    coastal channels, 207–209
    streams and rivers, 127
Organic contaminants
    bioavailability factors, 265–270
    biodegradability, 264–274
    bioremediation, 279–282
    co-solvents, 267
    example calculation, 273–274
    microorganisms, 280
    micropore exclusion, 269

Organic contaminants (*continued*)
  phytoremediation of, 281–282
  sorption on soils, 267–269
  water solubility, 265–267
Organic matter, oxidation sequence, 93
Oxbow, see also Billabong, 148
  creation, 149–155

Peatland restoration, 52, 163, 176–177
  treatment wetland, 235–237
Pennsylvania, 137–140, 235, 294
Pesticides, 78–79
Phenotypes, 88
Philippines, mangrove swamp loss,
  165, 216
Phosphate mine restoration, 172, 174,
  303–306
Phosphorus
  and biomanipulation, 119
  cycle, 65
  floculation, 114–115
  limitation, 106
  removal by wetlands, 153, 155, 174,
    226–227, 238, 241–242, 243,
    249, 251–252
Photosynthetic equation, 68–69
Phytoplankton diversity, 82–83
Phytoremediation, see also
  Bioremediation, 281–282
Planting techniques, wetlands, 189–191
Plants, see also Vegetation
  bioaccumulators, 282, 285–286
  cadmium uptake, 275–279
  chemical composition, 106, 269
  example calculation of
    contamination, 273–274
  exotic, 186–187
  phytoremediation, 281–282
  uptake of heavy metals, 274–279
  wetland introduction, 185–191
Point bars, 147–148
Poland, 299
Pollution
  control, 34
  flow, 12
  effects, 72–74
  organic, 78
  pesticides, 78–79
  secondary, 16–17

Population, human, 9, 14–15
Power
  conversion factors, 70
  maximum, 90
Preservationism, 18
Prigogine, 89–90, 92
Productivity
  aquatic, 108
  and biodiversity, 191
  conversion factors, 70
  gross primary, 67–68
  net primary, 67–68
  primary, 67–68
  secondary, 64
  wetlands, 191
Puerto Rico, 202
Pulse(s)
  flood, 159
  paradigm, 97
  stability, 97
Pulsing, 97, 359

Quanicassee River, Michigan, 224–227
Quebec, Canada, 176

Reclamation, 288
Reclamation ecology, 24
Recycling, 96–97
  in China, 315
Redfield ratio, 69–71, 96
Reefs, see also Coral reefs, 227–229
  prefabricated and designed, 229
Reference system, restoration, 161
Rehabilitation, 19
  definition, 288
Renewable resources, 14–15
Resilience, ecosystem, 81
Resistance, ecological, 81
Resource limitation, see also Limiting
  factors, 14–15
  bottom-up vs. top-down, 61
Resourcism, 18
Respiration, 67–69
Restoration
  abandoned river channel, 158, 160–
    162
  definition, 288
  ecosystem, 19
  forests, 296–303

indices, 161
lakes and reservoirs, 105–124
reference study, 161
streams and rivers, 125- 162
success, 159, 161–162, 191, 193–
    194
terrestrial ecosystems, 288–308
Rhine River, 79–80, 125
Rhône River, 159–162
Rice paddies, China, 331
Riparian ecosystems, 129–130
restoration, 158, 160–162
Riparian forest, see also Bottomland
    forest(s)
functions and values, 150
restoration, 54, 153–159
River, see also River restoration
basins, 126–127
channel, 127, 147–148
ecology, 130–133
geomorphic zones, 126–127
geomorphology, 128–129
meanders, 140–141
pollution control, 323–325
restoration, see River restoration
River Continuum Concept (RCC),
    131–132
River restoration, 3–6, 24, 54
classification, 133–134
channel restoration, 140–147
dam removal, 133–140
examples, 135
floodplain ecosystems, 147–160
measuring success, 159, 161–162
techniques, 133–162
Rock-reed filters, 254
Root-zone wetland, 53
r-selection, 81, 85
Ruhr River, 125
Russia, coal reserves, 292

Saginaw Bay, Michigan, 224–227
Salt marsh(es), 197–199
creation, 54
definition, 197
high marsh, 198
low marsh, 198
restoration, 5, 7, 48–49, 54, 202–
    215

restoration in China, 321–323
Sanitary engineering, 35
Scale
spatial, 62, 98–99
temporal, 62, 98–99
SCOPE (Scientific Committee on
    Problems in the Environment), 27
Sea grass beds, 200–201
Sediment fences, intertidal, 54
Seed bank, 183, 191
Selenium, 253
Self-design, 19, 28–32, 55, 98, 177,
    193, 359
coastal, 205, 212
definition, 28–29
mine restoration, 305
reef restoration, 229
urban forest reconstruction, 308
wetlands, 185
Self-organization, 28–29
definition, 28
Self-regulation, see also Self-design,
    177
Septage treatment, 52
Shannon–Weaver (Shannon–Weiner)
    index, 81
Shell game, 15–16
Silk production, 321–322
Skjern River, Denmark, 5–6, 145–147
Slough, 148
Slovenia, 114
Soil(s)
color, 182, 184–185
contamination, example problem,
    273–274
hydric, 182–185
microbial adaptations, 270
organic contamination, 264–274
pore size, effect on contaminant
    solubility, 269–270
porosity, 245–246, 270
restoration, see Bioremediation
treatment wetlands, 253–256
wetland creation and restoration,
    182–185
Soil–water partitioning coefficient, 267
Solar aquatics, 24, 41
Solar constant, 70
Spain, 238, 250

STELLA model
  Chinese fishpond, 325–330
  Chinese wetlands, 331–335
  diagrams, 353–354
  mine drainage wetland, 294–296
  nitrogen in wetland, 359–366
  reducing metals in agricultural land,
    275–279
  use for ecological engineering, 359
Stoplogs, wetlands, 181
Stormwater treatment, 237–239
Stream density, coastal, 207–208
Stream order, 127
Success
  ecological, 48–49
  river restoration, 159, 161–162
  wetland creation and restoration,
    191, 193–194
Succession, see also Development, 84–
  86
  properties, 85–86
  terrestrial restoration, 289
Suspended solids, wetland removal,
  251–252
Sustainability, 17–18, 41–42
Sustainable
  development, 18
  resource management, 50
Swamp, 164
Sweden, 26, 53, 113, 238, 250
Switzerland, 114
Synthetic ecology, 24
Systems approach, 33
Systems ecology, 56–61
  need for, 60
System theories
  ascendency, 93
  dissaptive structures, 89–90
  entropy minimization, 89–90
  exergy maximization, 90–92
  LeChâtelier's principle, 92–93
  maximum power, 90

Terraces, 148–149
Thailand, 243
  mangrove loss, 216
Thalweg, 140–141
Thames River, England, 125

Thermodynamics, laws of, 66–67
Threshold agents, 72–74
Tidal creeks, 8
Tile drains, 178–179
Toxic substances, 16
Trajectory, ecological, 46–49
Transformity, see Emergy
Treatment wetland(s), 5–6, 53, 166,
  230–262
  agricultural runoff, 238–242
  agricultural wastewater, 242–243
  basin morphology, 246–247
  chemical loading, 247–253
  economics, 259–260
  hydrology, 243–246
  institutional considerations, 261
  landfield leachate, 242
  liner, 253–254
  mine drainage, 237
  mosquito control, 258
  municipal wastewater, 235
  pathogens, 258–259
  soils, 253–256
  urban stormwater, 237–239
  vegetation, 256
Tropical
  mine restoration, 301–303
  rain forest, restoration, 301–303

United Kingdom, 176, 235, 243
U.S. Army Corps of Engineers
  dredging, 195
  Everglades restoration, 174
  greening its mission, 9
  wetland mitigation, 167
U.S. Department of Agriculture, 167–
  168
United States, wetland loss, 165

Vegetation, see also Plants
  classification-treatment wetlands,
    234
  introducing to wetlands, 185–191
  planting, urban environments, 306–
    308
  treatment wetlands, 256
Viet Nam, mangrove loss, 216
Vollenweider model, 109–110, 226

Washington, 111–112, 250
Washington, DC, 52
Water quality, created oxbow, 153, 155
Watershed, 126
    restoration, 174–176
Weir, 181–183
Wetland(s), see also Wetland creation
        and restoration
    biodiversity, 18
    creation, see Wetland creation and
        restoration
    cropped, 167
    definitions, 165–166
    designer, 185
    diversity, 82–8
    enhancement, 166
    fish production, 316–318
    global extent, 164–165
    hydric soil, 182–185
    losses, 164–165
    macrophyte richness, 83
    management in China, 331–335
    mitigation, 166–172, 186, 191, 193
    nitrogen removal, see also Nitrogen,
        153, 155, 359–366
    nutrient source, sink, transformer,
        230–231, 359–366
    phosphorus removal, see also
        Phosphorus, 153, 155
    plant biodiversity, 187, 191
    replacement, see also Wetland(s),
        mitigation, 166
    reserve program, 167
    restoration, see also Wetland
        creation and restoration, 24

seed bank, 183
self-design, 185
soil, see also Soil, 182–185
soil chemistry, 255–256
soil depth, 255
soil organic content, 255
soil reduction, 255–256
treatment, see Treatment wetlands
vegetation, 185–191
watershed restoration, 174–176
wildlife, 170, 173
Wetland creation and restoration, 163–
        194
    definition, 165–166
    goals, 177–178
    hydrology, 181–182
    measuring success, 191, 193–194
    placing in the landscape, 178–180
    principles, 177, 194, 215
    reasons for, 166–177
    site selection, 180
    success, see Success
    techniques, 177–191
Whole-ecosystem experiment, 187–189
Wildlife
    effect of plant introduction, 188–
        189, 192
    mitigation wetland, 170, 173
    treatment wetlands, 257–258

Yangtze River, 331–335
Yin and yang, 310–311

Zero discharge, 9
Zinc, 253, 286